MATLAB Lessons, Examples, and Exercises

Mehdi Rahmani-Andebili

MATLAB Lessons, Examples, and Exercises

A Tutorial for Beginners and Experts

 Springer

Mehdi Rahmani-Andebili
ECE Department
University of Alabama
Tuscaloosa, AL, USA

ISBN 978-3-031-76176-8   ISBN 978-3-031-76177-5   (eBook)
https://doi.org/10.1007/978-3-031-76177-5

This Springer imprint is published by the registered company Springer Nature Switzerland AG
The registered company address is: Gewerbestrasse 11, 6330 Cham, Switzerland

If disposing of this product, please recycle the paper.

Preface

MATLAB is a high-performance programming language for technical and mathematical computing. It integrates computation, visualization, and programming in a user-friendly environment where problems and solutions are expressed in familiar mathematical notation. At some universities around the worlds, MATLAB is taught as an independent course. On the other hand, at many universities worldwide, MATLAB is taught as an accompanying part of many courses to help students to perform their projects and homework since it is a standard instructional tool for introductory and advanced courses in mathematics, engineering, and science. Moreover, at research centres as well as in industry, MATLAB is a tool for high-productivity research, development, and analysis.

This textbook has been prepared for instructors as well as for engineers, technical professionals, and students taking MATLAB or MATLAB-based courses. In each section of each chapter of the textbook, the lessons along with several examples and exercises are presented. The exercises that include writing the codes, executing them, and achieving the results need to be done by students in the assigned places to master programming skills. To prepare the textbook as a reader-friendly one, the codes, outputs, and descriptions of programs have been typed in blue, black, and green colors, respectively.

The textbook has been prepared as a comprehensive book to cover all the reference functions of MATLAB needed for mathematics, engineering, and science. It provides fundamentals of mathematics behind each reference function to help readers comprehend them and apply them to perform specific operations in engineering and scientific problems. Therefore, the textbook can be used as a tutorial for beginners and experts. The subjects covered in the textbook include the MATLAB environment; arithmetic, relational, and logical operators; special characters; special variables and constants, complex numbers; arrays and matrices; descriptive statistics; trigonometric functions, basic functions; plotting features; plot types such as 3-D plot, stairstep graph, line plot with error bars, stacked plot, log-log scale plot, semilog scale plot for x and y-axes, area plot, polar coordinates plot, contour plot, vector plot (quiver), 2-D and 3-D streamline plots, surface plot, mesh surface plot, display image, bar graph, discrete sequence plot, histogram plot, 2-D and 3-D pie charts, and scatter plot; programming methods; numerical methods; and symbolic math.

Tuscaloosa, AL, USA   Mehdi Rahmani-Andebili

The Other Works Published by the Author

The author has already published the books and textbooks below with Springer Nature.

1. Power System Analysis - Comprehensive Lessons, *Springer Nature*, 2024.

2. Mathematics of Engineering and Science - Practice Problems, Methods, and Solutions, *Springer Nature*, 2024.

3. Power System Analysis - Practice Problems, Methods, and Solutions, *Springer Nature*, 2022.

4. Feedback Control Systems Analysis and Design- Practice Problems, Methods, and Solutions, *Springer Nature*, 2022.

5. AC Electric Machines- Practice Problems, Methods, and Solutions, *Springer Nature*, 2022.

6. DC Electric Machines, Electromechanical Energy Conversion Principles, and Magnetic Circuit Analysis- Practice Problems, Methods, and Solutions, *Springer Nature*, 2022.

7. Advanced Electrical Circuit Analysis - Practice Problems, Methods, and Solutions, *Springer Nature*, 2022.

8. AC Electrical Circuit Analysis - Practice Problems, Methods, and Solutions, *Springer Nature*, 2021.

9. DC Electrical Circuit Analysis - Practice Problems, Methods, and Solutions, *Springer Nature*, 2020.

10. Differential Equations- Practice Problems, Methods, and Solutions, *Springer Nature*, 2022.

11. Calculus III - Practice Problems, Methods, and Solutions, *Springer Nature*, 2023.

12. Calculus II - Practice Problems, Methods, and Solutions, *Springer Nature*, 2023.

13. Calculus I (2nd Ed.) - Practice Problems, Methods, and Solutions, *Springer Nature*, 2023.

14. Precalculus (2nd Ed.) - Practice Problems, Methods, and Solutions, *Springer Nature*, 2024.

15. Calculus - Practice Problems, Methods, and Solutions, *Springer Nature*, 2021.

16. Precalculus - Practice Problems, Methods, and Solutions, *Springer Nature*, 2021.

17. Planning and Operation of Electric Vehicles in Smart Grid, *Springer Nature*, 2023.

18. Applications of Artificial Intelligence in Planning and Operation of Smart Grid, *Springer Nature*, 2022.

19. Design, Control, and Operation of Microgrids in Smart Grids, *Springer Nature*, 2021.

20. Applications of Fuzzy Logic in Planning and Operation of Smart Grids, *Springer Nature*, 2021.

21. Operation of Smart Homes, *Springer Nature*, 2021.

22. Planning and Operation of Plug-in Electric Vehicles: Technical, Geographical, and Social Aspects, *Springer Nature*, 2019.

Contents

MATLAB Environment

Abstract

In this chapter, the reference functions concerned with the MATLAB environment are presented and described. In this regard, several examples and exercises for each section of the chapter are presented. The exercises that include writing the codes, executing them, and achieving the results need to be done by students to master programming skills. In this book, the codes, outputs, and descriptions are in blue, black, and green colors, respectively. To program in MATLAB, a script file can be created and saved with an appropriate name (e.g., untitled01) in the preferred directory of a computer. The program can be run by clicking on the "Run" available on the top toolbar of the script in MATLAB or calling the script by typing its name in Command Window or in the other scripts.

1.1 The Reference Function "clc": Text Clearance

- This reference function is used to clear all the text from the Command Window, resulting in a clear screen [1].
- However, you can use the up-arrow key "↑" in the Command Window to recall the statements from the command history.

Example

```
clc
```

Description

All the text will be cleared from the Command Window.

1.2 The Reference Function "clear": Variables Clearance

- This reference function is used to remove all variables or the selected variables from the current workspace and release them from system memory.

Example

```
clear
```

Description

All the variables will be removed from the current workspace.

Example

a = 1
b = 2
clear a

Description

Only the variable "a" will be removed from the current workspace.

1.3   The Reference Function "format": Numeric Formatting

- This reference function is used to change the output display format to the specified format. The numeric formats affect only how numbers appear in the display, not how MATLAB computes or saves them.

Example

format long
pi

Output

ans =
   3.141592653589793

Description

The code sets the output format to the long fixed-decimal format and displays the value of pi.

Example

format short
pi

Output

ans =
   3.1416

Description

The code sets the output format to the short fixed-decimal format and display the value of pi.

1.4 The Reference Functions "tic" and "toc": Time Measurement

- These reference functions are used to measure the elapsed time. The tic function records the current time, and the toc function uses the recorded value to calculate the elapsed time.

Example

```
tic
rand(12000,4400);
toc
```

Output

Elapsed time is 0.445531 seconds.

Exercise

Execute the codes below in your computer and write the result(s) in the assigned location(s).

```
tic
rand(12000,4400);
toc
```

Output

Elapsed time is (   ) seconds.

1.5 Simultaneously Clicking of "Ctrl" and "C": Execution Stopping

- It is used to stop the execution of a MATLAB command.

Reference

1. MATLAB 2023a.

Abstract

In this chapter, the reference functions concerned with the arithmetic operators in MATLAB are presented and described. MATLAB includes two types of arithmetic operations including matrix and array operations. Matrix operations are defined by the rules of linear algebra, while array operations are carried out element-by-element. The period character "." is used for array operations to distinguish them from the matrix operations. In this regard, several examples and exercises for each section of the chapter are presented. The exercises that include writing the codes, executing them, and achieving the results need to be done by students to master programming skills. In this book, the codes, outputs, and descriptions are in blue, black, and green colors, respectively. To program in MATLAB, a script file can be created and saved with an appropriate name (e.g., untitled01) in the preferred directory of a computer. The program can be run by clicking on the "Run" available on the top toolbar of the script in MATLAB or calling the script by typing its name in Command Window or in the other scripts.

2.1   The Arithmetic Operator "+": Addition

- This arithmetic operator is used to add two scalar quantities, one scalar and one matrix, or two matrices. Herein, the sizes of matrices must be the same or be compatible [1–9].
- Moreover, if the operator is applied to a scalar and a matrix, the scalar will be combined with each element of the matrix.
- Furthermore, the addition of two vectors with different orientations (one row vector and one column vector) will implicitly expand to form a matrix.
- Herein, it should be mentioned that plus(A,B) is an alternate way to execute A + B.
- Moreover, the character pair ".+" is not used since the matrix and array operations are the same for addition.

Example

```
A = 1
B = 2
C = A + B
```

Output

```
A =
   1
```

B =
  2

C =
  3

Description

Adding two scalar quantities.

Example

A = [0 1; 1 0]
C = A + 2

Output

A =
   0   1
   1   0

C =
   2   3
   3   2

Description

Adding one scalar and one matrix.

Example

A = [1 0; 2 4]
B = [5 9; 2 1]
C = A + B

Output

A =
   1   0
   2   4

B =
   5   9
   2   1

C =
   6   9
   4   5

Description

Adding two matrices.

Example

a = [1 2]
b = [1 2 3]'
c = a + b

Output

a =
     1   2

b =
     1
     2
     3

c =
     2   3
     3   4
     4   5

Description

Adding two vectors with different orientations.

Exercise

Execute the codes below in your computer and write the result(s) in the assigned location(s).

A = [9 1; 3 5]
B = [1 1; 2 2]
C = A + B

Output

A =

B =

C =

2.2   The Arithmetic Operator "−": Subtraction

- This arithmetic operator is used to subtract two scalar quantities, one scalar and one matrix, or two matrices from each other. Herein, the sizes of matrices must be the same or be compatible.
- Moreover, if the operator is applied to a scalar and a matrix, the scalar will be subtracted from each element of the matrix.
- In addition, the subtraction of two vectors with different orientations (one row vector and one column vector) will implicitly expand to form a matrix.
- Herein, it should be mentioned that minus(A,B) is an alternate way to execute A - B.
- Moreover, the character pair ".−" is not used since the matrix and array operations are the same for subtraction.

Example

A = 1
B = 2
C = A - B

Output

A =
   1

B =
   2

C =
   -1

Description

Subtracting two scalars.

Example

A = [0 1; 1 0]
C = A - 2

Output

A =
   0   1
   1   0

C =
   -2   -1
   -1   -2

Description

Subtracting one scalar and one matrix.

Example

A = [1 0; 2 4]
B = [5 9; 2 1]
C = A - B

Output

```
A =
   1     0
   2     4

B =
   5     9
   2     1

C =
  -4    -9
   0     3
```

Description

Subtracting two matrices.

Example

a = [1 2]
b = [1 2 3]'
c = a - b

Output

```
a =
   1     2

b =
   1
   2
   3

c =
   0     1
  -1     0
  -2    -1
```

Description

Subtracting two vectors with different orientations.

Example

A = [0 1; 1 0]
C = -A

Output

A =
   0   1
   1   0

C =
   0   -1
  -1   0

Description

Negating the elements of matrix A.

Exercise

Execute the codes below in your computer and write the result(s) in the assigned location(s).

A = [9 1; 3 5]
B = [1 1; 2 2]
C = A - B

Output

A =

B =

C =

2.3   The Arithmetic Operator "$*$": Multiplication

- This arithmetic operator is used to calculate the product of two scalar quantities, the product of one scalar and one matrix, and the matrix product of two matrices while the number of columns of the first matrix must be equal to the number of rows of the second matrix.
- If the dimension of the first matrix is m-by-p and the dimension of the second one is p-by-n matrix, then the resultant matrix is an m-by-n matrix.
- It should be mentioned that mtimes(A,B) is an alternative way to execute A*B.

Example

A = 5
B = 4
C = B*A

Output

A =
    5

B =
    4

C =
    20

Description

Negating two scalars.

Example

A = 3
B = [1; 2; 3; 4]
C = A*B

Output

A =
    3

B =
    1
    2
    3
    4

C =
    3
    6
    9
    12

Description

Negating one scalar and one vector.

Example

A = [1 2 1 3]
B = [1; 2; 3; 4]
C = A*B

Output

A =
   1   2   1   3

B =
   1
   2
   3
   4

C =
   20

Description

Negating two vectors.

Example

A = [1 2 1 3]
B = [1; 2; 3; 4]
C = B*A

Output

A =
   1   2   1   3

B =
   1
   2
   3
   4

C =
   1   2   1   3
   2   4   2   6
   3   6   3   9
   4   8   4   12

Description

Negating two vectors.

Exercise

Execute the codes below in your computer and write the result(s) in the assigned location(s).

A = [2 4; 5 11]
B = [3 6; 5 2]
C = A*B

Output

A =

B =

C =

2.4   The Arithmetic Operator ".∗": Element-Wise Multiplication

- This arithmetic operator, which is called element-wise multiplication, is used to multiply the corresponding elements of two matrices. The sizes of matrices must be the same or be compatible.
- Moreover, if the operator is applied to a scalar and a matrix, the scalar will be combined with each element of the matrix.
- Also, the multiplication of two vectors with different orientations (one row vector and one column vector) will implicitly expand to form a matrix.
- It should be mentioned that times(A,B) is an alternate way to execute A.*B.

Example

A = [1 0 3]
B = [2 3 7]
C = A.*B

Output

A =
   1   0   3

B =
   2   3   7

C =
   2   0   21

Description

Negating the corresponding elements of two vectors.

Example

A = [1 0 3; 5 3 8; 2 4 6]
B = [2 3 7; 9 1 5; 8 8 3]
C = A.*B

Output

A =
     1   0   3
     5   3   8
     2   4   6

B =
     2   3   7
     9   1   5
     8   8   3

C =
     2   0   21
    45   3   40
    16   32   18

Description

Negating the corresponding elements of two matrices.

Exercise

Execute the codes below in your computer and write the result(s) in the assigned location(s).

A = [1 11; 5 3]
B = [2 9; 1 5]
C = A.*B

Output

A =

B =

C =

2.5   The Arithmetic Operator "/": Right Division

- If A is a square matrix and B is a matrix with the same number of columns, then x = B/A is a solution to the equation x*A = B if it exists.
- Moreover, this arithmetic operator can be used to calculate the division of two scalar quantities and the division of a matrix by a scalar.
- It should be mentioned that mrdivide(B,A) is an alternative way to execute B/A.

Example

A = 3
B = 6
x = B/A

Output

A =
   3

B =
   6

x =
   2

Description

Division of two scalars.

Example

A = 3
B = [3 6 9]
x = B/A

Output

A =
   3

B =
   3   6   9

x =
   1   2   3

Description

Division a of vector by a scalar.

Example

A = [1 1 3; 2 0 4; -1 6 -1]
B = [2 19 8]
x = B/A

Output

A =
   1   1   3
   2   0   4
  -1   6   -1

B =
   2   19   8

x =
  1.0000   2.0000   3.0000

Description

Solving the equation x*A = B.

Exercise

Execute the codes below in your computer and write the result(s) in the assigned location(s).

A = [3 2 1; 1 1 2; -1 2 -1]
B = [2 4 8]
x = B/A

Output

A =

B =

x =

2.6   The Arithmetic Operator "./": Element-Wise Right Division

- This arithmetic operator, which is called element-wise right division, is used to divide the corresponding elements of two matrices. The sizes of matrices must be the same or be compatible.

- However, if one of the matrices is a scalar, the scalar is combined with each element of the other matrix.
- Also, vectors with different orientations (one row vector and one column vector) implicitly expand to form a matrix.
- It should be mentioned that rdivide(A,B) is an alternative way to execute A./B.

Example

A = [2 4 6 8; 3 5 7 9]
B = [10 10 10 10; 10 10 10 10]
x = A./B

Output

A =
     2   4   6   8
     3   5   7   9

B =
    10   10   10   10
    10   10   10   10

x =
    0.2000   0.4000   0.6000   0.8000
    0.3000   0.5000   0.7000   0.9000

Description

Dividing the corresponding elements of two matrices.

Example

A = 2
B = [4 2 1; 1 1 2; -1 2 -1]
x = A./B

Output

A =
     2

B =
     4   2   1
     1   1   2
    -1   2   -1

x =
    0.5000   1.0000   2.0000
    2.0000   2.0000   1.0000
   -2.0000   1.0000   -2.0000

Description

Dividing a scalar by a matrix.

Example

A = [1 2]
B = [1 2 3]'
C = A ./ B

Output

A =
    1   2

B =
    1
    2
    3

C =
    1.0000   2.0000
    0.5000   1.0000
    0.3333   0.6667

Description

Dividing two vectors with different orientations.

Exercise

Execute the codes below in your computer and write the result(s) in the assigned location(s).

A = [5 2 8; 2 4 12]
B = [5 1 4; 1 2 6]
x = A./B

Output

A =

B =

x =

2.7   The Arithmetic Operator "\": Left Division

- This arithmetic operator is used to calculate the division of two scalar quantities and the division of one matrix and one scalar.
- Moreover, if A is a square matrix and B is a matrix with the same number of rows, then x = A\B is a solution to the equation A*x = B, if it exists.
- It should be mentioned that x = mldivide(A,B) is an alternative way to execute x = A\B.

Example

A = 3
B = 6
x = B\A

Output

A =
   3

B =
   6

x =
   0.5000

Description

Division of two scalars.

Example

A = 3
B = [3 6 9]
x = A\B

Output

A =
   3

B =
   3   6   9

x =
   1   2   3

Description

Division of a vector by a scalar.

Example

```
A = [8 1 6; 3 5 7; 4 9 2]
B = [15; 15; 15]
x = A\B
```

Output

```
A =
    8    1    6
    3    5    7
    4    9    2

B =
   15
   15
   15

x =
   1.0000
   1.0000
   1.0000
```

Description

Solving the equation A*x = B.

Exercise

Execute the codes below in your computer and write the result(s) in the assigned location(s).

```
A = [8 2; 4 6]
B = [2; 8]
x = A\B
```

Output

```
A =

B =

x =
```

2.8   The Arithmetic Operator ".\": Element-Wise Left Division

- This arithmetic operator, which is called element-wise left division, is used to divide the corresponding elements of two matrices. The sizes of matrices must be the same or be compatible.
- However, if one of the matrices is a scalar, the scalar is combined with each element of the other matrix.
- Also, vectors with different orientations (one row vector and one column vector) implicitly expand to form a matrix.
- It should be mentioned that ldivide(B,A) is an alternative way to execute A./B.

Example

A = [2 4 6 8; 3 5 7 9]
B = [10 10 10 10; 10 10 10 10]
x = A.\B

Output

A =
```
   2    4    6    8
   3    5    7    9
```

B =
```
   10    10    10    10
   10    10    10    10
```

x =
```
   5.0000    2.5000    1.6667    1.2500
   3.3333    2.0000    1.4286    1.1111
```

Description

Dividing the corresponding elements of two matrices.

Example

A = 2
B = [4 2 1; 1 1 2; -1 2 -1]
x = B.\A

Output

A =
```
   2
```

B =
```
    4    2    1
    1    1    2
   -1    2   -1
```

x =
```
    0.5000      1.0000      2.0000
    2.0000      2.0000      1.0000
   -2.0000      1.0000     -2.0000
```

Description

Dividing a scalar by a matrix.

Example

```
A = [1 2]
B = [1 2 3]'
C = A .\ B
```

Output

A =
```
    1     2
```

B =
```
    1
    2
    3
```

C =
```
    1.0000      0.5000
    2.0000      1.0000
    3.0000      1.5000
```

Description

Dividing two vectors with different orientations.

Exercise

Execute the codes below in your computer and write the result(s) in the assigned location(s).

```
A = [5 2 8; 2 4 12]
B = [5 1 4; 1 2 6]
x = A.\B
```

Output

A =

B =

x =

2.9   The Arithmetic Operator "^": Power

- This arithmetic operator is used to raise a scalar or a matrix to the given power or root.
- Herein, if the exponent is -1, then the result will be the reciprocal of the given scalar and the inverse matrix of the given matrix, respectively.
- It should be mentioned that mpower(A,B) is an alternate way to execute A^B.

Example

```
A = 3
B = 4
C = A^B
```

Output

```
A =
   3

B =
   4

C =
   81
```

Description

Raising the value of a scalar to the given exponent.

Example

```
A = [1 2; 3 4]
B = 2
C = A^B
```

Output

```
A =
   1    2
   3    4

B =
   2

C =
    7    10
   15    22
```

Description

Raising the value of a matrix to the given exponent.

Example

A = 2
B = -1
C = A^B

Output

A =
   2

B =
   -1

C =
   0.5000

Description

Computing the reciprocal value of the given scalar.

Example

A = [1 2; 3 4]
B = -1
C = A^B

Output

A =
   1   2
   3   4

B =
   -1

C =
   -2.0000   1.0000
    1.5000   -0.5000

Computing the inverse matrix of the given matrix.

Exercise

Execute the codes below in your computer and write the result(s) in the assigned location(s).

A = [1 2 3; 4 5 6; 7 8 9]
B = 2
C = A^B

Output

A =

B =

C =

2.10   The Arithmetic Operator ".^": Element-Wise Power

- This arithmetic operator, which is called element-wise power, is used to raise each element of a matrix to the corresponding powers or roots in another matrix. The sizes of matrices must be the same or be compatible.
- However, if one of the matrices is a scalar, the scalar is combined with each element of the other matrix.
- Moreover, vectors with different orientations (one row vector and one column vector) implicitly expand to form a matrix.
- It should be mentioned that power(A,B) is an alternate way to execute A.^B.

Example

A = [1 2 3; 4 5 6; 7 8 9]
B = [1 1 1; 2 2 2; 3 3 3]
C = A.^B

Output

A =
    1   2   3
    4   5   6
    7   8   9

B =
    1   1   1
    2   2   2
    3   3   3

C =
```
      1        2        3
     16       25       36
    343      512      729
```

Description

Raising each element of a matrix to the corresponding powers in another matrix.

Example

A = [2 4; 16 9]
B = 1/2
C = A.^B

A =
```
      2        4
     16        9
```

B =
```
    0.5000
```

C =
```
    1.4142      2.0000
    4.0000      3.0000
```

Description

Computing the root of each element of a matrix by the corresponding roots in another matrix.

Example

A = [2 3]
B = [1; 2; 3]'
C = A.^B

A =
```
      2        3
```

B =
```
      1
      2
      3
```

C =
  2   3
  4   9
  8   27

Description

Applying the operator of element-wise power on two vectors with different orientations.

Exercise

Execute the codes below in your computer and write the result(s) in the assigned location(s).

A = [16 4; 9 81]
B = [0.5 1; 2 0.5]
C = A.^B

Output

A =

B =

C =

2.11   The Arithmetic Operator "'": Complex Conjugate Transpose

- This arithmetic operator, which is called complex conjugate transpose, is used to compute the complex conjugate transpose of a matrix.
- It should be mentioned that ctranspose(A) is an alternate way to execute A'.

Example

A = [2 1; 9 7; 2 8; 3 5]
B = A'

Output

A =
  2   1
  9   7
  2   8
  3   5

B =
  2   9   2   3
  1   7   8   5

Description

Computing the complex conjugate transpose of the given matrix.

Example

A = [-i 2+i; 4+2i -2i]
B = A'

Output

A =
    0.0000 − 1.0000i   2.0000 + 1.0000i
    4.0000 + 2.0000i   0.0000 − 2.0000i

B =
    0.0000 + 1.0000i   4.0000 − 2.0000i
    2.0000 − 1.0000i   0.0000 + 2.0000i

Description

Computing the complex conjugate transpose of the given matrix.

Exercise

Execute the codes below in your computer and write the result(s) in the assigned location(s).

A = [1 i; -i 1+i]
B = A'

Output

A =

B =

2.12   The Arithmetic Operator ".'": Transpose

- This arithmetic operator, which is called transpose, is used to compute the transpose of a matrix.
- It should be mentioned that transpose(A) is an alternate way to execute A.'.

Example

A = [-i 2+i; 4+2i -2i]
B = A.'

Output

A =

| 0.0000 − 1.0000i | 2.0000 + 1.0000i |
| 4.0000 + 2.0000i | 0.0000 − 2.0000i |

B =

| 0.0000 − 1.0000i | 4.0000 + 2.0000i |
| 2.0000 + 1.0000i | 0.0000 − 2.0000i |

Description

Computing the complex conjugate transpose of the given matrix.

Exercise

Execute the codes below in your computer and write the result(s) in the assigned location(s).

```
A = [1 i; -i 1+i]
B = A.'
```

Output

A =

B =

References

1. MATLAB 2023a.
2. Rahmani-Andebili, M. (2024). *Mathematics of engineering and science – Practice problems, methods, and solutions*. Springer Nature.
3. Rahmani-Andebili, M. (2022). *Differential equations – Practice problems, methods, and solutions*. Springer Nature.
4. Rahmani-Andebili, M. (2023). *Calculus III – Practice problems, methods, and solutions*. Springer Nature.
5. Rahmani-Andebili, M. (2023). *Calculus II – Practice problems, methods, and solutions*. Springer Nature.
6. Rahmani-Andebili, M. (2023). *Calculus I – Practice problems, methods, and solutions* (2nd ed.). Springer Nature.
7. Rahmani-Andebili, M. (2021). *Calculus – Practice problems, methods, and solutions*. Springer Nature.
8. Rahmani-Andebili, M. (2024). *Precalculus – Practice problems, methods, and solutions* (2nd ed.). Springer Nature.
9. Rahmani-Andebili, M. (2021). *Precalculus – Practice problems, methods, and solutions*. Springer Nature.

Abstract

In this chapter, the reference functions concerned with the relational operators in MATLAB are presented and described. In this regard, several examples and exercises for each section of the chapter are presented. The exercises that include writing the codes, executing them, and achieving the results need to be done by students to master programming skills. In this book, the codes, outputs, and descriptions are in blue, black, and green colors, respectively. To program in MATLAB, a script file can be created and saved with an appropriate name (e.g., untitled01) in the preferred directory of a computer. The program can be run by clicking on the "Run" available on the top toolbar of the script in MATLAB or calling the script by typing its name in Command Window or in the other scripts.

3.1 The Relational Operator "==": Logical Equality

- The operation A == B returns a logical array of logical values with elements set to logical 1 (true), where inputs A and B are equal; otherwise, the element is logical 0 (false). Herein, the comparison is done for both real and imaginary parts of numeric arrays [1–9].
- It should be mentioned that eq(A,B) is an alternative way to execute A == B.

Example

```
A = [1+i 3 2 4+i]
B = [1 3+i 2 4+i]
A == B
```

Output

```
A =
  1.0000 + 1.0000i   3.0000 + 0.0000i   2.0000 + 0.0000i   4.0000 + 1.0000i

B =
  1.0000 + 0.0000i   3.0000 + 1.0000i   2.0000 + 0.0000i   4.0000 + 1.0000i

ans =
  1×4 logical array
   0   0   1   1
```

Description

Comparing the elements of two vectors for equality.

Exercise

Execute the codes below in your computer and write the result(s) in the assigned location(s).

A = [i 1+i; -i 1-i]
B = [-i 1-i; -i -1+i]
A == B

Output

A =

B =

ans =

3.2   The Relational Operator "~=": Logical Inequality

- The operation A ~= B returns a logical array of logical values with elements set to logical 1 (true), where inputs A and B are not equal; otherwise, the element is logical 0 (false). Herein, the comparison is done for both real and imaginary parts of numeric arrays.
- It should be mentioned that ne(A,B) is an alternative way to execute A ~= B.

Example

A = [1+i 3 2 4+i]
B = [1 3+i 2 4+i]
A ~= B

Output

A =
  1.0000 + 1.0000i   3.0000 + 0.0000i   2.0000 + 0.0000i   4.0000 + 1.0000i

B =
  1.0000 + 0.0000i   3.0000 + 1.0000i   2.0000 + 0.0000i   4.0000 + 1.0000i

ans =
  1×4 logical array
   1   1   0   0

Description

Comparing the elements of two vectors for inequality.

Exercise

Execute the codes below in your computer and write the result(s) in the assigned location(s).

A = [i 1+i; -i 1-i]
B = [-i 1-i; -i -1+i]
A ~= B

Output

A =

B =

ans =

3.3 The Relational Operators ">" and "<": Greater/Less Than

- The operation A > B returns a logical array of logical values with elements set to logical 1 (true), where A is greater than B; otherwise, the element is logical 0 (false). Herein, the comparison is done only for the real part of numeric arrays.
- It should be mentioned that gt(A,B) is an alternative way to execute A > B.
- The operation A < B returns a logical array of logical values with elements set to logical 1 (true), where A is less than B; otherwise, the element is logical 0 (false). Herein, the comparison is done only for the real part of numeric arrays.
- It should be mentioned that lt(A,B) is an alternative way to execute A < B.

Example

A = [1 12 18 7 9 11 2 15]
A > 10

Output

A =
   1   12   18   7   9   11   2   15

ans =
  1×8 logical array
  0   1   1   0   0   1   0   1

Description

Applying the relational operator on the elements of the vector.

Example

A = [1+i 2-2i 1+3i 1-2i 5-i]
A > 2

Output

A =

   1.0000 + 1.0000i   2.0000 − 2.0000i   1.0000 + 3.0000i   1.0000 − 2.0000i   5.0000 − 1.0000i

ans =
  1×5 logical array
  0   0   0   0   1

Description

Applying the relational operator on the elements of the vector.

Exercise

Execute the codes below in your computer and write the result(s) in the assigned location(s).

A = [i 1 2 3 2i]
A > 2

Output

A =

ans =

3.4   The Relational Operators ">=" and "<=": Greater/Less Than or Equal to

- The operation A >= B returns a logical array of logical values with elements set to logical 1 (true), where A is greater than or equal to B; otherwise, the element is logical 0 (false). Herein, the comparison is done only for the real part of numeric arrays.
- It should be mentioned that ge(A,B) is an alternative way to execute A >= B.
- The operation A <= B returns a logical array of logical values with elements set to logical 1 (true), where A is less than or equal to B; otherwise, the element is logical 0 (false). Herein, the comparison is done only for the real part of numeric arrays.
- It should be mentioned that le(A,B) is an alternative way to execute A <= B.

Example

A = [1 12 18 7 9 11 2 15]
A >= 11

Output

A =

  1   12   18   7   9   11   2   15

ans =

  1×8 logical array

  0   1   1   0   0   1   0   1

Description

Applying the relational operator on the elements of the vector.

Example

A = [1+i 2-2i 1+3i 1-2i 5-i]
A >= 2

Output

A =

  1.0000 + 1.0000i   2.0000 − 2.0000i   1.0000 + 3.0000i   1.0000 − 2.0000i   5.0000 − 1.0000i

ans =

  1×5 logical array

  0   1   0   0   1

Description

Applying the relational operator on the elements of the vector.

Exercise

Execute the codes below in your computer and write the result(s) in the assigned location(s).

A = [i 1 2 3 2i]
A >= 2

Output

A =

ans =

References

1. MATLAB 2023a.
2. Rahmani-Andebili, M. (2024). *Mathematics of engineering and science – Practice problems, methods, and solutions*. Springer Nature.
3. Rahmani-Andebili, M. (2022). *Differential equations – Practice problems, methods, and solutions*. Springer Nature.
4. Rahmani-Andebili, M. (2023). *Calculus III – Practice problems, methods, and solutions*. Springer Nature.
5. Rahmani-Andebili, M. (2023). *Calculus II – Practice problems, methods, and solutions*. Springer Nature.
6. Rahmani-Andebili, M. (2023). *Calculus I – Practice problems, methods, and solutions* (2nd ed.). Springer Nature.
7. Rahmani-Andebili, M. (2021). *Calculus – Practice problems, methods, and solutions*. Springer Nature.
8. Rahmani-Andebili, M. (2024). *Precalculus – Practice problems, methods, and solutions* (2nd ed.). Springer Nature.
9. Rahmani-Andebili, M. (2021). *Precalculus – Practice problems, methods, and solutions*. Springer Nature.

Abstract

In this chapter, the reference functions concerned with the logical operators in MATLAB are presented and described. In this regard, several examples and exercises for each section of the chapter are presented. The exercises that include writing the codes, executing them, and achieving the results need to be done by students to master programming skills. In this book, the codes, outputs, and descriptions are in blue, black, and green colors, respectively. To program in MATLAB, a script file can be created and saved with an appropriate name (e.g., untitled01) in the preferred directory of a computer. The program can be run by clicking on the "Run" available on the top toolbar of the script in MATLAB or calling the script by typing its name in Command Window or in the other scripts.

4.1 The Logical Operator "&": Logical AND

- The operation A & B performs a logical AND of inputs A and B and returns an array containing elements set to either logical 1 (true) or logical 0 (false). An element of the output is set to logical 1 (true) if both A and B contain a nonzero element at that same location. Otherwise, the element is set to 0 [1–9].
- It should be mentioned that and(A,B) is an alternate way to execute A & B.

Example

```
A = [5 7 0; 0 2 9; 5 0 0]
B = [6 6 0; 1 3 5; -1 0 0]
A & B
```

Output

```
A =
   5   7   0
   0   2   9
   5   0   0

B =
    6   6   0
    1   3   5
   -1   0   0
```

ans =
  3×3 logical array
  1   1   0
  0   1   1
  1   0   0

Description

Applying the logical AND operator on the matrices.

Exercise

Execute the codes below in your computer and write the result(s) in the assigned location(s).

A = [0 0; 1 2]
B = [0 0; 2 1]
A & B

Output

A =

B =

ans =

4.2   The Logical Operator "|": Logical OR

- The operation A | B performs a logical OR of inputs A and B and returns an array containing elements set to either logical 1 (true) or logical 0 (false). An element of the output is set to logical 1 (true) if either A or B contain a nonzero element at that same location. Otherwise, the element is set to 0.
- It should be mentioned that or(A,B) is an alternate way to execute A | B.

Example

A = [5 7 0; 0 2 9; 5 0 0]
B = [6 6 0; 1 3 5; -1 0 0]
A | B

Output

A =
  5   7   0
  0   2   9
  5   0   0

B =
```
   6      6      0
   1      3      5
  -1      0      0
```

ans =
   3×3 logical array
```
   1      1      0
   1      1      1
   1      0      0
```

Description

Applying the logical OR operator on the matrices.

Exercise

Execute the codes below in your computer and write the result(s) in the assigned location(s).

```
A = [1 0; 1 2]
B = [0 0; 2 1]
A | B
```

Output

A =

B =

ans =

4.3   The Logical Operator "&&": Logical Short-Circuiting AND

- The operation expr1 && expr2 represents a logical AND operation that employs logical short-circuiting behavior. That is, expr2 is not evaluated if expr1 is logical 0 (false). Each expression must evaluate to a scalar logical result.

Example

```
b = 1;
a = 20;
x = (b ~= 0) && (a/b > 18.5)
```

Output

x =
   logical
   1

Description

Applying the logical short-circuiting AND operator on the mathematical expressions.

Example

```
b = 0;
a = 20;
x = (b ~= 0) && (a/b > 18.5)
```

Output

```
x =
  logical
   0
```

Description

Applying the logical short-circuiting AND operator on the mathematical expressions.

Exercise

Execute the codes below in your computer and write the result(s) in the assigned location(s).

```
b = 1;
a = 1;
x = (a == 1) && (a+b >= 2)
```

Output

```
x =
```

4.4   The Logical Operator "‖": Logical Short-Circuiting OR

- The operation expr1 && expr2 represents a logical OR operation that employs logical short-circuiting behavior. That is, expr2 is not evaluated if expr1 is logical 1 (true). Each expression must evaluate to a scalar logical result.

Example

```
b = 1;
a = 20;
x = (b ~= 0) || (a/b > 18.5)
```

Output

x =
  logical
   1

Description

Applying the logical short-circuiting OR operator on the mathematical expressions.

Example

b = 0;
a = 20;
x = (b ~= 0) || (a/b > 18.5)

Output

x =
  logical
   1

Description

Applying the logical short-circuiting OR operator on the mathematical expressions.

Exercise

Execute the codes below in your computer and write the result(s) in the assigned location(s).

b = 1;
a = 1;
x = (a == 1) || (a+b >= 2)

Output

x =

4.5   The Logical Operator "~": Logical NOT

- The operation ~A returns a logical array of logical values of the same size as A. The output contains logical 1 (true) values where A is zero and logical 0 (false) values where A is nonzero.
- It should be mentioned that not(A) is an alternate way to execute ~A.

Example

A = [1 0; 1 2]
B = ~A

Output

A =

   1   0
   1   2

B =

   2×2 logical array

   0   1
   0   0

Description

Applying the logical NOT operator on a matrix.

Exercise

Execute the codes below in your computer and write the result(s) in the assigned location(s).

A = [10 20; 0 0]
B = ~A

Output

B =

References

1. MATLAB 2023a.
2. Rahmani-Andebili, M. (2024). *Mathematics of engineering and science – Practice problems, methods, and solutions*. Springer Nature.
3. Rahmani-Andebili, M. (2022). *Differential equations – Practice problems, methods, and solutions*. Springer Nature.
4. Rahmani-Andebili, M. (2023). *Calculus III – Practice problems, methods, and solutions*. Springer Nature.
5. Rahmani-Andebili, M. (2023). *Calculus II – Practice problems, methods, and solutions*. Springer Nature.
6. Rahmani-Andebili, M. (2023). *Calculus I – Practice problems, methods, and solutions* (2nd ed.). Springer Nature.
7. Rahmani-Andebili, M. (2021). *Calculus – Practice problems, methods, and solutions*. Springer Nature.
8. Rahmani-Andebili, M. (2024). *Precalculus – Practice problems, methods, and solutions* (2nd ed.). Springer Nature.
9. Rahmani-Andebili, M. (2021). *Precalculus – Practice problems, methods, and solutions*. Springer Nature.

Abstract

In this chapter, the reference functions concerned with the special characters in MATLAB are presented and described. In this regard, several examples and exercises for each section of the chapter are presented. The exercises that include writing the codes, executing them, and achieving the results need to be done by students to master programming skills. In this book, the codes, outputs, and descriptions are in blue, black, and green colors, respectively. To program in MATLAB, a script file can be created and saved with an appropriate name (e.g., untitled01) in the preferred directory of a computer. The program can be run by clicking on the "Run" available on the top toolbar of the script in MATLAB or calling the script by typing its name in Command Window or in the other scripts.

5.1 The Special Character "...": Command Continuation

- This special operator, which is called "dot dot dot" or ellipsis, is used to indicate line continuation. In other words, MATLAB ignores the rest of the line and continues the current command on the next line if this special operator is used at the end of the line [1].

Example

```
y = 1 + ...
    2 + ...
    3 + ...
    4
```

Output

```
=
10
```

5.2 The Special Character ",": Elements Separation

- This special operator, which is called comma, is used to separate elements in a row, array subscripts, function input and output arguments, and commands entered on the same line.

Example

A = [12,13; 14,15]

Output

A =
   12   13
   14   15

Description

Separating row elements to create an array.

Example

A(1,2)

Output

ans =
   13

Description

Separating subscripts.

Example

command, command, command

Description

Separating multiple commands on a line.

5.3   The Special Character ":": Matrix Creation and Reshaping

- This special operator, which is called colon, is used to create vectors and reshape a matrix into a column vector.

Example

x = 1:10

Output

x =

| 1 | 2 | 3 | 4 | 5 | 6 | 7 | 8 | 9 | 10 |

Description

Creating a regularly spaced vector.

Example

x = 1:3:19

Output

x =

| 1 | 4 | 7 | 10 | 13 | 16 | 19 |

Description

Creating a vector that increments by a given quantity.

Example

A = [12,13; 14,15]
A(:)

Output

A =

| 12 | 13 |
| 14 | 15 |

ans =

12
14
13
15

Description

Reshaping a matrix into a column vector.

Exercise

Execute the codes below in your computer and write the result(s) in the assigned location(s).

x = -25:5:25

Output

x =

Exercise

Execute the codes below in your computer and write the result(s) in the assigned location(s).

B = [1,2; 3,4]
B(:)

Output

A =

ans =

5.4   The Special Character ";": Output Display Suppressing and Elements Separation

- This special operator, which is called semicolon, is used to separate rows in an array creation command and suppress the output display of a line of code.

Example

A = [12 13; 14 15]

Output

A =
    12   13
    14   15

Description

Separating rows to create an array.

Example

A = [12 13; 14 15];

Description

Suppressing the output display of a line of code.

Example

A = 12.5; B = 42.7, C = 1.25;

Output

B =
    42.7000

Description

Separating and suppressing multiple commands on a single line.

5.5   The Special Character "%": Command Nonexecution

- This special operator is used to indicate nonexecutable line of code within the body of a program. This text is normally used to include comments in the code.

Example

A = [12 13; 14 15]; % Note: This is a . . .

Output

A =
    12   13
    14   15

Description

Adding a comment for guidance.

5.6   The Special Character "( )": Parentheses

- This special operator, which is called parentheses, is used to call the elements of a matrix.

Example

A = [12 13; 14 15];
A(1,1)

Output

ans =
    12

Description

Calling the element in the first row and column of the matrix.

Example

A = [12 13; 14 15];
A(1,:)

Output

ans =
    12   13

Description

Calling the elements in the first row of the matrix.

Example

A = [12 13; 14 15];
A(:,2)

Output

ans =
    13
    15

Description

Calling the elements in the second column of the matrix.

Exercise

Execute the codes below in your computer and write the result(s) in the assigned location(s).

A = [1 2 3; 4 5 6; 7 8 9];
A(2,3)

Output

ans =

Exercise

Execute the codes below in your computer and write the result(s) in the assigned location(s).

A = [1 2 3; 4 5 6; 7 8 9];
A(:,3)

Output

ans =

5.7   The Special Character "[ ]": Square Brackets

- This special operator, which is called square brackets, is used to create a matrix as well as to add a new part to a matrix or delete part of it.

Example

A = [1 2; 3 4]

Output

A =
    1   2
    3   4

Description

Creating a matrix.

Example

A = [1 2; 3 4];
B = [A; 5 6]

Output

B =
    1   2
    3   4
    5   6

Description

Adding a new row to a matrix.

Example

A = [1 2; 3 4];
A(:,1) = []

Output

A =
   2
   4

Description

Deleting one of the columns of a matrix.

Exercise

Execute the codes below in your computer and write the result(s) in the assigned location(s).

A = [1 2 3; 4 5 6; 7 8 9];
B = [A; 10 11 12]

Output

B =

5.8   The Special Character "%{%}": Curly Brackets

• This special operator is used to enclose a block of comments that extend beyond one line.

Example

%{
The purpose of this routine is to compute
the value of ...
%}

Description

Enclosing multiline comments.

5.9 The Special Character " ' ": Single Quotes

- This special operator, which is called single quotes, is used to create character vectors that have class char.

Example

'Hello'

Output

ans =
   'Hello'

Description

Creating a character vector.

5.10 The Special Character """: Double Quotes

- This special operator, which is called double quotes, is used to create string scalars that have class string.

Example

"Hello"

Output

ans =
   "Hello"

Description

Creating a string scalar.

5.11 The Special Character: Newline

- This special operator is used to separate rows in an array construction statement. In that context, the newline character and semicolon are equivalent.

Example

A = [12   13
     14   15]

A =
  12   13
  14   15

Separating rows in an array creation command.

5.12   The Special Character "~": Logical NOT

- This special operator, which is called tilde, is used to represent logical NOT.

A = [0 1; -1 2]
~A

A =
   0   1
  -1   2

ans =
  2×2 logical array
  1   0
  0   0

Computing the logical NOT of a matrix.

A = [1 -1; 0 1]
B = [1 -2; 3 2]
A~=B

A =
  1   -1
  1   1

B =
     1   -2
     3   2

ans =
  2×2 logical array
     1   1
     1   1

Description

Determining if the elements of A are not equal to those of B.

Exercise

Execute the codes below in your computer and write the result(s) in the assigned location(s).

A = [1 2 3 4 5]
B = [5 4 3 2 1]
A~=B

Output

A =

B =

ans =

5.13 The Special Character "=": Equality Sign

- This special operator, which is called equal sign, is used to assign values to a variable. The syntax B = A stores the elements of A in variable B.

Example

A = [1 1; -1 0];
B = A

Output

B =
     1   1
    -1   0

Description

Storing the elements of matrix A in matrix B.

Reference

1. MATLAB 2023a.

Abstract

In this chapter, the reference functions concerned with the special variables and constants in MATLAB are presented and described. In this regard, several examples and exercises for each section of the chapter are presented. The exercises that include writing the codes, executing them, and achieving the results need to be done by students to master programming skills. In this book, the codes, outputs, and descriptions are in blue, black, and green colors, respectively. To program in MATLAB, a script file can be created and saved with an appropriate name (e.g., untitled01) in the preferred directory of a computer. The program can be run by clicking on the "Run" available on the top toolbar of the script in MATLAB or calling the script by typing its name in Command Window or in the other scripts.

6.1 The Special Variable "ans": Most Recent Answer

- This special operator, which is called "most recent answer," is the variable created when an output is returned without a specified output argument. MATLAB creates the ans variable and stores the output there. Using the value of ans in a script or function is not recommended because its value can change frequently [1–9].

Example

```
1 + 1
```

Output

```
ans =
   2
```

```
ans + 100
```

Output

```
ans =
   102
```

Description

Saving the output in the ans variable. As can be seen, the value of ans has changed from 2 to 102.

6.2   The Special Constant "pi": Mathematical Constant

- This special mathematical constant returns the floating-point number nearest to the value of π.

Example

pi

Output

ans =
   3.1416

Description

Returning the value of π.

Example

r = 2;
A = pi*r^2

Output

A =
   12.5664

Description

Computing the surface area of a circle.

Example

r = 2;
V = 4/3*pi*r^3

Output

V =
   33.5103

Description

Computing the volume of a sphere.

6.3   The Special Constant "i" or "j": Imaginary Unit

- These special constants return the basic imaginary unit which is equivalent to $\sqrt{-1}$. They can be used to create a complex scalar, vector, or matrix.

Example

z = 1+i

Output

z =
  1.0000 + 1.0000i

Description

A complex scalar.

Example

x = [1:4]';
y = [2:2:8]';
z = x+i*y

Output

z =
  1.0000 + 2.0000i
  2.0000 + 4.0000i
  3.0000 + 6.0000i
  4.0000 + 8.0000i

Description

A complex vector.

6.4   The Special Constant "NaN" or "nan": Not a Number

- This special constant returns the scalar representation of "not a number." Operations return NaN when they have undefined numeric results, such as 0/0 or 0*Inf.
- Also, NaN(n) returns an n-by-n matrix of NaN values.

Example

0/0

Output

ans =
   NaN

Description

An undefined numeric value.

Example

0*Inf

Output

ans =
   NaN

Description

An undefined numeric value.

Example

A = NaN(3)

Output

A =
   NaN   NaN   NaN
   NaN   NaN   NaN
   NaN   NaN   NaN

Description

Creating a 3-by-3 matrix of NaN values.

Exercise

Execute the codes below in your computer and write the result(s) in the assigned location(s).

a = Inf- Inf
b = Inf/Inf

Output

a =

b =

6.5 The Special Constant "Inf" or "inf": Infinite

- This special constant returns the scalar representation of positive infinity. Operations return Inf when their result is too large to represent as a floating point number, such as 1/0 or log(0).
- Also, Inf(n) returns an n-by-n matrix of Inf values.

Example

1/0

Output

ans =
    Inf

Description

Infinite value.

Example

exp(1000)

Output

ans =
    Inf

Description

A numeric value which is too large.

Example

log(0)

Output

```
ans =
   -Inf
```

Description

Infinite value.

Example

A = Inf(3)

Output

```
A =
   Inf   Inf   Inf
   Inf   Inf   Inf
   Inf   Inf   Inf
```

Description

Creating a 3-by-3 matrix of Inf values.

References

1. MATLAB 2023a.
2. Rahmani-Andebili, M. (2024). *Mathematics of engineering and science – Practice problems, methods, and solutions*. Springer Nature.
3. Rahmani-Andebili, M. (2022). *Differential equations – Practice problems, methods, and solutions*. Springer Nature.
4. Rahmani-Andebili, M. (2023). *Calculus III – Practice problems, methods, and solutions*. Springer Nature.
5. Rahmani-Andebili, M. (2023). *Calculus II – Practice problems, methods, and solutions*. Springer Nature.
6. Rahmani-Andebili, M. (2023). *Calculus I – Practice problems, methods, and solutions* (2nd ed.). Springer Nature.
7. Rahmani-Andebili, M. (2021). *Calculus – Practice problems, methods, and solutions*. Springer Nature.
8. Rahmani-Andebili, M. (2024). *Precalculus – Practice problems, methods, and solutions* (2nd ed.). Springer Nature.
9. Rahmani-Andebili, M. (2021). *Precalculus – Practice problems, methods, and solutions*. Springer Nature.

Abstract

In this chapter, the reference functions concerned with the complex numbers in MATLAB are presented and described. In this regard, several examples and exercises for each section of the chapter are presented. The exercises that include writing the codes, executing them, and achieving the results need to be done by students to master programming skills. In this book, the codes, outputs, and descriptions are in blue, black, and green colors, respectively. To program in MATLAB, a script file can be created and saved with an appropriate name (e.g., untitled01) in the preferred directory of a computer. The program can be run by clicking on the "Run" available on the top toolbar of the script in MATLAB or calling the script by typing its name in Command Window or in the other scripts.

7.1 The Reference Function "real": Real Quantity

- This reference function returns the real part of each element in an array [1–9].

Example

```
z = 20+14i;
x = real(z)
```

Output

```
x =
    20
```

Description

Real part of a complex scalar.

Example

```
z = [i −1+3i 0.3]
x = real(z)
```

Output

z =

  0.0000 + 1.0000i   −1.0000 + 3.0000i   0.3000 + 0.0000i

x =

  0   −1.0000   0.3000

Description

Real part of a complex vector.

Exercise

Execute the codes below in your computer and write the result(s) in the assigned location(s).

```
z = [2i  1+i;  −3  1−i];
x = real(z)
```

Output

x =

7.2   The Reference Function "imag": Imaginary Quantity

- This reference function returns the imaginary part of each element in an array.

Example

```
z = 20+14i;
x = imag(z)
```

Output

x =
  14

Description

Imaginary part of a complex scalar.

Example

```
z = [i  −1+3i  0.3]
x = imag(z)
```

Output

z =

0.0000 + 1.0000i   −1.0000 + 3.0000i   0.3000 + 0.0000i

x =

1   3   0

Description

Imaginary part of a complex vector.

Exercise

Execute the codes below in your computer and write the result(s) in the assigned location(s).

```
z = [2i  1+i;  −3  1−i];
x = imag(z)
```

Output

x =

7.3   The Reference Function "angle": Phase Angle

- This reference function returns the phase angle (radians) in the interval $[-\pi, \pi]$ for each element of a complex array.
- The phase angle of a complex quantity (radians) in polar form ($z = re^{i\theta}$) is θ. Moreover, the phase angle of a complex quantity (radians) in rectangular form ($z = x + iy$) can be calculated as follows:

$$
\theta_z = \begin{cases}
\tan^{-1}\left|\dfrac{y}{x}\right| & \text{if } x>0, y>0 \\[2mm]
\pi - \tan^{-1}\left|\dfrac{y}{x}\right| & \text{if } x<0, y>0 \\[2mm]
\pi + \tan^{-1}\left|\dfrac{y}{x}\right| & \text{if } x<0, y<0 \\[2mm]
-\tan^{-1}\left|\dfrac{y}{x}\right| & \text{if } x>0, y<0
\end{cases}
$$

Example

```
z = 2*exp(i*0.5);
angle(z)
```

Output

ans =

0.5000

Description

Computing the phase angle of a complex quantity (radians) which is in polar form.

Example

```
z = 1+i;
angle(z)
```

Output

```
ans =
   0.7854
```

Description

Computing the phase angle of a complex quantity (radians) which is in rectangular form.

Exercise

Execute the codes below in your computer and write the result(s) in the assigned location(s).

```
z = exp(i);
angle(z)
```

Output

```
ans =
```

Exercise

Execute the codes below in your computer and write the result(s) in the assigned location(s).

```
z = i;
angle(z)
```

Output

```
ans =
```

7.4   The Reference Function "conj": Complex Conjugate

- This reference function returns the complex conjugate of each element in a complex array.

Example

z = [−i 2+1i; 4+2i −2i]
conj(z)

Output

z =
     0.0000 − 1.0000i   2.0000 + 1.0000i
     4.0000 + 2.0000i   0.0000 − 2.0000i

ans =
     0.0000 + 1.0000i   2.0000 − 1.0000i
     4.0000 − 2.0000i   0.0000 + 2.0000i

Description

Computing the complex conjugate of each element in a complex matrix.

Exercise

Execute the codes below in your computer and write the result(s) in the assigned location(s).

z = [−i 2; 4 i];
conj(z)

Output

ans =

7.5   The Reference Function "isreal": Real Quantity Identifier

- This reference function returns the logical 1 (true) when numeric array A does not have an imaginary part, and logical 0 (false) otherwise.

Example

A = isreal(1)
B = isreal(i)
C = isreal(1+i)
D = isreal(1+0i)

Output

A =
  logical
  1

B =
  logical
  0

C =
  logical
  0

D =
  logical
  1

Description

Applying the reference function on several complex scalars.

Example

E = isreal([1 2 i])

Output

E =
  logical
  0

Description

Applying the reference function on a complex vector.

Exercise

Execute the codes below in your computer and write the result(s) in the assigned location(s).

A = [7 3+4i 2 5i];
isreal(A)

Output

ans =

References

1. MATLAB 2023a.
2. Rahmani-Andebili, M. (2024). *Mathematics of engineering and science – Practice problems, methods, and solutions*. Springer Nature.
3. Rahmani-Andebili, M. (2022). *Differential equations – Practice problems, methods, and solutions*. Springer Nature.
4. Rahmani-Andebili, M. (2023). *Calculus III – Practice problems, methods, and solutions*. Springer Nature.
5. Rahmani-Andebili, M. (2023). *Calculus II – Practice problems, methods, and solutions*. Springer Nature.
6. Rahmani-Andebili, M. (2023). *Calculus I – Practice problems, methods, and solutions* (2nd ed.). Springer Nature.
7. Rahmani-Andebili, M. (2021). *Calculus – Practice problems, methods, and solutions*. Springer Nature.
8. Rahmani-Andebili, M. (2024). *Precalculus – Practice problems, methods, and solutions* (2nd ed.). Springer Nature.
9. Rahmani-Andebili, M. (2021). *Precalculus – Practice problems, methods, and solutions*. Springer Nature.

Arrays and Matrices in MATLAB

8

Abstract

In this chapter, the reference functions concerned with the arrays and matrices in MATLAB are presented and described. In this regard, several examples and exercises for each section of the chapter are presented. The exercises that include writing the codes, executing them, and achieving the results need to be done by students to master programming skills. In this book, the codes, outputs, and descriptions are in blue, black, and green colors, respectively. To program in MATLAB, a script file can be created and saved with an appropriate name (e.g., untitled01) in the preferred directory of a computer. The program can be run by clicking on the "Run" available on the top toolbar of the script in MATLAB or calling the script by typing its name in Command Window or in the other scripts.

8.1 The Reference Function "size": Array Dimensions Identifier

- This reference function returns a row vector whose elements are the dimensions of the array [1–9].

Example

```
V = [1 2 3]
size(V)
```

Output

```
V =
   1   2   3

ans =
   1   3
```

Description

Determining the size of a vector.

Example

```
M = [1 2 3; 4 5 6]
size(M)
```

Output

```
M =
   1    2    3
   4    5    6

ans =
   2    3
```

Description

Determining the size of a matrix.

Example

```
V = []
size(V)
```

Output

```
V =
   []

ans =
   0    0
```

Description

Determining the size of an empty array.

Exercise

Execute the codes below in your computer and write the result(s) in the assigned location(s).

```
M = [1 2 3; 4 5 6]'
size(M)
```

Output

```
M =

ans =
```

8.2   The Reference Function "length": Array Length Identifier

- This reference function returns the length of the largest array dimension.
- For vectors, the length is simply the number of elements.
- The length of an empty array is zero.

Example

V = [1 2 3]
length(V)

Output

V =
   1   2   3

ans =
   3

Description

Determining the length of a vector.

Example

M = [1 2 3; 4 5 6]
length(M)

Output

M =
   1   2   3
   4   5   6

ans =
   3

Description

Determining the length of a matrix.

Example

V = []
length(V)

Output

V =
   []

ans =
   0

Description

Determining the length of an empty array.

Exercise

Execute the codes below in your computer and write the result(s) in the assigned location(s).

M = [1 2 3; 4 5 6]'
length(M)

Output

M =

ans =

8.3   The Reference Function "numel": Elements Number Identifier

- This reference function returns the number of elements in an array A.

Example

M = [1 2 3; 4 5 6]
numel(M)

Output

M =
   1   2   3
   4   5   6

ans =
   6

Description

Determining the number of elements of a matrix.

Exercise

Execute the codes below in your computer and write the result(s) in the assigned location(s).

M = [1 2 3; 4 5 6]'
numel(M)

Output

M =

ans =

8.4   The Reference Function "sort": Elements Sortation (Type 1)

- This reference function sorts the elements of an array in ascending order (by default).
- If the array is a vector, then it sorts the vector elements.
- If the array is a matrix, then it treats the columns of the array as vectors and sorts each column.

Example

A = [9 0 -7 5 3 8 -10 4 2]
sort(A)

Output

A =
    9   0   -7   5   3   8   -10   4   2

ans =
   -10   -7   0   2   3   4   5   8   9

Description

Sorting the elements of a vector in ascending order (by default).

Example

A = [3 6 5; 7 -2 4; 1 0 -9]
sort(A)

Output

```
A =
    3      6      5
    7     -2      4
    1      0     -9

ans =
    1     -2     -9
    3      0      4
    7      6      5
```

Description

Sorting the elements of each column of a matrix in ascending order (by default).

Example

```
A = [3 6 5; 7 -2 4; 1 0 -9]
sort(A,'descend')
```

Output

```
A =
    3      6      5
    7     -2      4
    1      0     -9

ans =
    7      6      5
    3      0      4
    1     -2     -9
```

Description

Sorting the elements of each column of a matrix in descending order.

Exercise

Execute the codes below in your computer and write the result(s) in the assigned location(s).

```
B = [3 6 5; 7 -2 4; 1 0 -9]'
sort(B,'descend')
```

B =

ans =

8.5   The Reference Function "sortrows": Elements Sortation (Type 2)

- This reference function sorts the rows of a matrix in ascending order (by default) based on the elements in the first column.
- When the first column contains repeated elements, it sorts according to the values in the next column and repeats this behavior for succeeding equal values.

A = [3 6 5; 7 -2 4; 1 0 -9]
sortrows(A)

A =
   3   6   5
   7   -2   4
   1   0   -9

ans =
   1   0   -9
   3   6   5
   7   -2   4

Sorting the rows of a matrix in ascending order (by default) based on the elements of the first column.

A = [3 6 5; 7 -2 4; 1 0 -9]
sortrows(A,'descend')

A =
   3   6   5
   7   -2   4
   1   0   -9

ans =
   7   -2   4
   3   6   5
   1   0   -9

Description

Sorting the rows of a matrix in descending order based on the elements of the first column.

Exercise

Execute the codes below in your computer and write the result(s) in the assigned location(s).

A = [9 8 7; 6 5 4; 3 2 1]
sortrows(A)

Output

A =

ans =

8.6   The Reference Function "flip": Elements Order Flipping

- This reference function returns an array of the same size, but with the order of the elements reversed.
- If the array is a vector, then it reverses the order of the elements along the length of the vector.
- If the array is a matrix, then it reverses the elements in each column.

Example

A = [1 3 2]
flip(A)

Output

A =
   1   3   2

ans =
   2   3   1

Description

Applying the reference function on a vector.

Example

A = [3 6 5; 7 -2 4; 1 0 -9]
flip(A)

Output

```
A =
    3     6     5
    7    -2     4
    1     0    -9

ans =
    1     0    -9
    7    -2     4
    3     6     5
```

Description

Applying the reference function on a matrix.

Exercise

Execute the codes below in your computer and write the result(s) in the assigned location(s).

A = [1 0 0; 0 2 0; 0 0 3]
flip(A)

Output

A =

ans =

8.7   The Reference Function "repmat": Repeated Matrix Creation

- This reference function returns an array containing n copies of an array in the row and column dimensions.
- For example, repmat(k,m,n) returns a m-by-n matrix whose elements are k.

Example

A = repmat(1,3,2)

Output

```
A =
    1     1
    1     1
    1     1
```

Description

Creating a 3-by-2 matrix whose elements are 1.

Example

A = [1 2; 3 4]
B = repmat(A,2,2)

Output

A =
   1   2
   3   4

B =
   1   2   1   2
   3   4   3   4
   1   2   1   2
   3   4   3   4

Description

Repeating the copies of a matrix into a 2-by-2 block arrangement.

Exercise

Execute the codes below in your computer and write the result(s) in the assigned location(s).

A = [1 0; 0 1]
B = repmat(A,2,2)

Output

A =

B =

8.8   The Reference Function "nnz": Nonzero Elements Number Identifier

- This reference function returns the number of nonzero elements in a matrix.
- Use it in conjunction with a relational operator to determine how many matrix elements meet a condition. Then, it will count the 1s, where the condition is true because relational operators produce logical matrices of 1s and 0s.

Example

A = [3 6 5; 7 -2 4; 1 0 -9]
nnz(A)

Output

A =
```
   3      6      5
   7     -2      4
   1      0     -9
```

ans =
    8

Description

Determining the number of nonzero elements in a matrix.

Example

A = [3 6 5; 7 -2 4; 1 0 -9];
nnz(A>5)

Output

ans =
    2

Description

Determining the number of elements that are greater than 5.

Exercise

Execute the codes below in your computer and write the result(s) in the assigned location(s).

A = [3 6 5; 7 -2 4; 1 0 -9];
nnz(A==5)

Output

ans =

8.9   The Reference Function "find": Elements' Linear Index Identifier

- This reference function returns a vector containing the linear indices of each nonzero element in an array.
- If the array is a vector, then it returns a vector with the same orientation as the array.
- If the array is a matrix, then it returns a column vector of the linear indices of the result.

Example

A = [1 0 2 4]
find(A)

Output

A =
   1   0   2   4

ans =
   1   3   4

Description

Finding the linear indices of nonzero elements of a vector.

Example

A = [1 0 2; 0 1 1; 0 0 4]
find(A)

Output

A =
   1   0   2
   0   1   1
   0   0   4

ans =
   1
   5
   7
   8
   9

Description

Finding the linear indices of nonzero elements of a matrix.

Example

A = [1 0 2; 0 1 1; 0 0 4]
find(~A)

Output

A =
   1   0   2
   0   1   1
   0   0   4

ans =
   2
   3
   4
   6

Description

Finding the linear indices of zero elements of a matrix.

Exercise

Execute the codes below in your computer and write the result(s) in the assigned location(s).

A = [0 22 2; 11 1 1; 3 5 0]
find(~A)

Output

A =

ans =

8.10   The Reference Function "rank": Matrix Rank Identifier

- This reference function returns the rank of a matrix.
- The number of linearly independent columns (or rows) in a matrix is the rank of the matrix.
- A matrix is called "full rank" if its rank is the highest possible for a matrix of the same size and "rank deficient" if it does not have full rank.
- The rank gives a measure of the dimension of the range or column space of the matrix, which is the collection of all linear combinations of the columns.

Example

A = [3 2 4; -1 1 2; 9 5 10]
rank(A)

Output

A =
```
   3     2     4
  -1     1     2
   9     5    10
```

ans =
   2

Description

Determining the rank of a matrix.

Exercise

Execute the codes below in your computer and write the result(s) in the assigned location(s).

A = [1 2 3; 2 4 6; 3 6 9];
rank(A)

Output

ans =

Exercise

Execute the codes below in your computer and write the result(s) in the assigned location(s).

A = [1 2 3; 2 5 4; 1 3 2];
rank(A)

Output

ans =

8.11   The Reference Function "trace": Diagonal Elements Adder

- This reference function calculates the sum of the diagonal elements of a matrix.

Example

A = [1 2 3; 2 5 4; 1 3 2]
trace(A)

Output

A =
1	2	3
2	5	4
1	3	2

ans =
   8

Description

Computing the trace of the matrix.

Example

A = [1 2 3; 2 5 4; 1 3 2]'
trace(A)

Output

A =
1	2	1
2	5	3
3	4	2

ans =
   8

Description

Computing the trace of the transposed matrix.

Exercise

Execute the codes below in your computer and write the result(s) in the assigned location(s).

A = [0 1 11; 1 0 3; 5 6 0]
trace(A)

8.12   The Reference Function "det": Matrix Determinant

- This reference function returns the determinant of a square matrix.

Example

A = [1 -2 4; -5 2 0; 1 0 3]
det(A)

Output

A =
     1   -2   4
    -5   2   0
     1   0   3

ans =
   -32

Description

Computing the determinant of a square matrix.

Exercise

Execute the codes below in your computer and write the result(s) in the assigned location(s).

A = [1 2 3; 10 20 30; 2 4 6]
det(A)

Output

A =
     1   2   3
    10   20   30
     2   4   6

ans =
    0

8.13   The Reference Function "eig": Matrix Eigenvalues

- This reference function returns a column vector containing the eigenvalues of a square matrix.

Example

A = [1 0.5 0.3 0.2; 0.5 1 0.6 0.5; 0.3 0.6 1 0.7; 0.2 0.5 0.7 1]
eig(A)

Output

A =
    1.0000   0.5000   0.3000   0.2000
    0.5000   1.0000   0.6000   0.5000
    0.3000   0.6000   1.0000   0.7000
    0.2000   0.5000   0.7000   1.0000

ans =
    0.2827
    0.3865
    0.8961
    2.4347

Description

Computing the eigenvalues of a square matrix.

Exercise

Execute the codes below in your computer and write the result(s) in the assigned location(s).

A = [1 0.5 0.3; 0.5 1 0.6; 0.3 0.6 1];
eig(A)

Output

ans =

8.14   The Reference Function "inv": Inverse Matrix

- This reference function computes the inverse of a square matrix.
- Alternatively, A^(-1) can be used to compute the inverse of a square matrix A.

Example

```
A = [1 2; -2 1]
inv(A)
```

Output

```
A =
    1     2
   -2     1

ans =
    0.2000    -0.4000
    0.4000     0.2000
```

Description

Computing the inverse of a square matrix.

Example

```
A = [1 0 2; -1 5 0; 0 3 -9]
inv(A)
```

Output

```
A =
    1     0     2
   -1     5     0
    0     3    -9

ans =
    0.8824    -0.1176     0.1961
    0.1765     0.1765     0.0392
    0.0588     0.0588    -0.0980
```

Description

Computing the inverse of a square matrix.

Exercise

Execute the codes below in your computer and write the result(s) in the assigned location(s).

```
A = [1 1 2; -1 2 -1; 3 3 4]
inv(A)
```

Output

ans =

8.15   The Reference Function "norm": Vector Norm

- This reference function returns the p-norm of a vector v by using norm(v,p), where p is the given norm.
- The general definition for the p-norm of a vector v that has N elements is as follows.

$$\|v\|_p = \left(\sum_{k=1}^{N} |v_k|^p \right)^{\frac{1}{p}}$$

- The 2-norm is also called norm, magnitude, or Euclidean length. It can be calculated for a vector v that has N elements, as follows. Herein, the format of the reference function can be norm(v) or norm(v,2).

$$\|v\| = \sqrt{\sum_{k=1}^{N} |v_k|^2}$$

Example

v = [1 -2 3]
norm(v)

Output

ans =
   3.7417

Description

Computing the 2-norm of a vector.

Example

v = [1 -2 3]
norm(v,1)

Output

ans =
   6

Description

Computing the 1-norm of a vector which is sum of the element magnitudes.

Exercise

Execute the codes below in your computer and write the result(s) in the assigned location(s).

v = [0 -1 2 -2]
norm(v)

Output

ans =

8.16   The Reference Function "exp": Exponential Value

• This reference function returns the element-by-element exponential value of a matrix.

Example

A = [1 1 0; 0 0 2; 0 0 -1]
exp(A)

Output

A =
   1   1   0
   0   0   2
   0   0   -1

ans =
   2.7183   2.7183   1.0000
   1.0000   1.0000   7.3891
   1.0000   1.0000   0.3679

Description

Computing the element-by-element exponential value of a matrix.

Exercise

Execute the codes below in your computer and write the result(s) in the assigned location(s).

A = [1 1 1; 0 1 1; 0 0 1]
exp(A)

Output

A =

ans =

8.17   The Reference Function "log": Natural Logarithmic Value

- This reference function returns the element-by-element natural logarithm value of a matrix.

Example

A = [1 0; 2.7183 Inf]
log(A)

Output

A =
   1.0000   0
   2.7183   Inf

ans =
   0   -Inf
   1.0000   Inf

Description

Computing the element-by-element natural logarithm value of a matrix.

Exercise

Execute the codes below in your computer and write the result(s) in the assigned location(s).

A = [1 2; 2.71828 3];
log(A)

Output

ans =

8.18   The Reference Function "cross": Matrices Cross Product

- This reference function returns the cross product of two arrays.
- If the arrays are vectors, then they must have a length of 3. The result is a vector that is perpendicular to both vectors.
- If the arrays are matrices, then they must have the same size. In this case, the function treats them as collections of three-element vectors.

Example

A = [4 -2 1];
B = [1 -1 3];
C = cross(A,B)

Output

C =
    -5   -11   -2

Description

The cross product of two vectors.

Example

A = [1 1 1; 0 1 1; 0 0 1]
B = [0 1 11; 1 0 3; 5 6 0]
C = cross(A,B)

Output

A =
    1   1   1
    0   1   1
    0   0   1

B =
    0   1   11
    1   0   3
    5   6   0

C =
    0   6   -3
   -5   -6   11
    1   -1   -8

Description

The cross product of two matrices.

Exercise

Execute the codes below in your computer and write the result(s) in the assigned location(s).

A = [1 1 1; 0 1 1; 0 0 1];
B = [0 1 11; 1 0 3; 5 6 0];
C = cross(B,A)

8.19   The Reference Function "dot": Matrices Dot Product

- This reference function returns the dot product of two arrays.
- If the arrays are vectors, then they must have the same length. The result is a scalar.
- If the arrays are matrices, then they must have the same size. In this case, the function treats them as collections of vectors.

Example

A = [4 -2 1];
B = [1 -1 3];
C = dot(A,B)

Output

C =
   9

Description

The dot product of two vectors.

Example

A = [1+i 1-i -1+i -1-i];
B = [3-4i 6-2i 1+2i 4+3i];
C = dot(A,B)

Output

C =
   1.0000 − 5.0000i

Description

The dot product of two vectors including complex quantities.

Example

A = [1 1 1; 0 1 1; 0 0 1]
B = [0 1 11; 1 0 3; 5 6 0]
C = dot(A,B)

Output

A =
```
   1    1    1
   0    1    1
   0    0    1
```

B =
```
   0    1   11
   1    0    3
   5    6    0
```

C =
```
   0    1   14
```

Description

The dot product of two matrices.

Exercise

Execute the codes below in your computer and write the result(s) in the assigned location(s).

A = [1 1 1; 0 1 1; 0 0 1];
B = [0 1 11; 1 0 3; 5 6 0];
C = dot(B,A)

8.20   The Reference Function "kron": Kronecker Tensor Product

- This reference function returns the Kronecker tensor product of two matrices.
- In kron(A,B), if A is an m-by-n matrix and B is a p-by-q matrix, then kron(A,B) is an m*p-by-n*q matrix formed by taking all possible products between the elements of A and the matrix B.

Example

A = [1 0 0; 0 1 0; 0 0 1]
B = [1 -1;-1 1]
kron(A,B)

Output

A =
```
   1    0    0
   0    1    0
   0    0    1
```

B =
```
   1   -1
  -1    1
```

ans =
```
   1   -1    0    0    0    0
  -1    1    0    0    0    0
   0    0    1   -1    0    0
   0    0   -1    1    0    0
   0    0    0    0    1   -1
   0    0    0    0   -1    1
```

Description

Creating a block diagonal matrix by using the reference function.

Example

A = [1 2 3; 4 5 6]
B = ones(2)
kron(A,B)

Output

A =
```
   1    2    3
   4    5    6
```

B =
```
   1    1
   1    1
```

ans =
```
   1    1    2    2    3    3
   1    1    2    2    3    3
   4    4    5    5    6    6
   4    4    5    5    6    6
```

Description

Expanding the size of a matrix by using the reference function.

8.21   The Reference Function "tril": Matrix's Lower Triangular Identifier

- This reference function in the format tril(A) returns the lower triangular portion of a matrix.
- This reference function in the format tril(A,k) returns the elements on and below the kth diagonal of a matrix.

Example

A = [1 1 1 1; 1 1 1 1; 1 1 1 1; 1 1 1 1]
tril(A)

Output

A =
1	1	1	1
1	1	1	1
1	1	1	1
1	1	1	1

ans =
1	0	0	0
1	1	0	0
1	1	1	0
1	1	1	1

Description

Extracting the lower triangular portion of the matrix.

Example

A = [1 1 1 1; 1 1 1 1; 1 1 1 1; 1 1 1 1]
tril(A,2)

Output

A =
1	1	1	1
1	1	1	1
1	1	1	1
1	1	1	1

ans =
1	1	1	0
1	1	1	1
1	1	1	1
1	1	1	1

Description

Extracting the elements on and below the 2nd diagonal of the matrix.

Example

A = [1 1 1 1; 1 1 1 1; 1 1 1 1; 1 1 1 1]
tril(A,-2)

Output

A =
1	1	1	1
1	1	1	1
1	1	1	1
1	1	1	1

ans =
0	0	0	0
0	0	0	0
1	0	0	0
1	1	0	0

Description

Extracting the elements on and below the -2'th diagonal of the matrix.

Exercise

Execute the codes below in your computer and write the result(s) in the assigned location(s).

A = [1 2 3; 4 5 6; 7 8 9]
tril(A)

Output

A =

ans =

8.22   The Reference Function "triu": Matrix's Upper Triangular Identifier

- This reference function in the format tril(A) returns the upper triangular portion of a matrix.
- This reference function in the format tril(A,k) returns the elements on and above the kth diagonal of a matrix.

Example

A = [1 1 1 1; 1 1 1 1; 1 1 1 1; 1 1 1 1]
triu(A)

Output

```
A =
    1    1    1    1
    1    1    1    1
    1    1    1    1
    1    1    1    1

ans =
    1    1    1    1
    0    1    1    1
    0    0    1    1
    0    0    0    1
```

Description

Extracting the upper triangular portion of the matrix.

Example

```
A = [1 1 1 1; 1 1 1 1; 1 1 1 1; 1 1 1 1]
triu(A,2)
```

Output

```
A =
    1    1    1    1
    1    1    1    1
    1    1    1    1
    1    1    1    1

ans =
    0    0    1    1
    0    0    0    1
    0    0    0    0
    0    0    0    0
```

Description

Extracting the elements on and above the 2nd diagonal of the matrix.

Example

```
A = [1 1 1 1; 1 1 1 1; 1 1 1 1; 1 1 1 1]
triu(A,-2)
```

Output

```
A =
   1   1   1   1
   1   1   1   1
   1   1   1   1
   1   1   1   1


ans =
   1   1   1   1
   1   1   1   1
   1   1   1   1
   0   1   1   1
```

Description

Extracting the elements on and above the -2'th diagonal of the matrix.

Exercise

Execute the codes below in your computer and write the result(s) in the assigned location(s).

A = [1 2 3; 4 5 6; 7 8 9]
triu(A)

Output

A =

ans =

8.23   The Reference Function "linsolve": Matrix Equation Solver

• This reference function solves the matrix equation AX = B.

Example

A = [2 1 1; -1 1 -1;1 2 3]
B = [2; 3; -10]
X = linsolve(A,B)

Output

```
A =
    2    1    1
   -1    1   -1
    1    2    3
```

B =
     2
     3
   -10

X =
     3
     1
    -5

Description

Solving the system of linear equations.

Exercise

Execute the codes below in your computer and write the result(s) in the assigned location(s).

A = [2 1; -1 1]
B = [2; 3]
X = linsolve(A,B)

Output

A =

B =

X =

8.24   The Reference Function "zeros": Array of All Zeros

- This reference function in the format X = zeros returns the scalar 0.
- This reference function in the format X = zeros(n) returns an n-by-n matrix of zeros.
- This reference function in the format X = zeros(sz) returns an array of zeros, where the size vector, sz, defines size(X).

Example

X = zeros(3)

Output

X =
     0   0   0
     0   0   0
     0   0   0

Description

Creating a 3-by-3 matrix of zeros.

Example

X = zeros([2 3])

Output

X =

    0   0   0
    0   0   0

Description

Creating a 2-by-3 matrix of zeros.

Exercise

Execute the codes below in your computer and write the result(s) in the assigned location(s).

Y = zeros(2)

Output

Y =

8.25   The Reference Function "ones": Array of All Ones

- This reference function in the format X = ones returns the scalar 1.
- This reference function in the format X = ones(n) returns an n-by-n matrix of ones.
- This reference function in the format X = ones(sz) returns an array of ones, where the size vector, sz, defines size(X).

Example

X = ones(3)

Output

X =

    1   1   1
    1   1   1
    1   1   1

Description

Creating a 3-by-3 matrix of ones.

Example

X = ones([2 3])

Output

X =
   1   1   1
   1   1   1

Description

Creating a 2-by-3 matrix of ones.

Exercise

Execute the codes below in your computer and write the result(s) in the assigned location(s).

Y = ones(2)

Output

Y =

8.26 The Reference Function "rand": Uniformly Distributed Random Numbers

- This reference function in the format X = rand returns a random scalar drawn from the uniform distribution in the interval (0,1).
- This reference function in the format X = rand(n) returns an n-by-n matrix of uniformly distributed random numbers.
- This reference function in the format X = rand(sz) returns an array of random numbers, where the size vector, sz, defines size(X).
- This reference function in the format

Example

X = rand

Output

X =
   0.8147

Description

Creating a random number drawn from the uniform distribution in the interval (0,1).

Example

X = rand(3)

Output

X =
   0.9058   0.6324   0.5469
   0.1270   0.0975   0.9575
   0.9134   0.2785   0.9649

Description

Creating a 3-by-3 matrix of random numbers drawn from the uniform distribution in the interval (0,1).

Example

X = rand([2 3])

Output

X =
   0.1576   0.9572   0.8003
   0.9706   0.4854   0.1419

Description

Creating a 2-by-3 matrix of random numbers drawn from the uniform distribution in the interval (0,1).

Exercise

Execute the codes below in your computer and write the result(s) in the assigned location(s).

Y = rand(2)

Output

Y =

8.27   The Reference Function "meshgrid": 2-D and 3-D Grids

- This reference function in the format [X,Y] = meshgrid(x,y) returns 2-D grid coordinates based on the coordinates contained in vectors x and y. X is a matrix, where each row is a copy of x, and Y is a matrix, where each column is a copy of y. The grid represented by the coordinates X and Y has length(y) rows and length(x) columns.
- This reference function in the format [X,Y] = meshgrid(x) is the same as [X,Y] = meshgrid(x,x), returning square grid coordinates with grid size length(x)-by-length(x).
- This reference function in the format [X,Y,Z] = meshgrid(x,y,z) returns 3-D grid coordinates defined by the vectors x, y, and z. The grid represented by X, Y, and Z has size length(y)-by-length(x)-by-length(z).
- This reference function in the format [X,Y,Z] = meshgrid(x) is the same as [X,Y,Z] = meshgrid(x,x,x), returning 3-D grid coordinates with grid size length(x)-by-length(x)-by-length(x).

Example

```
x = 1:3;
y = 1:5;
[X,Y] = meshgrid(x,y)
```

Output

X =
```
   1    2    3
   1    2    3
   1    2    3
   1    2    3
   1    2    3
```

Y =
```
   1    1    1
   2    2    2
   3    3    3
   4    4    4
   5    5    5
```

Description

Creating 2-D grid coordinates with x-coordinates defined by the vector x, and y-coordinates defined by the vector y.

Example

```
x = 1:3;
[X,Y] = meshgrid(x)
```

Output

X =

1	2	3
1	2	3
1	2	3

Y =

1	1	1
2	2	2
3	3	3

Description

Creating 2-D grid coordinates with x-coordinates defined by the vector x.

Example

```
x = 1:3;
y = 1:5;
z = 7:8;
[X,Y,Z] = meshgrid(x,y,z)
```

Output

X(:,:,1) =

1	2	3
1	2	3
1	2	3
1	2	3
1	2	3

X(:,:,2) =

1	2	3
1	2	3
1	2	3
1	2	3
1	2	3

Y(:,:,1) =

1	1	1
2	2	2
3	3	3
4	4	4
5	5	5

Y(:,:,2) =

1	1	1
2	2	2
3	3	3
4	4	4
5	5	5

Z(:,:,1) =

7	7	7
7	7	7
7	7	7
7	7	7
7	7	7

Z(:,:,2) =

8	8	8
8	8	8
8	8	8
8	8	8
8	8	8

Description

Creating 3-D grid coordinates with x-coordinates defined by the vector x, y-coordinates defined by the vector y, and z-coordinates defined by the vector z.

Exercise

Execute the codes below in your computer and write the result(s).

```
x = 1:3;
y = 1:5;
z = 7:8;
[X,Y,Z] = meshgrid(x)
```

References

1. MATLAB 2023a.
2. Rahmani-Andebili, M. (2024). *Mathematics of engineering and science – Practice problems, methods, and solutions*. Springer Nature.
3. Rahmani-Andebili, M. (2022). *Differential equations – Practice problems, methods, and solutions*. Springer Nature.
4. Rahmani-Andebili, M. (2023). *Calculus III – Practice problems, methods, and solutions*. Springer Nature.
5. Rahmani-Andebili, M. (2023). *Calculus II – Practice problems, methods, and solutions*. Springer Nature.
6. Rahmani-Andebili, M. (2023). *Calculus I – Practice problems, methods, and solutions* (2nd ed.). Springer Nature.
7. Rahmani-Andebili, M. (2021). *Calculus – Practice problems, methods, and solutions*. Springer Nature.
8. Rahmani-Andebili, M. (2024). *Precalculus – Practice problems, methods, and solutions* (2nd ed.). Springer Nature.
9. Rahmani-Andebili, M. (2021). *Precalculus – Practice problems, methods, and solutions*. Springer Nature.

Abstract

In this chapter, the reference functions concerned with the descriptive statistics in MATLAB are presented and described. In this regard, several examples and exercises for each section of the chapter are presented. The exercises that include writing the codes, executing them, and achieving the results need to be done by students to master programming skills. In this book, the codes, outputs, and descriptions are in blue, black, and green colors, respectively. To program in MATLAB, a script file can be created and saved with an appropriate name (e.g., untitled01) in the preferred directory of a computer. The program can be run by clicking on the "Run" available on the top toolbar of the script in MATLAB or calling the script by typing its name in Command Window or in the other scripts.

9.1 The Reference Function "sum": Matrix Elements Adder

- The reference function in the format sum(A) returns sum of the elements of A if A is a vector [1].
- The reference function in the format sum(A) returns a row vector containing the sum of each column if A is a matrix.
- The reference function in the format sum(A,"all") returns the sum of all elements of A.

Example

```
A = [1 2 3];
sum(A)
```

Output

```
ans =
   6
```

Description

Applying the reference function on a vector.

Example

A = [1 2 3; 4 5 6; 7 8 9]
B = sum(A)
C = sum(A,"all")

Output

A =

1	2	3
4	5	6
7	8	9

B =
   12   15   18

C =
  45

Description

Applying the reference function on a matrix.

Exercise

Execute the codes below in your computer and write the result(s) in the assigned location(s).

A = [1 2 3; 4 5 6; 7 8 9]'
B = sum(A)
C = sum(A,"all")

Output

A =

B =

C =

9.2   The Reference Function "prod": Matrix Elements Multiplier

- The reference function in the format prod(A) returns product of the elements of A if A is a vector.
- The reference function in the format prod(A) returns a row vector of the products of each column if A is a matrix.
- The reference function in the format prod(A,"all") returns the product of all elements of A.

Example

```
A = [1 2 3];
prod(A)
```

Output

```
ans =
    6
```

Description

Applying the reference function on a vector.

Example

```
A = [1 2 3; 4 5 6; 7 8 9]
B = prod(A)
C = prod(A,"all")
```

Output

```
A =
    1    2    3
    4    5    6
    7    8    9

B =
    28    80    162

C =
    362880
```

Description

Applying the reference function on a matrix.

Exercise

Execute the codes below in your computer and write the result(s) in the assigned location(s).

```
A = [1 2 3; 4 5 6; 7 8 9]'
B = prod(A)
C = prod(A,"all")
```

Output

A =

B =

C =

9.3   The Reference Function "max": Maximum Element Identifier

- If this reference function is applied on a vector, it returns the maximum element of the vector.
- If this reference function is applied on a matrix, it returns a row vector containing the maximum value of each column of A.

Example

A = [1 2 3];
max(A)

Output

ans =
   3

Description

Applying the reference function on a vector.

Example

A = [-2+2i 4+i -1-3i];
max(A)

Output

ans =
   4.0000 + 1.0000i

Description

Applying the reference function on a vector including complex quantities. As can be seen, the element with the maximum magnitude is selected.

Example

A = [1 2 3; 4 5 6; 7 8 9]
max(A)

Output

A =
 1   2   3
 4   5   6
 7   8   9

ans =
 7   8   9

Description

Applying the reference function on a matrix.

Exercise

Execute the codes below in your computer and write the result(s) in the assigned location(s).

A = [23 5 3; 4 11 6];
max(A)

Output

ans =

9.4   The Reference Function "min": Minimum Element Identifier

- If this reference function is applied on a vector, it returns the minimum element of the vector.
- If this reference function is applied on a matrix, it returns a row vector containing the minimum value of each column of A.

Example

A = [1 2 3];
min(A)

Output

ans =
 1

Description

Applying the reference function on a vector.

Example

```
A = [-2+2i 4+i -1-3i];
min(A)
```

Output

```
ans =
   -2.0000 + 2.0000i
```

Description

Applying the reference function on a vector including complex quantities. As can be seen, the element with the minimum magnitude is selected.

Example

```
A = [1 2 3; 4 5 6; 7 8 9]
min(A)
```

Output

```
A =
   1   2   3
   4   5   6
   7   8   9

ans =
   1   2   3
```

Description

Applying the reference function on a matrix.

Exercise

Execute the codes below in your computer and write the result(s) in the assigned location(s).

```
A = [23 5 3; 4 11 6];
min(A)
```

Output

ans =

9.5   The Reference Function "bounds": Maximum and Minimum Elements Identifier

- The reference function in the format [a,b]=bounds(A) returns the minimum and maximum values of A if it is a vector.
- The reference function in the format [a,b]=bounds(A) returns two row vectors containing the minimum and maximum values of columns of A if it is a matrix.
- The reference function in the format [a,b]= bounds(A,"all") returns the minimum and maximum values of all elements of A.

Example

A = [1 5 -1 6 3 7 0 8];
[a,b]=bounds(A)

Output

a =
  -1

b =
  8

Description

Applying the first format of the reference function on a vector.

Example

A = [1 5 -1; 6 3 7; 12 0 8]
[a,b]=bounds(A)

Output

A =
   1   5   -1
   6   3   7
  12   0   8

a =
   1   0   -1

b =
  12   5   8

Description

Applying the first format of the reference function on a matrix.

Example

A = [1 5 -1; 6 3 7; 12 0 8]
[a,b]= bounds(A,"all")

Output

```
A =
     1     5    -1
     6     3     7
    12     0     8

a =
    -1

b =
    12
```

Description

Applying the second format of the reference function on a matrix.

Exercise

Execute the codes below in your computer and write the result(s) in the assigned location(s).

A = [1 5 -1 6; 3 7 12 0; 11 45 -2 8; -5 -1 6 0]
[a,b]= bounds(A)
[c,d]= bounds(A,"all")

Output

A =

a =

b =

c =

d =

9.6   The Reference Function "mean": Mean

- This reference function in the format mean(A) returns the mean of the elements of A if it is a vector.
- This reference function in the format mean(A) returns a row vector containing the mean of each column of A if it is a matrix.
- This reference function in the format mean(A,"all") returns the mean over all elements of A.

Example

A = [1 5 -1 6 3 7 0 8];
mean(A)

Output

ans =
   3.6250

Description

Applying the first format of the reference function on a vector.

Example

A = [1 5 -1; 6 3 7; 12 0 8]
mean(A)

Output

A =
    1   5   -1
    6   3   7
   12   0   8

ans =
   6.3333   2.6667   4.6667

Description

Applying the first format of the reference function on a matrix.

Example

A = [1 5 -1; 6 3 7; 12 0 8]
mean(A,"all")

Output

A =
```
    1     5    -1
    6     3     7
   12     0     8
```

ans =
```
  4.5556
```

Description

Applying the second format of the reference function on a matrix.

Exercise

Execute the codes below in your computer and write the result(s) in the assigned location(s).

```
A = [1 5 -1 6; 3 7 12 0; 11 45 -2 8; -5 -1 6 0];
a = mean(A)
b = mean(A,"all")
```

Output

a =

b =

9.7   The Reference Function "median": Median Identifier

- This reference function in the format mean(A) returns the median of the elements of A if it is a vector.
- This reference function in the format mean(A) returns a row vector containing the median of each column of A if it is a matrix.
- This reference function in the format mean(A,"all") returns the median over all elements of A.

Example

```
A = [1 5 -1 6 3 7 0 8];
median(A)
```

Output

ans =
```
   4
```

Description

Applying the first format of the reference function on a vector.

Example

A = [1 5 -1; 6 3 7; 12 0 8]
median(A)

Output

A =
```
    1     5    -1
    6     3     7
   12     0     8
```

ans =
```
    6     3     7
```

Description

Applying the first format of the reference function on a matrix.

Example

A = [1 5 -1; 6 3 7; 12 0 8]
median(A,"all")

Output

A =
```
    1     5    -1
    6     3     7
   12     0     8
```

ans =
```
    5
```

Description

Applying the second format of the reference function on a matrix.

Exercise

Execute the codes below in your computer and write the result(s) in the assigned location(s).

A = [1 5 -1 6; 3 7 12 0; 11 45 -2 8; -5 -1 6 0];
a = median(A)
b = median(A,"all")

Output

a =

b =

9.8   The Reference Function "mode": Mode Identifier

- This reference function in the format mode(A) returns the mode (the most frequently occurring value) of the elements of A if it is a vector. Herein, when there are multiple values occurring equally frequently, mode returns the smallest of those values.
- This reference function in the format mode(A) returns a row vector containing the mode of each column of A if it is a matrix.
- This reference function in the format mode(A,"all") returns the mode over all elements of A.

Example

A = [3 3 1 4 3]
mode(A)

Output

ans =
   3

Description

Applying the first format of the reference function on a vector.

Example

A = [3 3 1 4; 0 0 1 1; 0 1 2 4]
mode(A)

Output

A =

3	3	1	4
0	0	1	1
0	1	2	4

ans =

0	0	1	4

Description

Applying the first format of the reference function on a matrix.

Example

A = [3 3 1 4; 0 0 1 1; 0 1 2 4]
mode(A,"all")

Output

A =

3	3	1	4
0	0	1	1
0	1	2	4

ans =

1

Description

Applying the second format of the reference function on a matrix.

Exercise

Execute the codes below in your computer and write the result(s) in the assigned location(s).

A = [1 5 -1 6; 3 7 6 0; 11 45 6 8; 1 -1 6 0]
a = mode(A)
b = mode(A,"all")

Output

A =

a =

b =

9.9   The Reference Function "std": Standard Deviation

- If this reference function is applied on a scalar, it returns 0.
- This reference function returns the standard deviation of the elements of a vector.
- This reference function returns a row vector containing the standard deviation of each column of a matrix.

Example

A = 2;
std(A)

Output

ans =
    0

Description

Applying the reference function on a scalar.

Example

A = [4 -5 1];
std(A)

Output

ans =
    4.5826

Description

Applying the reference function on a vector.

Example

A = [4 -5 1; 2 3 5; -9 1 7]
std(A)

Output

A =
     4   -5   1
     2   3   5
    -9   1   7

ans =
    7.0000   4.1633   3.0551

Description

Applying the reference function on a matrix.

Exercise

Execute the codes below in your computer and write the result(s) in the assigned location(s).

A = [1 5; 3 7; -9 2]
std(A)

Output

A =

ans =

9.10   The Reference Function "var": Variance

- If this reference function is applied on a scalar, it returns 0.
- This reference function returns the variance of the elements of a vector.
- This reference function returns a row vector containing the variance of each column of a matrix.

Example

A = 5;
var(A)

Output

ans =
   0

Description

Applying the reference function on a scalar.

Example

A = [4 -5 1];
var(A)

Output

ans =
   21

Description

Applying the reference function on a vector.

Example

A = [4 -5 1; 2 3 5; -9 1 7]
var(A)

Output

A =
```
    4    -5     1
    2     3     5
   -9     1     7
```

ans =
```
   49.0000    17.3333     9.3333
```

Description

Applying the reference function on a matrix.

Exercise

Execute the codes below in your computer and write the result(s) in the assigned location(s).

A = [1 5; 3 7; -9 2]
var(A)

Output

A =

ans =

9.11   The Reference Function "cov": Covariance

- If this reference function in the format cov(A) is applied on a scalar, it returns 0.
- If this reference function in the format cov(A) is applied on a vector of observations, then the result is the scalar-valued variance.
- If this reference function in the format cov(A) is applied on a matrix whose columns represent random variables and whose rows represent observations, then the result is the covariance matrix with the corresponding column variances along the diagonal.
- If this reference function in the format cov(A,B) is applied on two scalars, it returns a 2-by-2 block of zeros.
- If this reference function in the format cov(A,B) is applied on two vectors of observations with equal length, then the result is the 2-by-2 covariance matrix.
- If this reference function in the format cov(A,B) is applied on two matrices (with the same size), then it treats the matrices as vectors and is equivalent to cov(A(:),B(:)).

Example

A = 6;
cov(A)

Output

ans =
  0

Description

Applying the reference function on a scalar.

Example

A = [4 -5 1];
cov(A)

Output

ans =
  21

Description

Applying the reference function on a vector.

Example

A = [4 -5 1; 2 3 5; -9 1 7]
cov(A)

Output

A =
   4   -5   1
   2   3   5
  -9   1   7

ans =
   49.0000   -12.0000   -18.0000
  -12.0000   17.3333   10.6667
  -18.0000   10.6667   9.3333

Description

Applying the reference function on a matrix.

Example

A = 5;
B = 6;
cov(A,B)

Output

ans =
   0   0
   0   0

Description

Applying the reference function on two scalars.

Example

A = [4 -5 1];
B = [1 2 3];
cov(A,B)

Output

ans =
   21.0000   -1.5000
   -1.5000   1.0000

Description

Applying the reference function on two vectors.

Example

A = [4 -5 1; 2 3 5; -9 1 7]
B = [7 2 5; 6 4 2; -1 7 8]
cov(A,B)

Output

```
A =
    4    -5    1
    2     3    5
   -9     1    7

B =
    7     2    5
    6     4    2
   -1     7    8

ans =
   25.2500    11.1250
   11.1250     8.7778
```

Description

Applying the reference function on two matrices.

Exercise

Execute the codes below in your computer and write the result(s) in the assigned location(s).

```
A = [2 0 -9; 3 4 1];
B = [5 2 6; -4 4 9];
cov(A,B)
```

Output

```
ans =
```

9.12   The Reference Function "corrcoef": Correlation Coefficients

- If this reference function in the format corrcoef(A) is applied on a matrix A, it returns the matrix of correlation coefficients for A, where the columns of A represent random variables, and the rows represent observations.
- If this reference function in the format corrcoef(A,B) is applied on two matrices A and B, it returns the correlation coefficients between them.

Example

```
A = [4 -5 1; 2 3 5; -9 1 7]
corrcoef(A)
```

Output

```
A =
    4     -5      1
    2      3      5
   -9      1      7


ans =
    1.0000     -0.4118     -0.8417
   -0.4118      1.0000      0.8386
   -0.8417      0.8386      1.0000
```

Description

Applying the reference function on a matrix.

Example

A = [4 -5 1; 2 3 5; -9 1 7]
B = [1 2 3; 4 5 6; 7 8 9]
corrcoef(A,B)

Output

```
A =
    4     -5      1
    2      3      5
   -9      1      7


B =
    1      2      3
    4      5      6
    7      8      9


ans =
    1.0000     0.0462
    0.0462     1.0000
```

Description

Applying the reference function on two matrices.

Exercise

Execute the codes below in your computer and write the result(s) in the assigned location(s).
A = [4 -5 1; 2 3 5; -9 1 7]
B = [1 2 3; 4 5 6; 7 8 9]
corrcoef(B,A)

Output

ans =

9.13   The Reference Function "cumsum": Cumulative Adder

- If this reference function is applied on a vector, then the result is a vector of the same size containing the cumulative sum of the vector.
- If this reference function is applied on a matrix, then the result is a matrix of the same size containing the cumulative sum in each column of the matrix.

Example

A = [4 -5 1];
cumsum(A)

Output

ans =
    4   -1   0

Description

Applying the reference function on a vector.

Example

A = [4 -5 1; 2 3 5; -9 1 7]
cumsum(A)

Output

A =
    4   -5   1
    2   3   5
   -9   1   7

ans =
    4   -5   1
    6   -2   6
   -3   -1   13

Description

Applying the reference function on a matrix.

Exercise

Execute the codes below in your computer and write the result(s) in the assigned location(s).

A = [1 5; 3 7; -9 2]
cumsum(A)

Output

A =

ans =

9.14   The Reference Function "cumprod": Cumulative Multiplier

- If this reference function is applied on a vector, then the result is a vector of the same size containing the cumulative product of the vector.
- If this reference function is applied on a matrix, then the result is a matrix of the same size containing the cumulative product in each column of the matrix.

Example

A = [4 -5 1];
cumprod(A)

Output

ans =
    4   -20   -20

Description

Applying the reference function on a vector.

Example

A = [4 -5 1; 2 3 5; -9 1 7]
cumprod(A)

Output

```
A =
    4    -5    1
    2     3    5
   -9     1    7

ans =
     4     -5     1
     8    -15     5
   -72    -15    35
```

Description

Applying the reference function on a matrix.

Exercise

Execute the codes below in your computer and write the result(s) in the assigned location(s).

A = [1 5; 3 7; -9 2]
cumprod(A)

Output

A =

ans =

9.15   The Reference Function "cummax": Cumulative Maxima Identifier

- If this reference function is applied on a vector, then the result is a vector of the same size containing the cumulative maxima of the vector.
- If this reference function is applied on a matrix, then the result is a matrix of the same size containing the cumulative maxima in each column of the matrix.

Example

A = [4 -5 1 5 11];
cummax(A)

Output

```
ans =
    4    4    4    5    11
```

Description

Applying the reference function on a vector.

Example

A = [-5 4 6; 2 5 3; 1 7 -9]
cummax(A)

Output

A =
```
   -5    4    6
    2    5    3
    1    7   -9
```

ans =
```
   -5    4    6
    2    5    6
    2    7    6
```

Description

Applying the reference function on a matrix.

Exercise

Execute the codes below in your computer and write the result(s) in the assigned location(s).

A = [1 -1 5; 3 7 6; -9 2 9]
cummax(A)

Output

A =

ans =

9.16   The Reference Function "cummin": Cumulative Minima Identifier

- If this reference function is applied on a vector, then the result is a vector of the same size containing the cumulative minima of the vector.
- If this reference function is applied on a matrix, then the result is a matrix of the same size containing the cumulative minima in each column of the matrix.

Example

A = [4 -5 1 -11 11];
cummin(A)

Output

ans =
    4   -5   -5   -11   -11

Description

Applying the reference function on a vector.

Example

A = [-5 4 6; -20 5 3; 1 7 -9]
cummin(A)

Output

A =
    -5   4   6
   -20   5   3
     1   7   -9

ans =
    -5   4   6
   -20   4   3
   -20   4   -9

Description

Applying the reference function on a matrix.

Exercise

Execute the codes below in your computer and write the result(s) in the assigned location(s).

A = [1 -1 5; 3 7 4; -9 2 9]
cummin(A)

Output

A =

ans =

9.17   The Reference Function "smoothdata": Noisy Data Smoother

- This reference function returns a moving average of the elements of a vector using a fixed window length that is determined heuristically. The window slides down the length of the vector, computing an average over the elements within each window.
- If the reference function is applied on a matrix, then it computes the moving average down each column of the matrix.

Example

```
x = 1:100;
A = cos(2*pi*0.05*x+2*pi*rand) + 0.5*randn(1,100);
B = smoothdata(A);
plot(x,A)
hold on
plot(x,B)
legend("Input Data","Smoothed Data")
```

Description

Creating a vector containing noisy data, and then smoothing the data with a moving average. See Fig. 9.1.

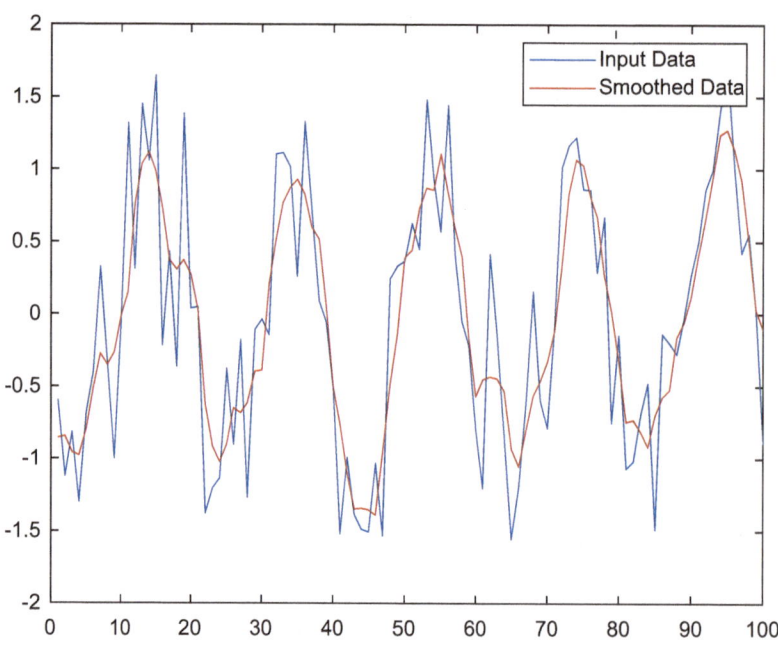

Fig. 9.1 The noisy data and the data smoothed with a moving average

9.18   The Reference Function "detrend": Best Straight-Fit Line Remover

- This reference function removes the best straight-fit line from the data and returns the remaining data.
- If the reference function is applied on a vector, then detrend subtracts the trend from the elements of the vector.
- If the reference function is applied on a matrix, then detrend operates on each column separately, subtracting each trend from the corresponding column of the matrix.

Example

```
t = 0:20;
A = 3*sin(t) + t;
D = detrend(A);
plot(t,A)
hold on
plot(t,D)
plot(t,A-D,":k")
legend("Input Data","Detrended Data","Trend")
```

Description

Creating a vector of data and removing the continuous linear trend. See Fig. 9.2.

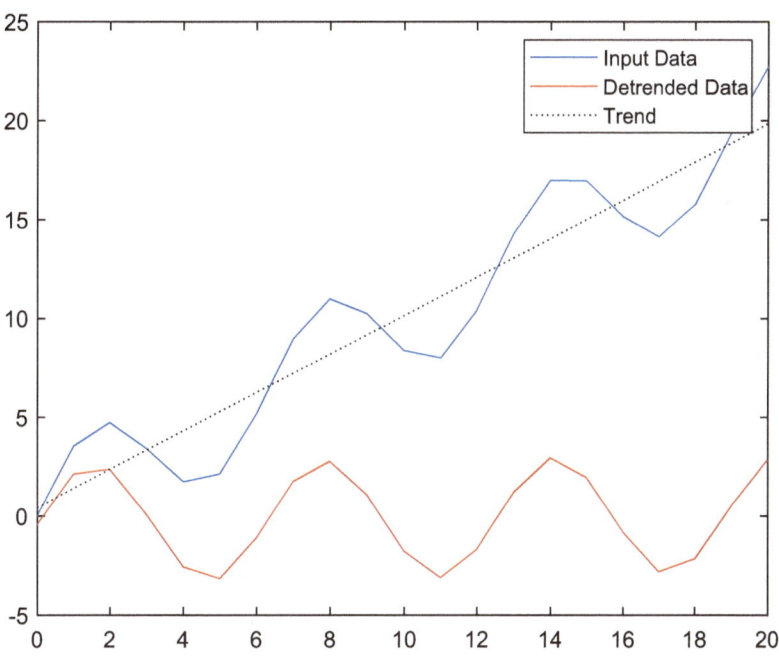

Fig. 9.2 The input data, detrended data, and trend line

9.19   The Reference Function "isoutlier": Outlier Identifier

- If this reference function in the format isoutlier(A) is applied on a vector A, then it returns a logical array whose elements are true when an outlier is detected in the corresponding element. By default, an outlier is a value that is more than three scaled median absolute deviations (MAD) from the median.
- If this reference function in the format isoutlier(A) is applied on a matrix A, then it operates on each column of A separately.
- If this reference function in the format isoutlier(A,method) is applied on a vector or a matrix A, then it specifies a method for detecting outliers. For example, isoutlier(A,"mean") returns true for all elements more than three standard deviations from the mean.

Example

A = [57 59 60 100 59 58 57 58 300 61 62 60 62 58 57];
isoutlier(A)

Output

ans =
  1×15 logical array
  0 0 0 1 0 0 0 0 1 0 0 0 0 0 0

Description

Finding the outliers in a vector of data by using the first format of the reference function. A logical 1 in the output indicates the location of an outlier.

Example

A = [57 59 60 100 59 58 57 58 300 61 62 60 62 58 57];
isoutlier(A,"mean")

Output

ans =
  1×15 logical array
  0 0 0 0 0 0 0 0 1 0 0 0 0 0 0

Description

Finding the outliers in a vector of data by using the second format of the reference function. A logical 1 in the output indicates the location of an outlier.

Exercise

Execute the codes below in your computer and write the result(s) in the assigned location(s).

A = [3 2 4 6 7 2 29 4 6 4 1 9 7 1 1 6 4 9 33];
isoutlier(A)

Output

ans =

Reference

1. MATLAB 2023a.

Abstract

In this chapter, the reference functions concerned with the trigonometric functions in MATLAB are presented and described. In this regard, several examples and exercises for each section of the chapter are presented. The exercises that include writing the codes, executing them, and achieving the results need to be done by students to master programming skills. In this book, the codes, outputs, and descriptions are in blue, black, and green colors, respectively. To program in MATLAB, a script file can be created and saved with an appropriate name (e.g. untitled01) in the preferred directory of a computer. The program can be run by clicking on the "Run" available on the top toolbar of the script in MATLAB or calling the script by typing its name in Command Window or in the other scripts.

10.1 The Reference Function "sin": Sine (Radian)

- This reference function in the format sin(X) returns the sine of the elements of X. The quantities in X are in radians [1–9].
- The function operates element-wise on arrays.
- For real values of X, sin(X) returns real values in the interval $[-1, 1]$.
- For complex values of X, sin(X) returns complex values.

Example

```
x = -pi:0.01:pi;
plot(x,sin(x))
grid on
```

Description

Plotting the sine function over the domain $-\pi \leq x \leq \pi$ radians. See Fig. 10.1.

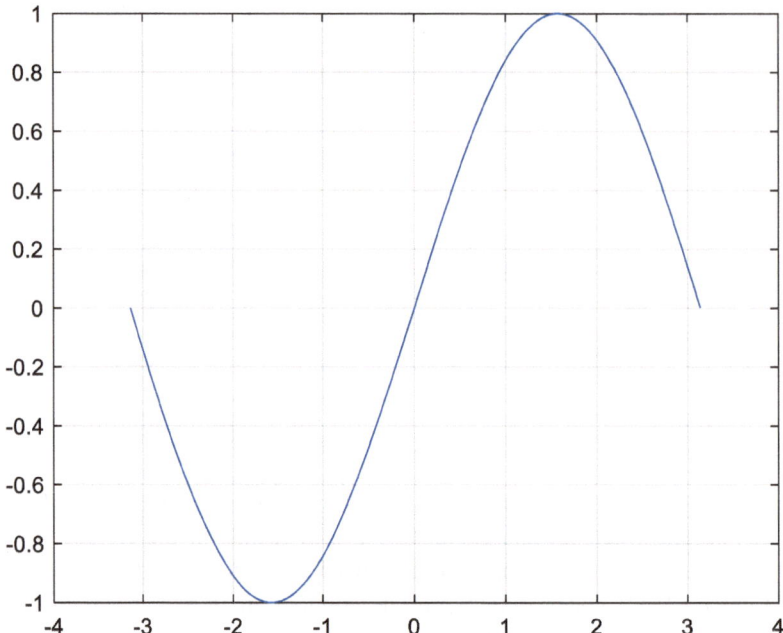

Fig. 10.1 The plot of the sine function over the domain $-\pi \leq x \leq \pi$ radians

Example

x = [-i pi+i*pi/2 -1+i*4];
y = sin(x)

Output

y =

   0.0000 - 1.1752i   0.0000 - 2.3013i   -22.9791 +14.7448i

Description

Computing the sine of the complex values in a vector.

Exercise

Execute the codes below in your computer and write the result(s) in the assigned location(s).

x = [-pi -pi/2 -pi/3 -pi/4 -pi/6 0 pi/6 pi/4 pi/3 pi/2 pi];
sin(x)

Output

ans =

10.2   The Reference Function "asin": Inverse Sine (Radian)

- This reference function in the format asin(X) returns the inverse sine of the elements of X in radians.
- For real values of X in the interval $[-1, 1]$, asin(X) returns values in the interval $[-\pi/2, \pi/2]$.
- For real values of X outside the interval $[-1, 1]$ and for complex values of X, asin(X) returns complex values.

Example

```
x = -1:0.01:1;
plot(x,asin(x))
grid on
```

Description

Plotting the inverse sine function over the intervals $-1 \leq x \leq 1$. See Fig. 10.2.

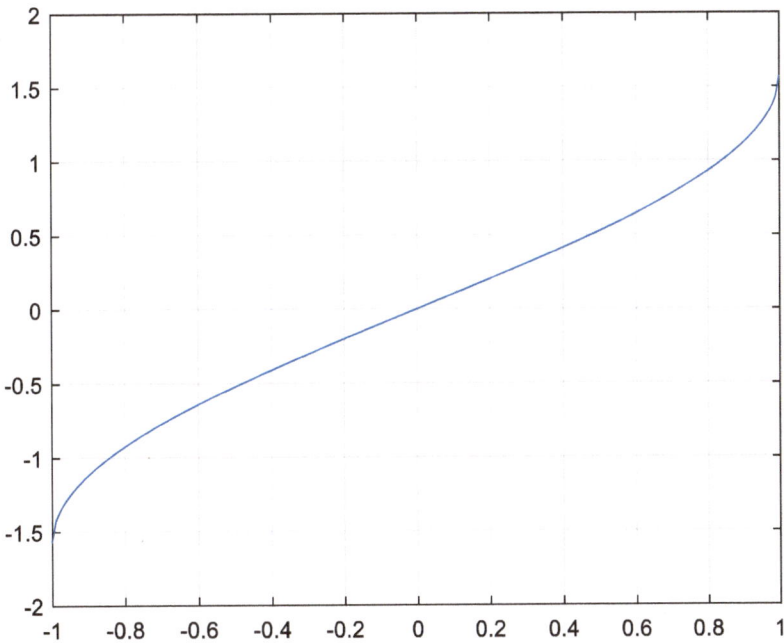

Fig. 10.2   The plot of the inverse sine function over the intervals $-1 \leq x \leq 1$

Example

```
x = [0.5i 1+3i -2.2+i];
y = asin(x)
```

Output

y =

   0.0000 + 0.4812i   0.3076 + 1.8642i   -1.1091 + 1.5480i

Description

Computing the inverse sine of the elements of a vector.

Exercise

Execute the codes below in your computer and write the result(s) in the assigned location(s).

```
x=[0 0.5 2^0.5/2 3^0.5/2 1];
asin(x)
```

Output

ans =

10.3   The Reference Function "sind": Sine (Degree)

- This reference function in the format sind(X) returns the sine of the elements of X. The quantities in X are in degrees.
- The function operates element-wise on arrays.
- For real values of X, sin(X) returns real values in the interval $[-1, 1]$.
- For complex values of X, sin(X) returns complex values.

Example

```
x = -180:180;
plot(x,sind(x))
grid on
```

Description

Plotting the sine function over the domain $-180° \leq x \leq 180°$. See Fig. 10.3.

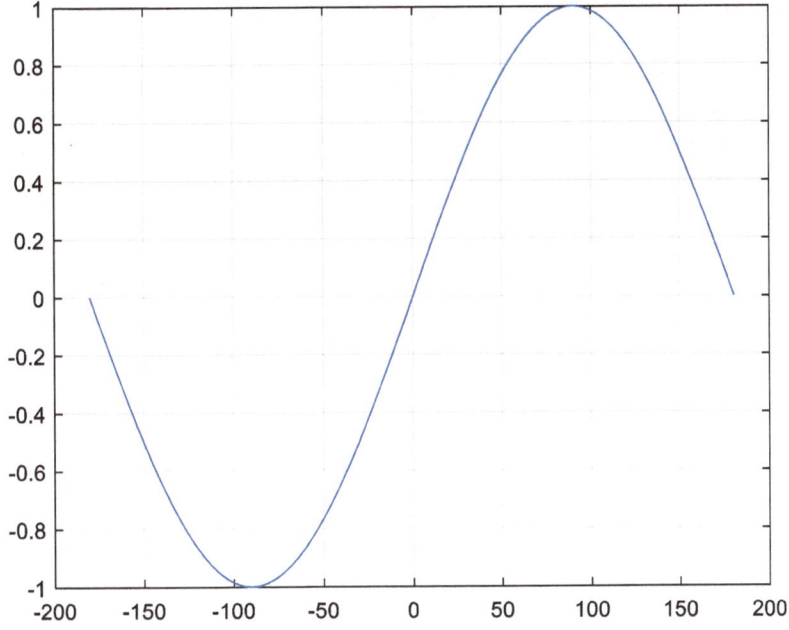

Fig. 10.3 The plot of sine function over the domain $-180° \leq x \leq 180°$

Example

```
x = [90+i 15+2i 10+3i];
y = sind(x)
```

Output

```
y =
   1.0002 + 0.0000i    0.2590 + 0.0337i    0.1739 + 0.0516i
```

Description

Computing the sine of the complex values in a vector.

Exercise

Execute the codes below in your computer and write the result(s) in the assigned location(s).

```
x = [-180 -90 -60 -45 -30 0 30 45 60 90 180];
sind(x)
```

Output

```
ans =
```

10.4   The Reference Function "asind": Inverse Sine (Degree)

- This reference function in the format asind(X) returns the inverse sine of the elements of X in degrees.
- For real values of X in the interval $[-1, 1]$, asind(X) returns values in the interval $[-90, 90]$.
- For real values of X outside the interval $[-1, 1]$ and for complex values of X, asind(X) returns complex values.

Example

x = -1:0.01:1;
plot(x,asind(x))
grid on

Description

Plotting the inverse sine function over the intervals $-1 \le x \le 1$. See Fig. 10.4.

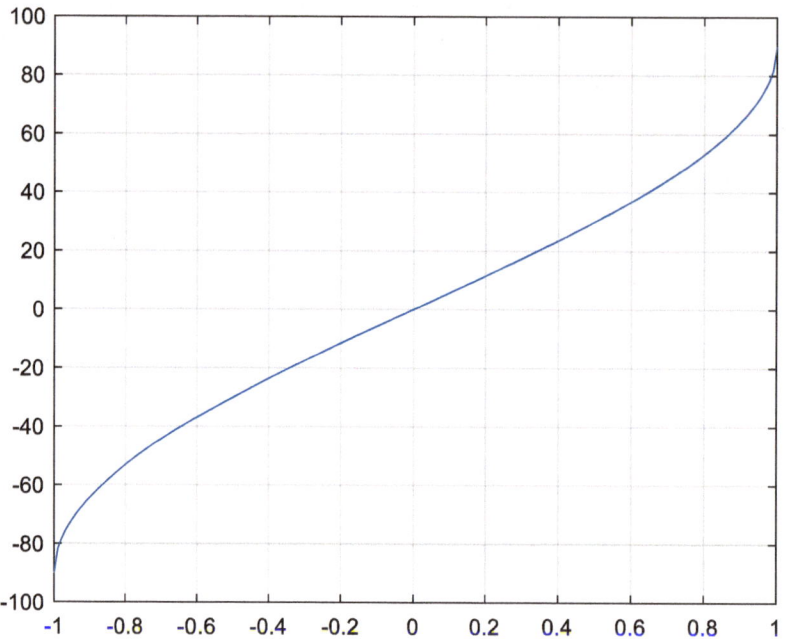

Fig. 10.4 The plot of inverse sine function over the intervals $-1 \le x \le 1$

Example

a = asind(-1)
b = asind(i)

Output

a =
  -90

b =
  0.0000 +50.4990i

Description

Computing the inverse sine of an element.

Exercise

Execute the codes below in your computer and write the result(s) in the assigned location(s).

```
x = [0 0.5 2^0.5/2 3^0.5/2 1];
asind(x)
```

Output

ans =

10.5   The Reference Function "sinh": Sine Hyperbolic (Radian)

- This reference function in the format sinh(X) returns the hyperbolic sine of the elements of X. The quantities in X are in radians.
- The function operates element-wise on arrays.
- The function accepts both real and complex inputs.

Example

```
x = -5:0.01:5;
y = sinh(x);
plot(x,y)
grid on
```

Description

Plotting the hyperbolic sine function over the domain $-5 \leq x \leq 5$ radians. See Fig. 10.5.

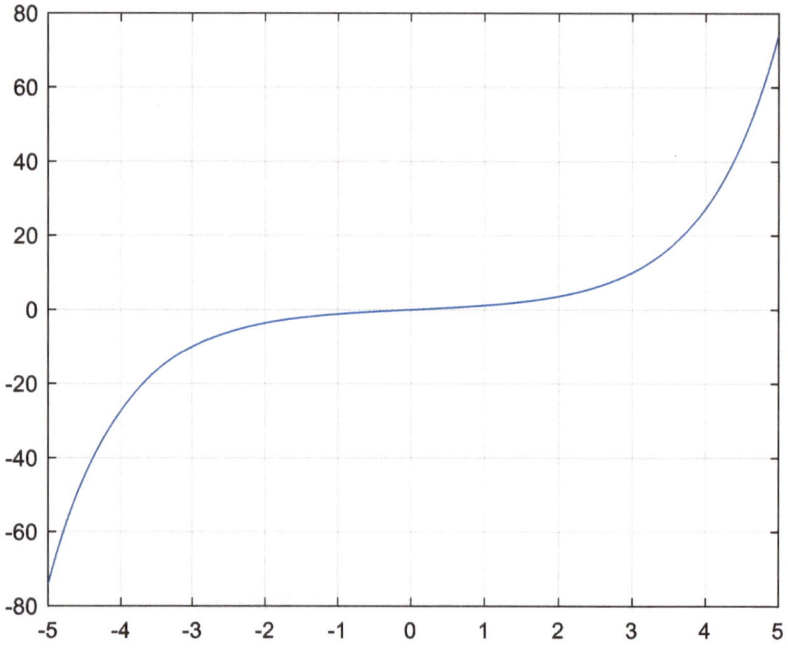

Fig. 10.5 The plot of hyperbolic sine function over the domain $-5 \leq x \leq 5$ radians

Example

```
x = [0 pi 2*pi 3*pi];
y = sinh(x)
```

Output

```
y =
   1.0e+03 *
        0    0.0115    0.2677    6.1958
```

Description

Computing the hyperbolic sine of the angles in a vector.

Example

```
sinh(i)
```

Output

```
ans =
   0.0000 + 0.8415i
```

Description

Computing the hyperbolic sine of a complex quantity.

Execute the codes below in your computer and write the result(s) in the assigned location(s).

x = [pi/6 pi/4 pi/3 pi/2];
sinh(x)

Output

ans =

10.6   The Reference Function "asinh": Inverse Sine Hyperbolic (Radian)

- This reference function in the format asinh(X) returns the inverse hyperbolic sine of the elements of X in radians.
- The function accepts both real and complex inputs.

Example

x = -5:.01:5;
plot(x,asinh(x))
grid on

Description

Plotting the inverse sine hyperbolic function over the intervals $-5 \leq x \leq 5$. See Fig. 10.6.

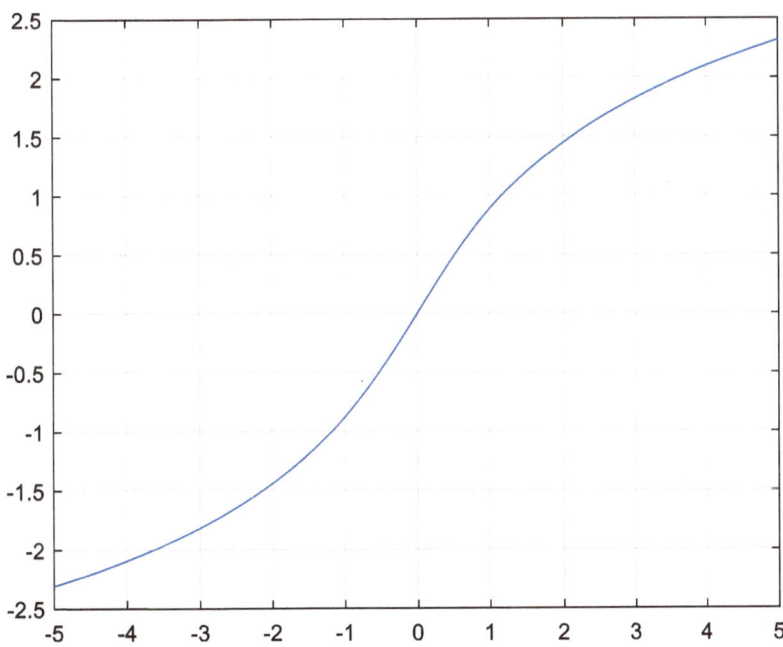

Fig. 10.6   The plot of inverse sine hyperbolic function over the intervals $-5 \leq x \leq 5$

Example

```
x = [2 -3 1+2i];
y = asinh(x)
```

Output

```
y =
   1.4436 + 0.0000i    -1.8184 + 0.0000i    1.4694 + 1.0634i
```

Description

Computing the inverse hyperbolic sine of the elements of a vector.

Exercise

Execute the codes below in your computer and write the result(s) in the assigned location(s).

```
x=[0 0.5 2^0.5/2 3^0.5/2 1];
asinh(x)
```

Output

```
ans =
```

10.7   The Reference Function "cos": Cosine (Radian)

- This reference function in the format cos(X) returns the cosine of the elements of X. The quantities in X are in radians.
- The function operates element-wise on arrays.
- For real values of X, cos(X) returns real values in the interval $[-1, 1]$.
- For complex values of X, cos(X) returns complex values.

Example

```
x = -pi:0.01:pi;
plot(x, cos(x))
grid on
```

Description

Plotting the cosine function over the domain $-\pi \leq x \leq \pi$ radians. See Fig. 10.7.

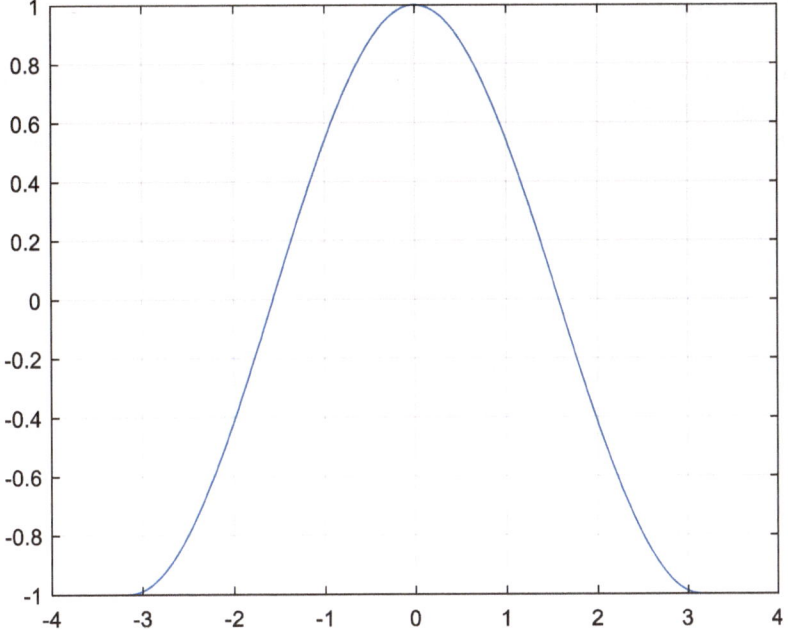

Fig. 10.7   The plot of cosine function over the domain $-\pi \leq x \leq \pi$ radians

Example

```
x = [-i pi+i*pi/2 -1+i*4];
y = cos(x)
```

Output

```
y =
   1.5431 + 0.0000i    -2.5092 - 0.0000i    14.7547 +22.9637i
```

Description

Computing the cosine of the complex values in a vector.

Exercise

Execute the codes below in your computer and write the result(s) in the assigned location(s).

```
x = [-pi -pi/2 -pi/3 -pi/4 -pi/6 0 pi/6 pi/4 pi/3 pi/2 pi];
cos(x)
```

Output

ans =

10.8   The Reference Function "acos": Inverse Cosine (Radian)

- This reference function in the format acos(X) returns the inverse cosine of the elements of X in radians.
- For real values of X in the interval $[-1, 1]$, acos(X) returns values in the interval $[0, \pi]$.
- For real values of X outside the interval $[-1, 1]$ and for complex values of X, acos(X) returns complex values.

Example

```
x = -1:0.01:1;
plot(x,acos(x))
grid on
```

Description

Plotting the inverse cosine function over the intervals $-1 \le x \le 1$. See Fig. 10.8.

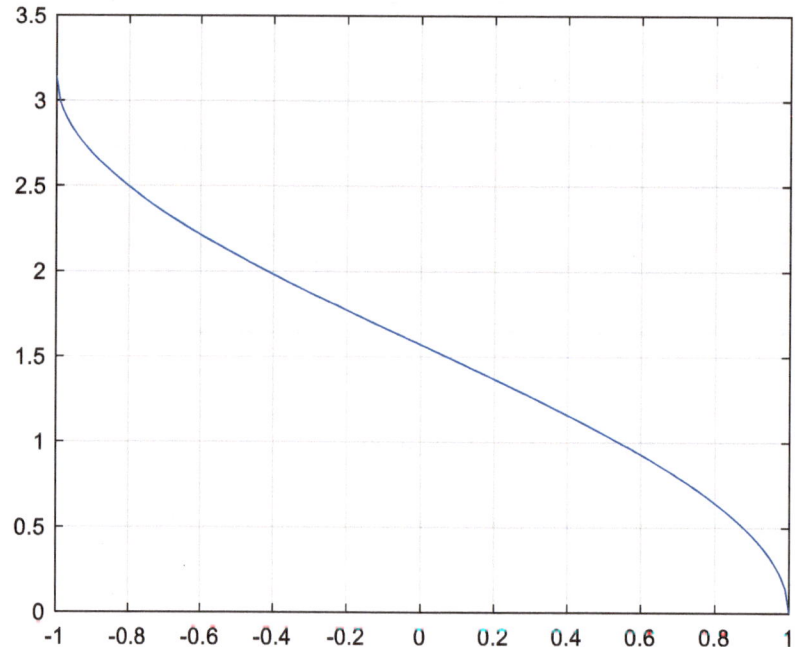

Fig. 10.8   The plot of inverse cosine function over the intervals $-1 \le x \le 1$

Example

```
x = [0.5i 1+3i -2.2+i];
y = acos(x)
```

Output

y =

  1.5708 - 0.4812i   1.2632 - 1.8642i   2.6799 - 1.5480i

Description

Computing the inverse cosine of the elements of a vector.

Exercise

Execute the codes below in your computer and write the result(s) in the assigned location(s).

x = [0 0.5 2^0.5/2 3^0.5/2 1];
acos(x)

Output

ans =

10.9   The Reference Function "cosd": Cosine (Degree)

- This reference function in the format cosd(X) returns the cosine of the elements of X. The quantities in X are in degrees.
- The function operates element-wise on arrays.
- For real values of X, cosd(X) returns real values in the interval $[-1, 1]$.
- For complex values of X, cosd(X) returns complex values.

Example

x = -180:180;
plot(x,cosd(x))
grid on

Description

Plotting the cosine function over the domain $-180° \le x \le 180°$. See Fig. 10.9.

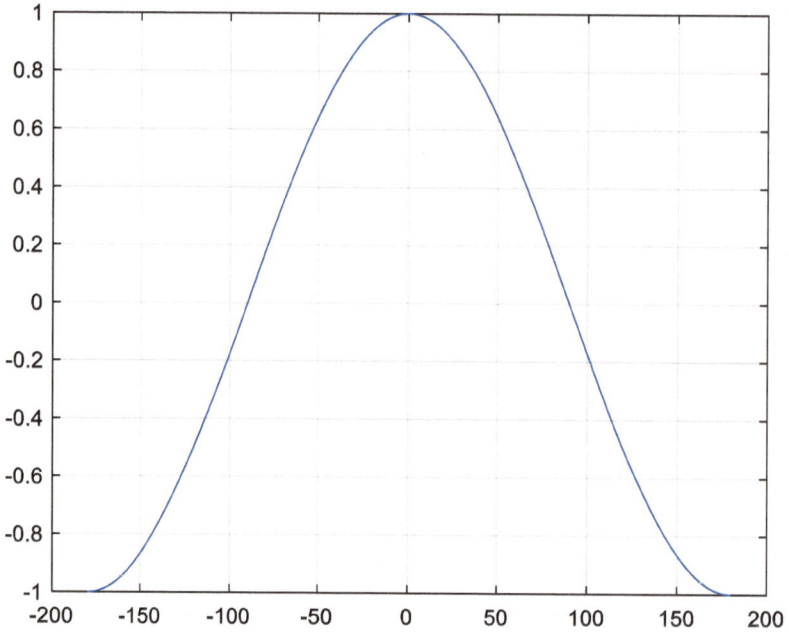

Fig. 10.9 The plot of cosine function over the domain $-180\,^\circ \le x \le 180^\circ$

Example

```
x = [180+i 45+2i 10+3i];
y = cosd(x)
```

Output

```
y =
  -1.0002 + 0.0000i    0.7075 - 0.0247i    0.9862 - 0.0091i
```

Description

Computing the cosine of the complex values in a vector.

Exercise

Execute the codes below in your computer and write the result(s) in the assigned location(s).

```
x = [-180 -90 -60 -45 -30 0 30 45 60 90 180];
cosd(x)
```

Output

```
ans =
```

10.10 The Reference Function "acosd": Inverse Cosine (Degree)

- This reference function in the format acosd(X) returns the inverse cosine of the elements of X in degrees.
- For real values of X in the interval [−1, 1], acosd(X) returns values in the interval [0, 180].
- For real values of X outside the interval [−1, 1] and for complex values of X, acosd(X) returns complex values.

Example

```
x = -1:0.01:1;
plot(x,acosd(x))
grid on
```

Description

Plotting the inverse cosine function over the intervals $-1 \leq x \leq 1$. See Fig. 10.10.

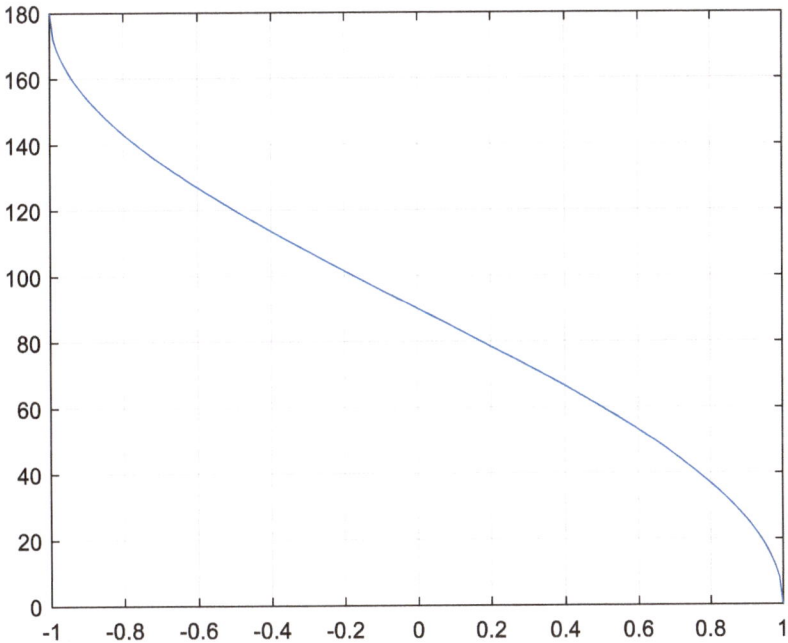

Fig. 10.10 The plot of inverse cosine function over the intervals $-1 \leq x \leq 1$

Example

```
acosd(-1)
```

Output

```
ans =
    180
```

Description

Computing the inverse cosine of a real quantity.

Exercise

Execute the codes below in your computer and write the result(s) in the assigned location(s).

```
x = [0 0.5 2^0.5/2 3^0.5/2 1];
acosd(x)
```

Output

ans =

10.11 The Reference Function "cosh": Cosine Hyperbolic (Radian)

- This reference function in the format cosh(X) returns the hyperbolic cosine of the elements of X. The quantities in X are in radians.
- The function operates element-wise on arrays.
- The function accepts both real and complex inputs.

Example

```
x = -5:0.01:5;
y = cosh(x);
plot(x,y)
grid on
```

Description

Plotting the hyperbolic cosine function over the domain $-5 \leq x \leq 5$ radians. See Fig. 10.11.

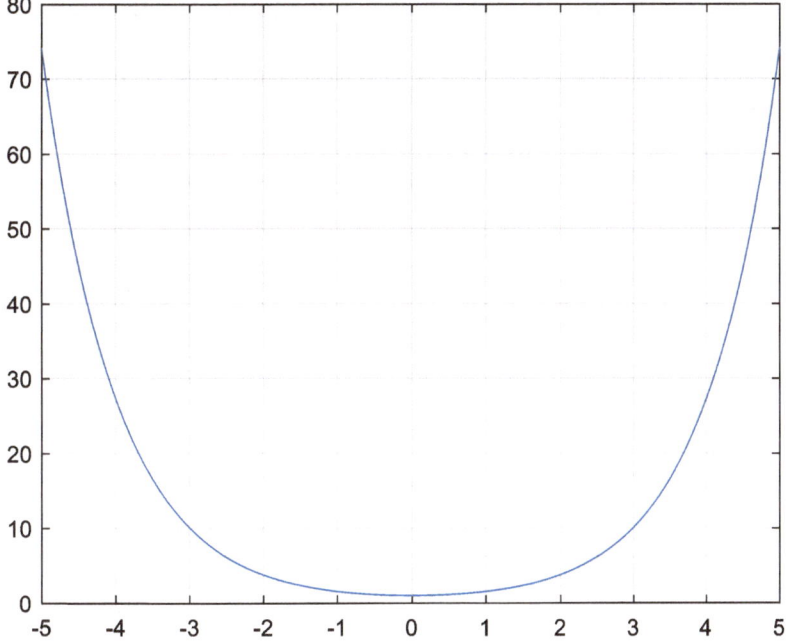

Fig. 10.11 The plot of hyperbolic cosine function over the domain $-5 \leq x \leq 5$ radians

Example

```
x = [0 pi 2*pi 3*pi];
y = cosh(x)
```

Output

```
y =
   1.0e+03 *
   0.0010    0.0116    0.2677    6.1958
```

Description

Computing the hyperbolic cosine of the angles in a vector.

Exercise

Execute the codes below in your computer and write the result(s) in the assigned location(s).

```
x = [pi/6 pi/4 pi/3 pi/2];
cosh(x)
```

Output

ans =

10.12 The Reference Function "acosh": Inverse Cosine Hyperbolic (Radian)

- This reference function in the format acosh(X) returns the inverse hyperbolic cosine of the elements of X in radians.
- The function accepts both real and complex inputs.

x = 1:0.01:5;
plot(x,acosh(x))
grid on

Plotting the inverse hyperbolic cosine function over the intervals $1 \leq x \leq 5$. See Fig. 10.12.

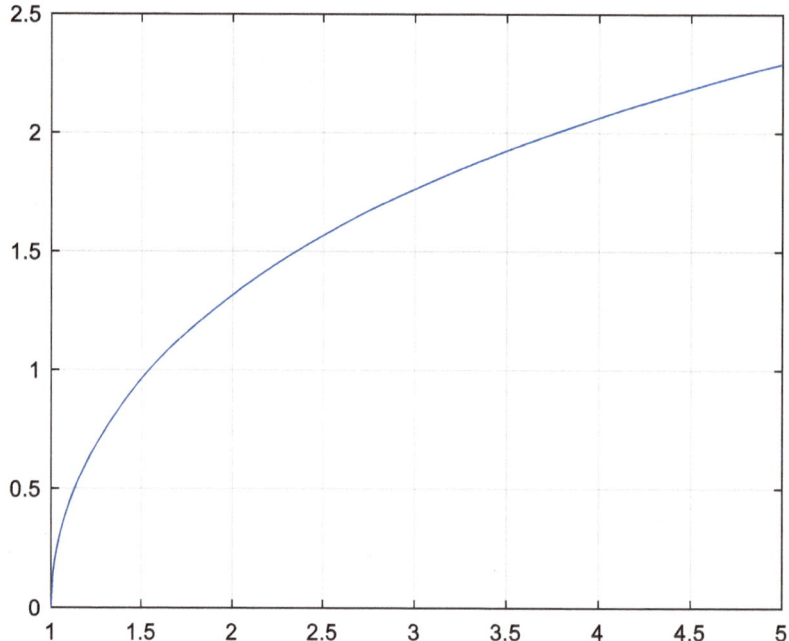

Fig. 10.12 The plot of inverse hyperbolic cosine function over the intervals $1 \leq x \leq 5$

x = [2 -3 1+2i];
y = acosh(x)

y =
    1.3170 + 0.0000i   1.7627 + 3.1416i   1.5286 + 1.1437i

Computing the inverse hyperbolic cosine of the elements of a vector.

Execute the codes below in your computer and write the result(s) in the assigned location(s).

```
x = [0 0.5 2^0.5/2 3^0.5/2 1];
acosh(x)
```

Output

ans =

10.13 The Reference Function "tan": Tangent (Radian)

- This reference function in the format tan(X) returns the tangent of the elements of X. The quantities in X are in radians.
- The function operates element-wise on arrays.
- For real values of X, tan(X) returns real values in the interval $(-\infty, \infty)$.
- For complex values of X, tan(X) returns complex values.

Example

```
x = (-pi/2)+0.01:0.01:(pi/2)-0.01;
plot(x,tan(x))
grid on
```

Description

Plotting the tangent function over the given domain in radians. See Fig. 10.13.

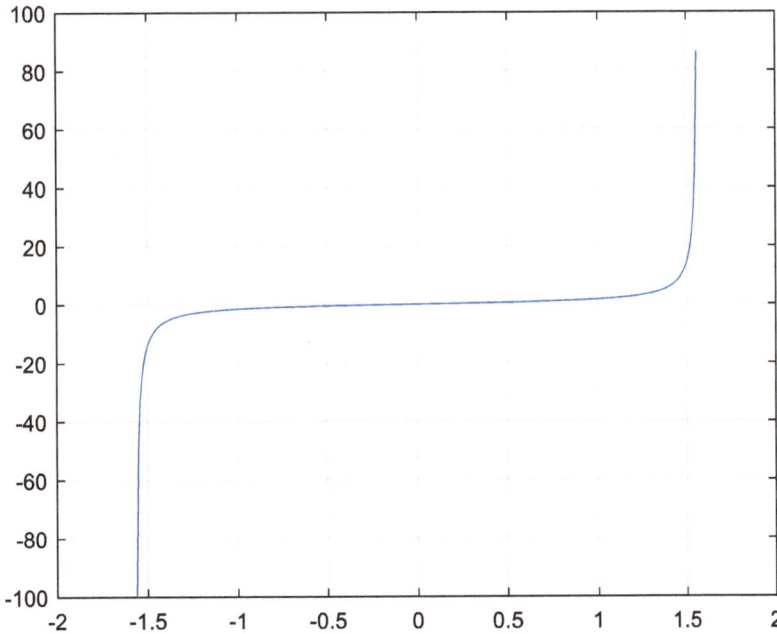

Fig. 10.13 The plot of tangent function over the given domain in radians

Example

```
x = [-i pi+i*pi/2 -1+i*4];
y = tan(x)
```

Output

```
y =
   0.0000 - 0.7616i     -0.0000 + 0.9172i     -0.0006 + 1.0003i
```

Description

Computing the tangent of the complex angles in a vector.

Exercise

Execute the codes below in your computer and write the result(s) in the assigned location(s).

```
x = [-pi/3 -pi/4 -pi/6 0 pi/6 pi/4 pi/3];
tan(x)
```

Output

```
ans =
```

10.14 The Reference Function "atan": Inverse Tangent (Radian)

- This reference function in the format atan(X) returns the inverse tangent of the elements of X in radians.
- For real values of X, atan(X) returns values in the interval $[-\pi/2, \pi/2]$.
- For complex values of X, atan(X) returns complex values.

Example

```
x = -20:0.01:20;
plot(x,atan(x))
grid on
```

Description

Plotting the inverse tangent function over the intervals $-20 \leq x \leq 20$. See Fig. 10.14.

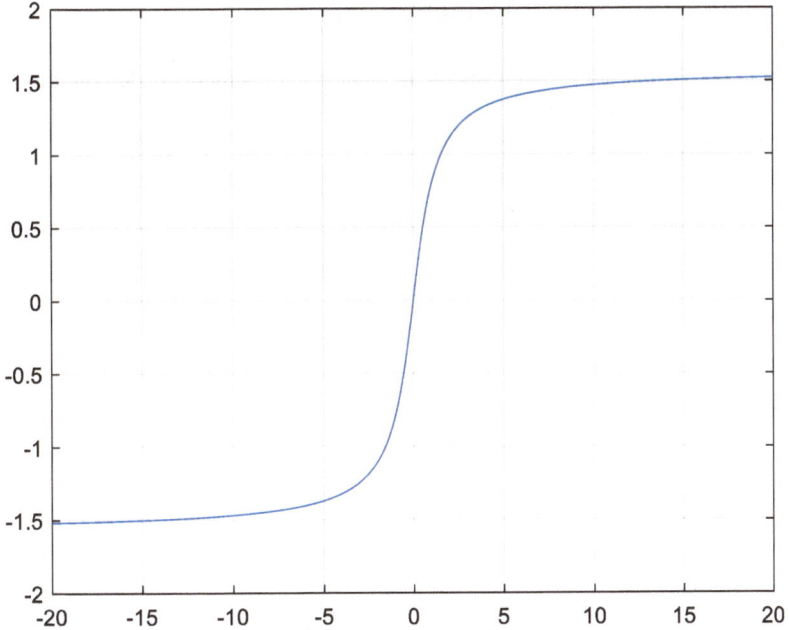

Fig. 10.14   The plot of inverse tangent function over the intervals $-20 \leq x \leq 20$

Example

x = [0.5i 1+3i -2.2+i];
y = atan(x)

Output

y =
    0.0000 + 0.5493i   1.4615 + 0.3059i   -1.2019 + 0.1506i

Description

Computing the inverse tangent of the elements of a vector.

Exercise

Execute the codes below in your computer and write the result(s) in the assigned location(s).

x = [0 3^0.5/3 1 3^0.5 Inf];
atan(x)

Output

ans =

10.15 The Reference Function "tand": Tangent (Degree)

- This reference function in the format tand(X) returns the tangent of the elements of X. The quantities in X are in degrees.
- The function operates element-wise on arrays.
- The function accepts both real and complex inputs.

```
x = -89:89;
plot(x, tand(x))
grid on
```

Plotting the tangent function over the given domain in degrees. See Fig. 10.15.

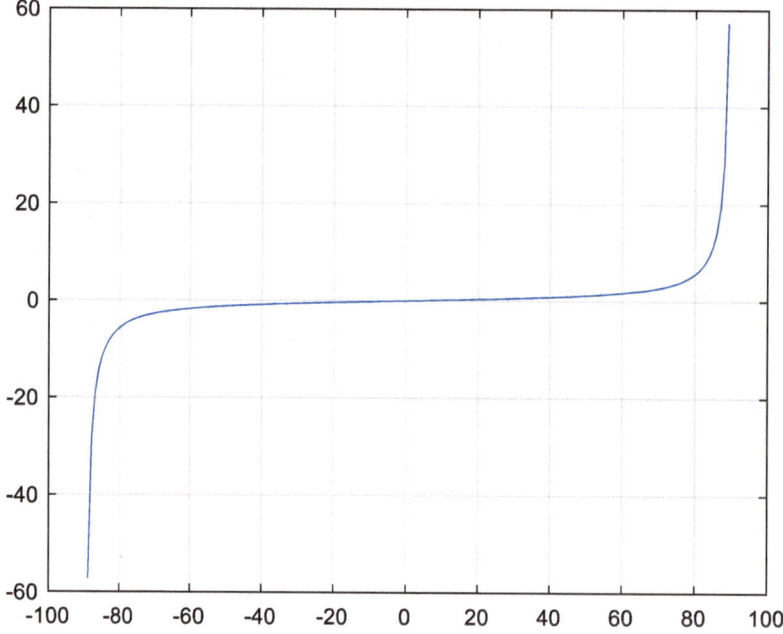

Fig. 10.15 The plot of tangent function over the given domain in degrees

```
x = [180+i 15+2i 10+3i];
y = tand(x)
```

Output

y =

   0.0000 + 0.0175i   0.2676 + 0.0374i   0.1758 + 0.0539i

Description

Computing the tangent of the complex angles in a vector.

Exercise

Execute the codes below in your computer and write the result(s) in the assigned location(s).

x = [-60 -45 -30 0 30 45 60];
tand(x)

Output

ans =

10.16 The Reference Function "atand": Inverse Tangent (Degree)

- This reference function in the format atand(X) returns the inverse tangent of the elements of X in degrees.
- For real values of X, atand(X) returns values in the interval $[-90, 90]$.
- For the complex values of X, atand(X) returns complex values.

Example

x = -20:0.01:20;
plot(x, atand(x))
grid on

Description

Plotting the inverse tangent function over the intervals $-20 \leq x \leq 20$. See Fig. 10.16.

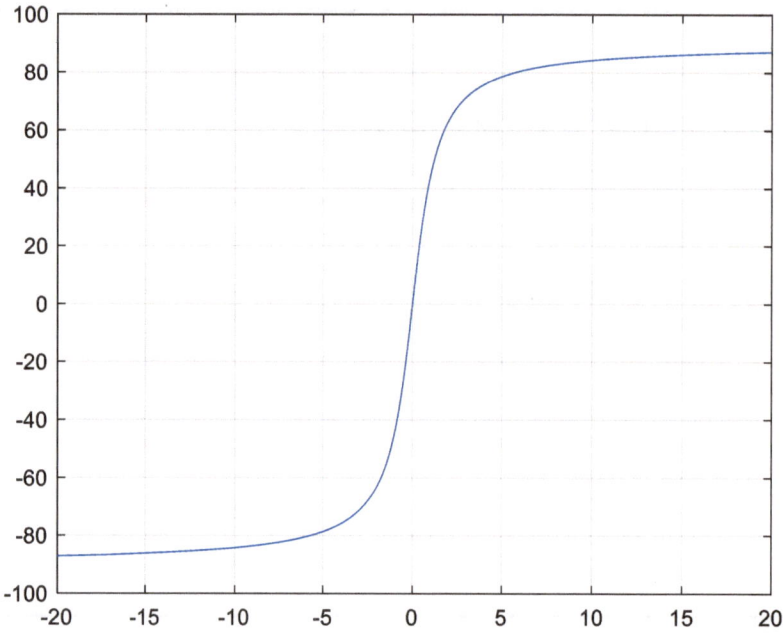

Fig. 10.16   The plot of inverse tangent function over the intervals $-20 \leq x \leq 20$

Example

```
a = atand(-1)
b = atand(10+i)
```

Output

```
a =
   -45

b =
   84.3450 + 0.5618i
```

Description

Computing the inverse tangent of an element.

Exercise

Execute the codes below in your computer and write the result(s) in the assigned location(s).

```
x = [0 3^0.5/3 1 3^0.5 Inf];
atand(x)
```

Output

```
ans =
```

10.17 The Reference Function "tanh": Tangent Hyperbolic (Radian)

- This reference function in the format tanh(X) returns the hyperbolic tangent of the elements of X. The quantities in X are in radians.
- The function operates element-wise on arrays.
- The function accepts both real and complex inputs.

Description

Plotting the hyperbolic tangent function over the domain $-5 \leq x \leq 5$ radians. See Fig. 10.17.

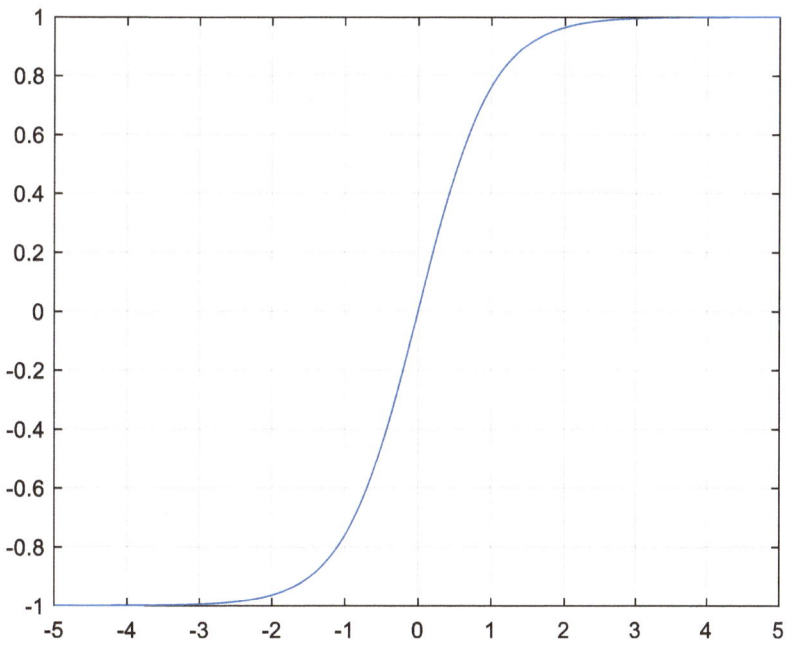

Fig. 10.17   The plot of hyperbolic tangent function over the domain $-5 \leq x \leq 5$ radians

Output

y =

  0   0.9963   1.0000   1.0000

Description

Computing the hyperbolic tangent of the angles in a vector.

Exercise

Execute the codes below in your computer and write the result(s) in the assigned location(s).

x = [pi/6 pi/4 pi/3 pi/2];
tanh(x)

Output

ans =

10.18 The Reference Function "atanh": Inverse Tangent Hyperbolic (Radian)

- This reference function in the format atanh(X) returns the inverse hyperbolic tangent of the elements of X in radians.
- The function accepts both real and complex inputs.

Example

x = -0.99:0.01:0.99;
plot(x,atanh(x))
grid on

Description

Plotting the inverse tangent hyperbolic function over the interval $-1 < x < 1$. See Fig. 10.18.

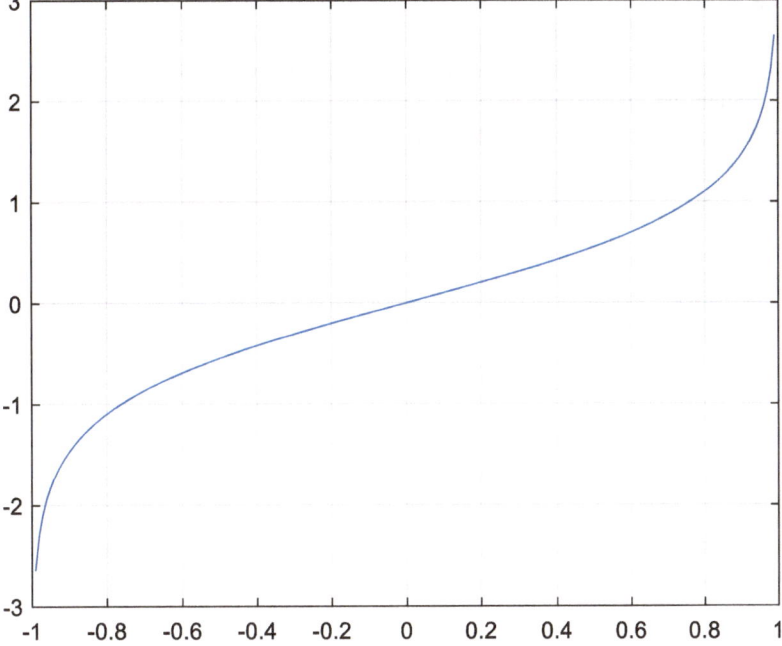

Fig. 10.18 The plot of inverse tangent hyperbolic function over the interval $-1 < x < 1$

Example

x = [2 -3 1+2i];
y = atanh(x)

Output

y =
   0.5493 + 1.5708i   -0.3466 - 1.5708i   0.1733 + 1.1781i

Description

Computing the inverse hyperbolic tangent of the elements of a vector.

Exercise

Execute the codes below in your computer and write the result(s) in the assigned location(s).

x = [0.2 0.4 0.6 0.8];
atanh(x)

Output

ans =

10.19 The Reference Function "cot": Cotangent (Radian)

- This reference function in the format cot(X) returns the cotangent of the elements of X. The quantities in X are in radians.
- The function operates element-wise on arrays.
- For real values of X, cot(X) returns real values in the interval $(-\infty, \infty)$.
- For complex values of X, cot(X) returns complex values.

Example

```
x = -pi+0.01:0.01:-0.01;
plot(x, cot(x))
grid on
```

Description

Plotting the cotangent function over the domain $-\pi < x < 0$ radians. See Fig. 10.19.

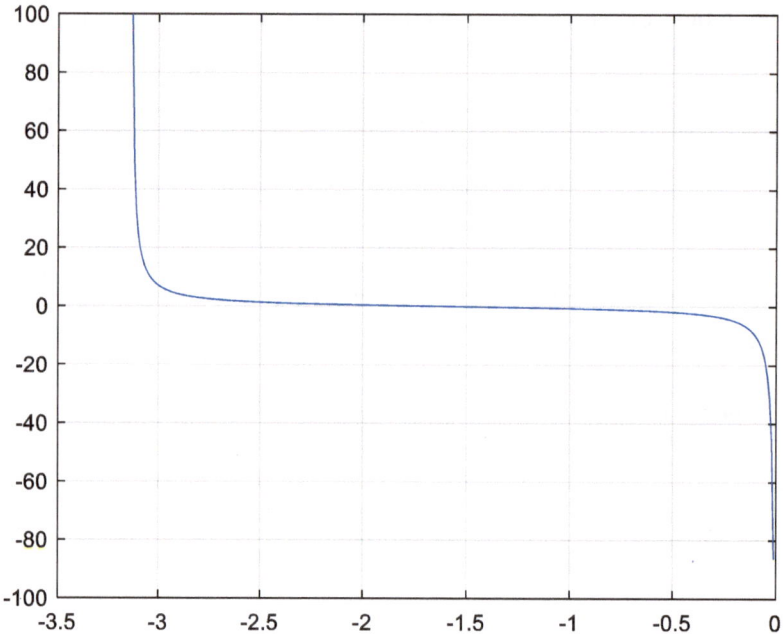

Fig. 10.19   The plot of cotangent function over the domain $-\pi < x < 0$ radians

Example

```
x = [-i pi+i*pi/2 -1+i*4];
y = cot(x)
```

Output

y =

   0.0000 + 1.3130i   -0.0000 - 1.0903i   -0.0006 - 0.9997i

Description

Computing the cotangent of the complex angles in a vector.

Exercise

Execute the codes below in your computer and write the result(s) in the assigned location(s).

```
x = [-pi/3 -pi/4 -pi/6 0 pi/6 pi/4 pi/3];
cot(x)
```

Output

ans =

10.20 The Reference Function "acot": Inverse Cotangent (Radian)

- This reference function in the format acot(X) returns the inverse cotangent of the elements of X in radians.
- For real values of X, acot(X) returns values in the interval $[-\pi/2, \pi/2]$.
- For complex values of X, acot(X) returns complex values.

Example

```
x = -20:0.01:20;
plot(x, acot(x))
grid on
```

Description

Plotting the inverse cotangent function over the intervals $-20 \leq x \leq 20$. See Fig. 10.20.

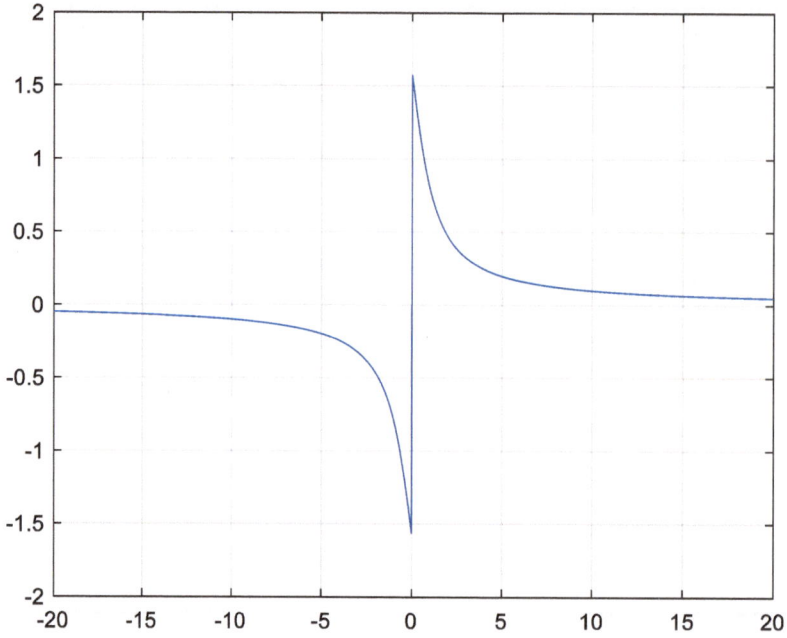

Fig. 10.20   The plot of inverse cotangent function over the intervals $-20 \leq x \leq 20$

x = [0.5i 1+3i -2.2+i];
y = acot(x)

y =
   1.5708 - 0.5493i   0.1093 - 0.3059i   -0.3689 - 0.1506i

Computing the inverse cotangent of the elements of a vector.

Execute the codes below in your computer and write the result(s) in the assigned location(s).

x = [0 3^0.5/3 1 3^0.5 Inf];
acot(x)

ans =

10.21 The Reference Function "cotd": Cotangent (Degree)

- This reference function in the format cotd(X) returns the cotangent of the elements of X. The quantities in X are in degrees
- The function operates element-wise on arrays
- The function accepts both real and complex inputs

Example

```
x = -179:-1;
plot(x, cotd(x))
grid on
```

Description

Plotting the cotangent function over the given domain in degrees. See Fig. 10.21.

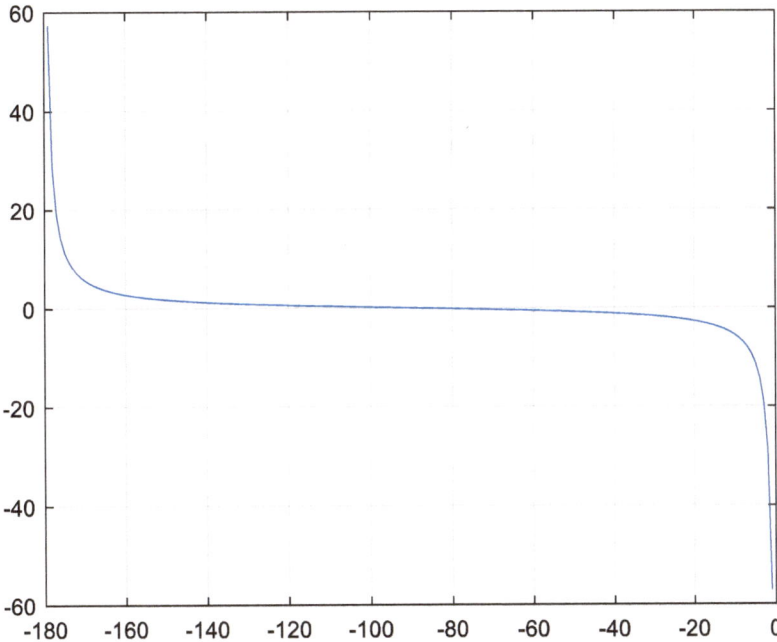

Fig. 10.21 The plot of cotangent function over the given domain in degrees

Example

```
x = [180+i 15+2i 10+3i];
y = cotd(x)
```

Output

y =

   0.0000 - 57.3016i   3.6654 - 0.5122i   5.1982 - 1.5945i

Description

Computing the cotangent of the complex angles in a vector.

Exercise

Execute the codes below in your computer and write the result(s) in the assigned location(s).

x = [-60 -45 -30 0 30 45 60];
cotd(x)

Output

ans =

10.22   The Reference Function "acotd": Inverse Cotangent (Degree)

- This reference function in the format acotd(X) returns the inverse cotangent of the elements of X in degrees.
- For real values of X, acotd(X) returns values in the interval $[-90, 90]$.
- For the complex values of X, acotd(X) returns complex values.

Example

x = -20:0.01:20;
plot(x, acotd(x))
grid on

Description

Plotting the inverse cotangent function over the intervals $-20 \le x \le 20$. See Fig. 10.22.

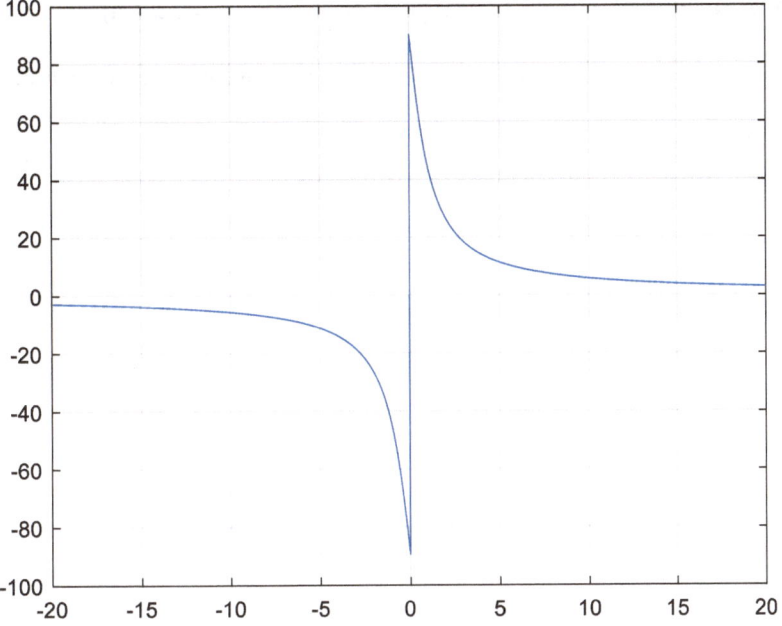

Fig. 10.22   The plot of inverse cotangent function over the intervals $-20 \leq x \leq 20$

a = acotd(-1)
b = acotd(10+i)

Output

a =
   -45

b =
   5.6550 - 0.5618i

Description

Computing the inverse cotangent of an element.

Exercise

Execute the codes below in your computer and write the result(s) in the assigned location(s).

x = [0 3^0.5/3 1 3^0.5 Inf];
acotd(x)

Output

ans =

10.23 The Reference Function "coth": Cotangent Hyperbolic (Radian)

- This reference function in the format coth(X) returns the hyperbolic cotangent of the elements of X. The quantities in X are in radians.
- The function operates element-wise on arrays.
- The function accepts both real and complex inputs.

Description

Plotting the hyperbolic cotangent function over the domain $-5 \leq x \leq 5$ radians. See Fig. 10.23.

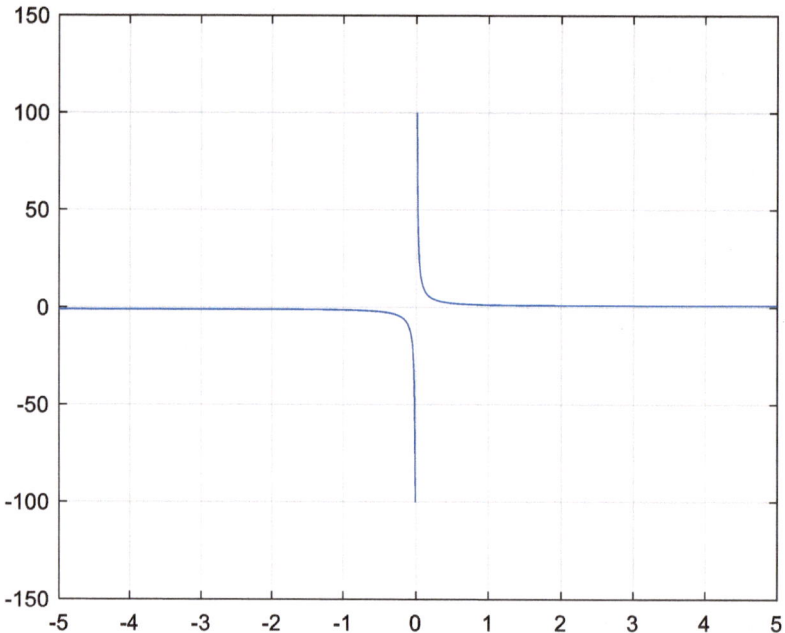

Fig. 10.23   The plot of hyperbolic cotangent function over the domain $-5 \leq x \leq 5$ radians

Example

x = [0 pi 2*pi 3*pi];
y = coth(x)

Output

y =

   Inf   1.0037   1.0000   1.0000

Description

Computing the hyperbolic cotangent of the angles in a vector.

Exercise

Execute the codes below in your computer and write the result(s) in the assigned location(s).

```
x = [pi/6 pi/4 pi/3 pi/2];
coth(x)
```

Output

ans =

10.24 The Reference Function "acoth": Inverse Cotangent Hyperbolic (Radian)

- This reference function in the format acoth(X) returns the inverse hyperbolic cotangent of the elements of X in radians.
- The function accepts both real and complex inputs.

Example

```
x1 = -30:0.1:-1.1;
x2 = 1.1:0.1:30;
plot(x1,acoth(x1),x2,acoth(x2))
grid on
```

Description

Plotting the inverse cotangent hyperbolic function over the given intervals. See Fig. 10.24.

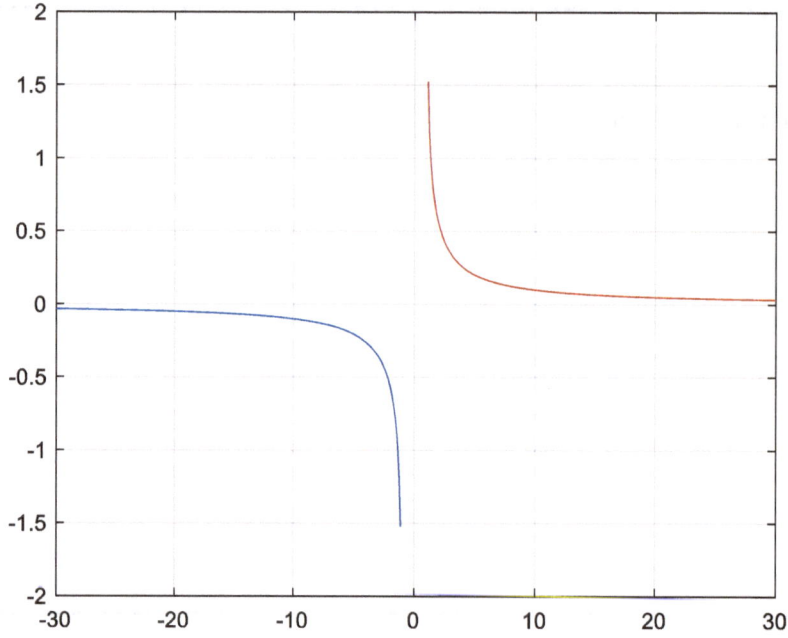

Fig. 10.24 The plot of inverse cotangent hyperbolic function over the given intervals

---

Example

x = [2 -3 1+2i];
y = acoth(x)

Output

y =
    0.5493 + 0.0000i   -0.3466 + 0.0000i   0.1733 - 0.3927i

Description

Computing the inverse hyperbolic cotangent of the elements of a vector.

Exercise

Execute the codes below in your computer and write the result(s) in the assigned location(s).

x = [2 4 6 8];
acoth(x)

Output

ans =

10.25 The Reference Function "sec": Secant (Radian)

- This reference function in the format sec(X) returns the secant of the elements of X. The quantities in X are in radians.
- The function operates element-wise on arrays.
- For real values of X, sec(X) returns real values in the interval $(-\infty, 1]$.
- For complex values of X, sec(X) returns complex values.

Example

x = -pi/2+0.01:0.01:pi/2-0.01;
plot(x,sec(x))
grid on

Description

Plotting the secant function over the given domain in radians. See Fig. 10.25.

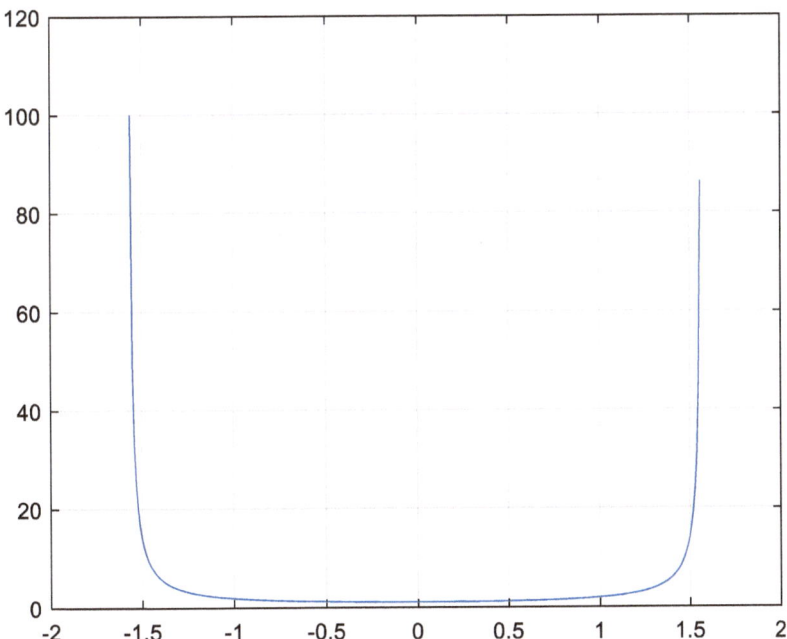

Fig. 10.25 The plot of secant function over the given domain in radians

Example

x = [-i pi+i*pi/2 -1+i*4];
y = sec(x)

y =

  0.6481 + 0.0000i   -0.3985 + 0.0000i   0.0198 - 0.0308i

Computing the secant of the complex angles in a vector.

Execute the codes below in your computer and write the result(s) in the assigned location(s).

```
x = [-pi -pi/3 -pi/4 -pi/6 0 pi/6 pi/4 pi/3 pi];
sec(x)
```

ans =

10.26 The Reference Function "asec": Inverse Secant (Radian)

- This reference function in the format asec(X) returns the inverse secant of the elements of X in radians.
- For real values of X in the intervals $(-\infty, -1]$ and $[1, \infty)$, asec(X) returns values in the interval $[0, \pi]$.
- For real values of X in the interval $(-1, 1)$ and for complex values of X, asec(X) returns complex values.

```
x1 = -5:0.01:-1;
x2 = 1:0.01:5;
plot(x1,asec(x1))
hold on
plot(x2,asec(x2))
grid on
```

Plotting the inverse secant function over the given intervals. See Fig. 10.26.

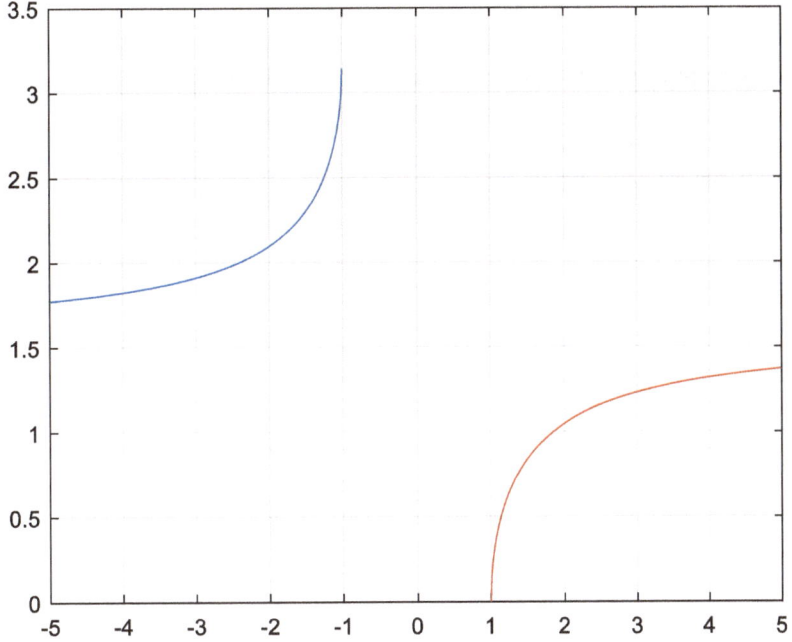

Fig. 10.26 The plot of inverse secant function over the given intervals

x = [0.5i 1+3i -2.2+i];
y = asec(x)

Output

y =
    1.5708 + 1.4436i   1.4749 + 0.2970i   1.9503 + 0.1833i

Description

Computing the inverse secant of the elements of a vector.

Exercise

Execute the codes below in your computer and write the result(s) in the assigned location(s).

x = [1 2 3 Inf];
asec(x)

Output

ans =

10.27 The Reference Function "secd": Secant (Degree)

- This reference function in the format secd(X) returns the secant of the elements of X. The quantities in X are expressed in degrees.
- The function operates element-wise on arrays.
- The function accepts both real and complex inputs.

Description

Plotting the secant function over the given domain in degrees. See Fig. 10.27.

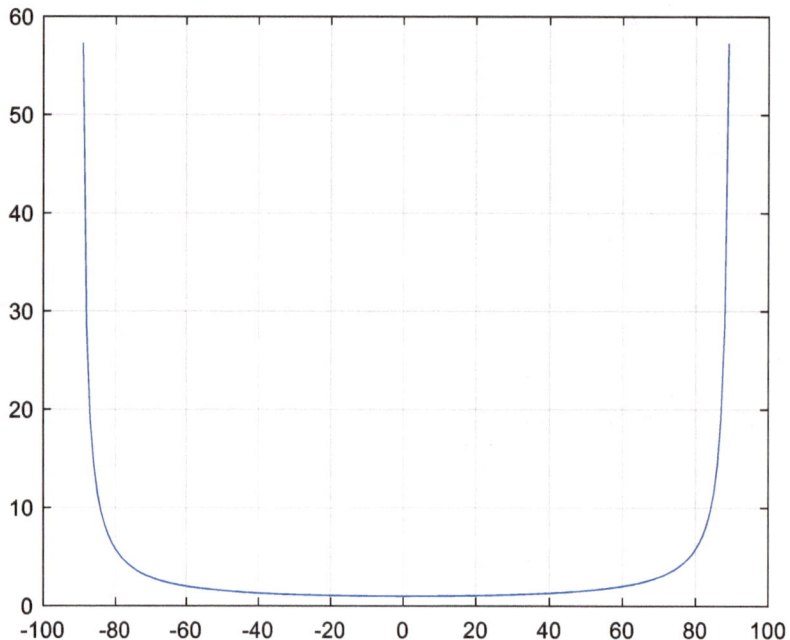

Fig. 10.27   The plot of secant function over the given domain in degrees

Example

x = [35+i 15+2i 10+3i];
y = secd(x)

Output

y =

   1.2204 + 0.0149i   1.0346 + 0.0097i   1.0140 + 0.0094i

Description

Computing the secant of the complex angles in a vector.

Exercise

Execute the codes below in your computer and write the result(s) in the assigned location(s).

```
x = [-180 -60 -45 -30 0 30 45 60 180];
secd(x)
```

Output

ans =

10.28 The Reference Function "asecd": Inverse Secant (Degree)

- This reference function in the format asecd(X) returns the inverse secant of the elements of X in degrees.
- For real values of X in the intervals $(-\infty, -1]$ and $[1, \infty)$, asecd(X) returns values in the interval $[0, 180]$.
- For real values of X in the interval $(-1, 1)$ and for complex values of X, asecd(X) returns complex values.

Example

```
x1 = -5:0.01:-1;
x2 = 1:0.01:5;
plot(x1,asecd(x1))
hold on
plot(x2,asecd(x2))
grid on
```

Description

Plotting the inverse secant function over the given intervals. See Fig. 10.28.

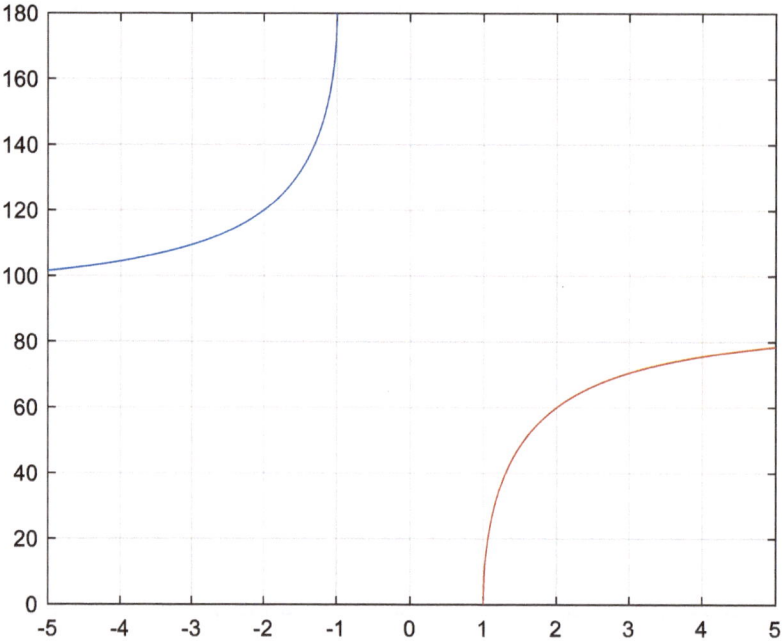

Fig. 10.28 The plot of inverse secant function over the given intervals

Example

asecd(-1)

Output

ans =
   180

Description

Computing the inverse secant of an element.

Exercise

Execute the codes below in your computer and write the result(s) in the assigned location(s).

x = [1 2 3 Inf];
asecd(x)

Output

ans =

10.29 The Reference Function "sech": Secant Hyperbolic (Radian)

- This reference function in the format sech(X) returns the hyperbolic secant of the elements of X. The quantities in X are in radians.
- The function operates element-wise on arrays.
- The function accepts both real and complex inputs.

Description

Plotting the hyperbolic secant function over the given domain in radians. See Fig. 10.29.

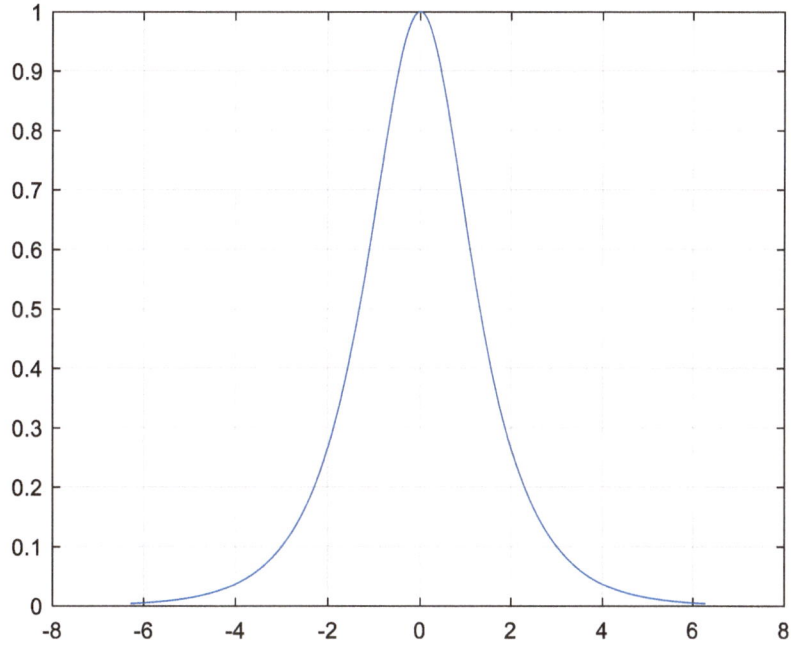

Fig. 10.29 The plot of hyperbolic secant function over the given domain in radians

Example

x = [0 pi 2*pi 3*pi];
y = sech(x)

Output

y =

   1.0e+03 *

   1.0000   0.0863   0.0037   0.0002

Description

Computing the hyperbolic secant of the angles in a vector.

Exercise

Execute the codes below in your computer and write the result(s) in the assigned location(s).

```
x = [pi/6 pi/4 pi/3 pi/2];
sech(x)
```

Output

ans =

10.30 The Reference Function "asech": Inverse Secant Hyperbolic (Radian)

- This reference function in the format asech(X) returns the inverse hyperbolic secant of the elements of X in radians.
- The function accepts both real and complex inputs.

Example

```
x = 0.01:0.001:1;
plot(x,asech(x))
grid on
```

Description

Plotting the inverse hyperbolic secant function over the given interval. See Fig. 10.30.

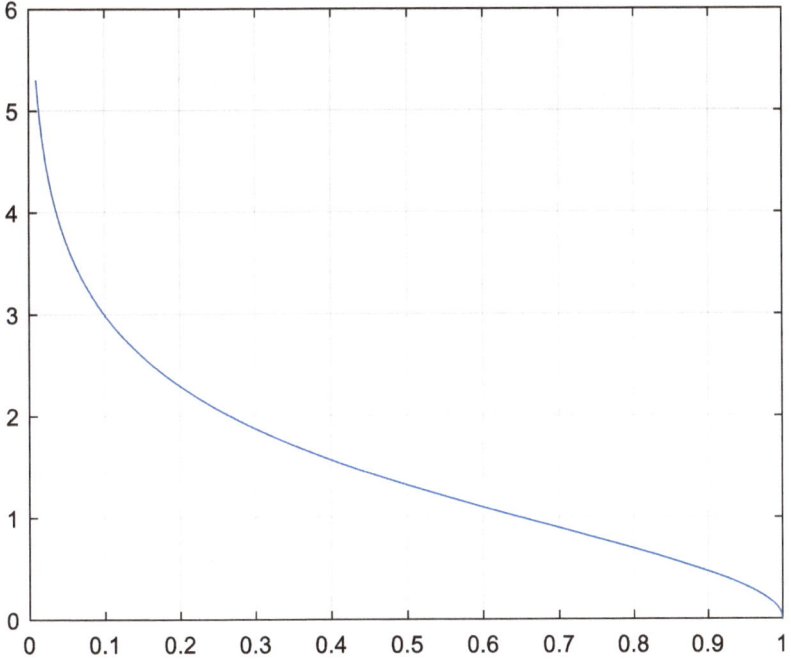

Fig. 10.30 The plot of inverse hyperbolic secant function over the given interval

Example

x = [2 -3 1+2i];
y = asech(x)

Output

y =
   0.0000 + 1.0472i   0.0000 + 1.9106i   0.3966 - 1.3845i

Description

Computing the inverse hyperbolic secant of the elements of a vector.

Exercise

Execute the codes below in your computer and write the result(s) in the assigned location(s).

x = [0 0.5 2^0.5/2 3^0.5/2 1];
asech(x)

Output

ans =

10.31 The Reference Function "csc": Cosecant (Radian)

- This reference function in the format csc(X) returns the cosecant of the elements of X. The quantities in X are in radians.
- The function operates element-wise on arrays.
- For real values of X, csc(X) returns real values in the interval $(-\infty, 1]$.
- For complex values of X, csc(X) returns complex values.

Example

x1 = -pi+0.01:0.01:-0.01;
x2 = 0.01:0.01:pi-0.01;
plot(x1,csc(x1),x2,csc(x2))
grid on

Description

Plotting the cosecant function over the given domain in radians. See Fig. 10.31.

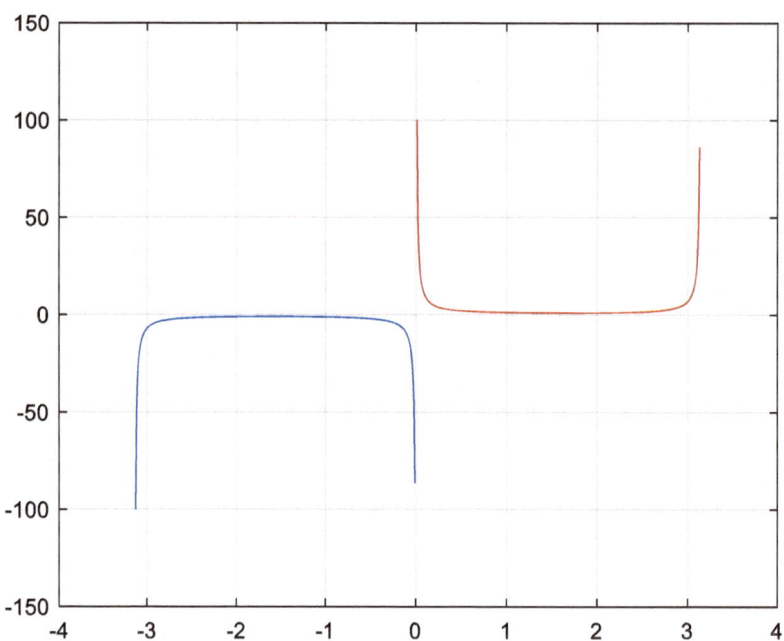

Fig. 10.31   The plot of cosecant function over the given domain in radians

Example

x = [-i pi+i*pi/2 -1+i*4];
y = csc(x)

Output

y =

  0.0000 + 0.8509i   0.0000 + 0.4345i   -0.0308 - 0.0198i

Description

Computing the cosecant of the complex angles in a vector.

Exercise

Execute the codes below in your computer and write the result(s) in the assigned location(s).

x = [-pi/2 -pi/3 -pi/4 -pi/6 pi/6 pi/4 pi/3 pi/2];
csc(x)

Output

ans =

10.32 The Reference Function "acsc": Inverse Cosecant (Radian)

- This reference function in the format acsc(X) returns the inverse cosecant of the elements of X in radians.
- For real values of X in the intervals $(-\infty, -1]$ and $[1, \infty)$, acsc(X) returns values in the interval $[-\pi/2, \pi/2]$.
- For real values of X in the interval $(-1, 1)$ and for complex values of X, acsc(X) returns complex values.

Example

x1 = -10:0.01:-1.01;
x2 = 1.01:0.01:10;
plot(x1,acsc(x1),'b')
hold on
plot(x2,acsc(x2),'b')
grid on

Description

Plotting the inverse cosecant function over the given intervals. See Fig. 10.32.

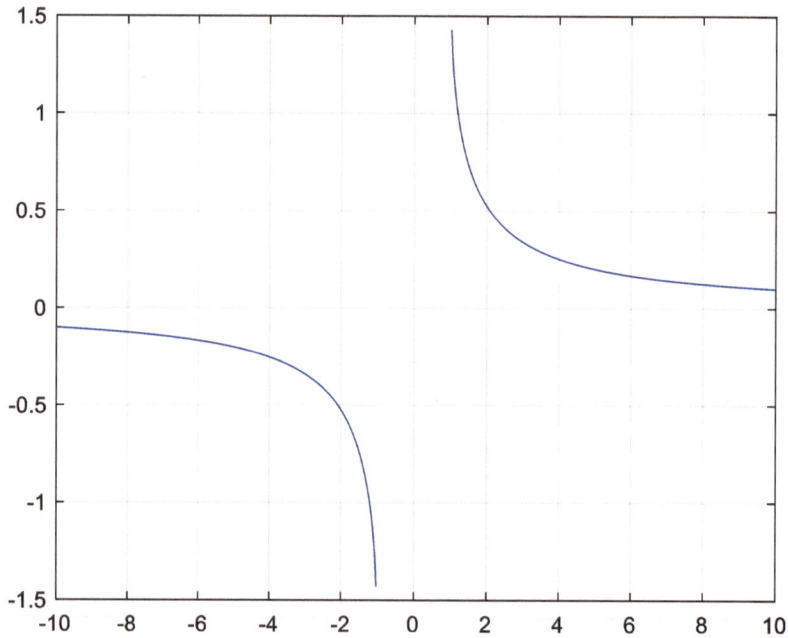

Fig. 10.32 The plot of inverse cosecant function over the given intervals

Example

x = [0.5i 1+3i -2.2+i];
y = acsc(x)

Output

y =
   0.0000 - 1.4436i   0.0959 - 0.2970i   -0.3795 - 0.1833i

Description

Computing the inverse cosecant of the elements of a vector.

Exercise

Execute the codes below in your computer and write the result(s) in the assigned location(s).

x = [1 2 3 Inf];
acsc(x)

Output

ans =

10.33 The Reference Function "cscd": Cosecant (Degree)

- This reference function in the format cscd(X) returns the cosecant of the elements of X. The quantities in X are expressed in degrees.
- The function operates element-wise on arrays.
- The function accepts both real and complex inputs.

```
x1 = -179:-1;
x2 = 1:179;
plot(x1,cscd(x1),x2,cscd(x2))
grid on
```

Description

Plotting the cosecant function over the given domain in degrees. See Fig. 10.33.

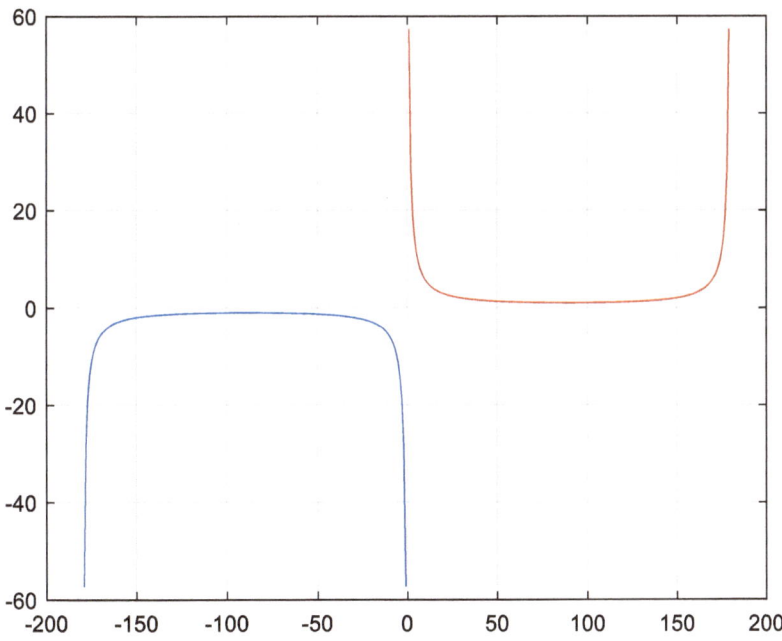

Fig. 10.33 The plot of cosecant function over the given domain in degrees

Example

```
x = [35+i 15+2i 10+3i];
y = cscd(x)
```

Output

y =

   1.7421 - 0.0434i   3.7970 - 0.4944i   5.2857 - 1.5681i

Description

Computing the cosecant of the complex angles in a vector.

Exercise

Execute the codes below in your computer and write the result(s) in the assigned location(s).

x = [-60 -45 -30 30 45 60];
cscd(x)

Output

ans =

10.34 The Reference Function "acscd": Inverse Cosecant (Degree)

- This reference function in the format acscd(X) returns the inverse cosecant of the elements of X in degrees.
- For real values of X in the intervals $(-\infty, -1]$ and $[1, \infty)$, acscd(X) returns values in the interval $[-90, 90]$.
- For real values of X in the interval $(-1, 1)$ and for complex values of X, acscd(X) returns complex values.

Example

x1 = -10:0.01:-1.01;
x2 = 1.01:0.01:10;
plot(x1,acscd(x1),'b')
hold on
plot(x2,acscd(x2),'b')
grid on

Description

Plotting the inverse cosecant function over the given intervals. See Fig. 10.34.

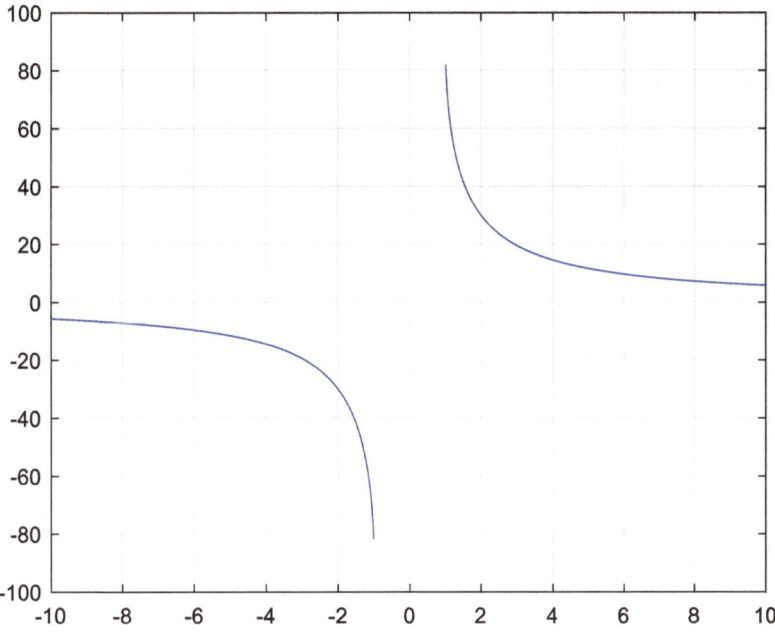

Fig. 10.34   The plot of inverse cosecant function over the given intervals

Example

acscd(-1)

Output

ans =
   -90

Description

Computing the inverse cosecant of an element.

Exercise

Execute the codes below in your computer and write the result(s) in the assigned location(s).

x = [1 2 3 Inf];
acscd(x)

Output

ans =

10.35 The Reference Function "csch": Cosecant Hyperbolic (Radian)

- This reference function in the format csch(X) returns the hyperbolic cosecant of the elements of X. The quantities in X are in radians.
- The function operates element-wise on arrays.
- The function accepts both real and complex inputs.

```
x1 = -pi+0.01:0.01:-0.01;
x2 = 0.01:0.01:pi-0.01;
y1 = csch(x1);
y2 = csch(x2);
plot(x1,y1,x2,y2)
grid on
```

Plotting the hyperbolic cosecant function over the given domain in radians. See Fig. 10.35.

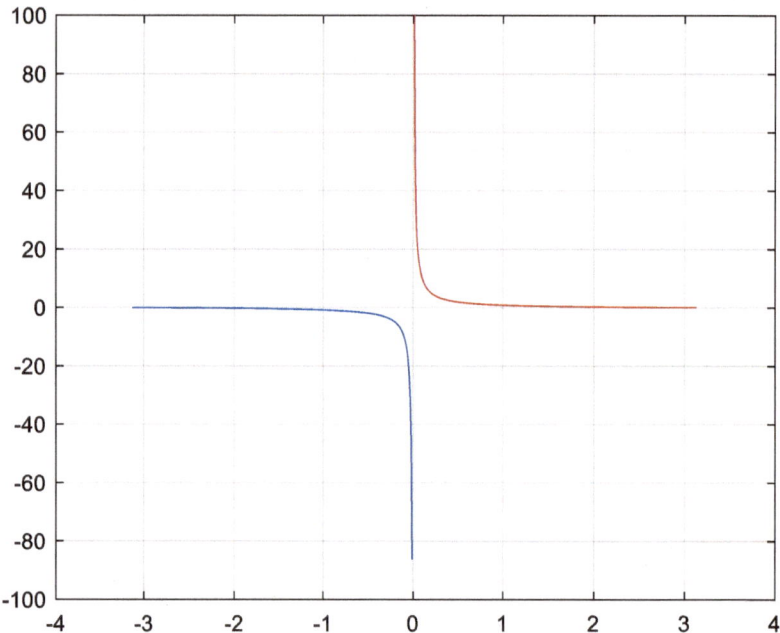

Fig. 10.35 The plot of hyperbolic cosecant function over the given domain in radians

```
x = [0 pi 2*pi 3*pi];
y = csch(x)
```

Output

y =
   Inf   0.0866   0.0037   0.0002

Description

Computing the hyperbolic cosecant of the angles in a vector.

Exercise

Execute the codes below in your computer and write the result(s) in the assigned location(s).

```
x = [pi/6 pi/4 pi/3 pi/2];
csch(x)
```

Output

ans =

10.36 The Reference Function "acsch": Inverse Cosecant Hyperbolic (Radian)

- This reference function in the format acsch(X) returns the inverse hyperbolic cosecant of the elements of X in radians.
- The function accepts both real and complex inputs.

Example

```
x1 = -20:0.01:-1;
x2 = 1:0.01:20;
plot(x1,acsch(x1),x2,acsch(x2))
grid on
```

Description

Plotting the inverse hyperbolic cosecant function over the given interval. See Fig. 10.36.

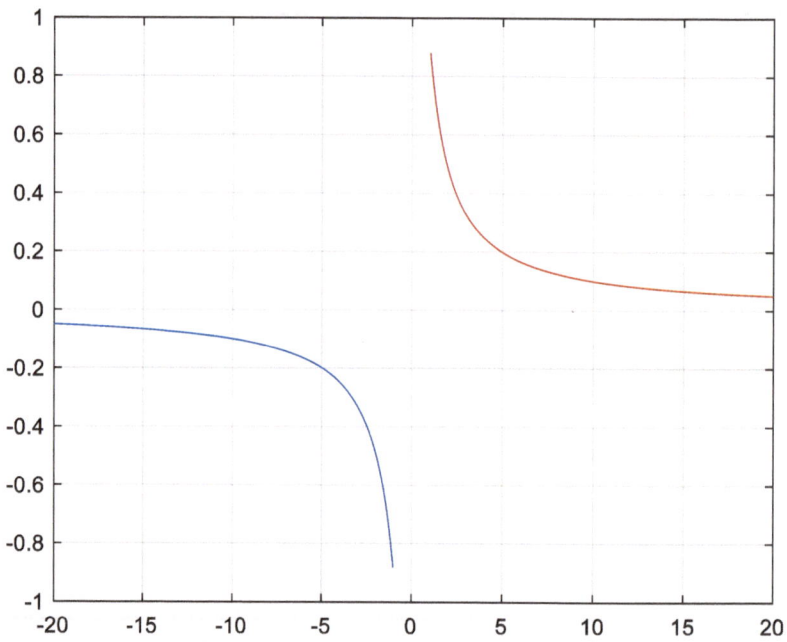

Fig. 10.36 The plot of inverse hyperbolic cosecant function over the given interval

```
x = [2 -3 1+2i];
y = acsch(x)
```

```
y =
   0.4812 + 0.0000i    -0.3275 + 0.0000i    0.2156 - 0.4016i
```

Computing the inverse hyperbolic cosecant of the elements of a vector.

Execute the codes below in your computer and write the result(s) in the assigned location(s).

```
x = [0 0.5 2^0.5/2 3^0.5/2 1];
acsch(x)
```

ans =

References

1. MATLAB 2023a.
2. Rahmani-Andebili, M. (2024). *Mathematics of engineering and science – Practice problems, methods, and solutions*. Springer Nature.
3. Rahmani-Andebili, M. (2022). *Differential equations – Practice problems, methods, and solutions*. Springer Nature.
4. Rahmani-Andebili, M. (2023). *Calculus III – Practice problems, methods, and solutions*. Springer Nature.
5. Rahmani-Andebili, M. (2023). *Calculus II – Practice problems, methods, and solutions*. Springer Nature.
6. Rahmani-Andebili, M. (2023). *Calculus I – Practice problems, methods, and solutions* (2nd ed.). Springer Nature.
7. Rahmani-Andebili, M. (2021). *Calculus – Practice problems, methods, and solutions*. Springer Nature.
8. Rahmani-Andebili, M. (2024). *Precalculus – Practice problems, methods, and solutions* (2nd ed.). Springer Nature.
9. Rahmani-Andebili, M. (2021). *Precalculus – Practice Problems, Methods, And Solutions*. Springer Nature.

Abstract

In this chapter, the reference functions concerned with the basic functions in MATLAB are presented and described. In this regard, several examples and exercises for each section of the chapter are presented. The exercises that include writing the codes, executing them, and achieving the results need to be done by students to master programming skills. In this book, the codes, outputs, and descriptions are in blue, black, and green colors, respectively. To program in MATLAB, a script file can be created and saved with an appropriate name (e.g., untitled01) in the preferred directory of a computer. The program can be run by clicking on the "Run" available on the top toolbar of the script in MATLAB or by calling the script by typing its name in the Command Window or in the other scripts.

11.1 The Reference Function "abs": Absolute Value and Complex Magnitude

- This reference function in the format abs(X) returns the absolute value of each element in input X [1–9].
- If X is complex, the function returns the complex magnitude.

Example

```
x = [-1 0 4];
abs(x)
```

Output

```
ans =
    1    0    4
```

Description

Computing the absolute value of the real elements of a vector.

Example

```
x = 6+8i;
abs(x)
```

M. Rahmani-Andebili, *MATLAB Lessons, Examples, and Exercises*, https://doi.org/10.1007/978-3-031-76177-5_11

Output

ans =
   10

Description

Computing the magnitude of a complex quantity.

Exercise

Execute the codes below in your computer and write the result(s) in the assigned location(s).

```
x = [3+4i 3 4i -3-4i];
abs(x)
```

Output

ans =

11.2   The Reference Function "exp": Complex Exponential Value

- This reference function in the format exp(X) returns the exponential value of each element in array X.
- If the element x is a real value, it computes e^x.
- If the element is a complex quantity, it computes the complex exponential value based on Euler's identity as follows.

$$e^z = e^{x+iy} = e^x(\cos y + i \sin y)$$

Example

```
x = 1;
exp(x)
```

Output

ans =
   2.7183

Description

Computing the exponential value of a real element.

Example

x = i*pi;
exp(x)

Output

ans =
   -1.0000 + 0.0000i

Description

Computing the exponential value of a complex element.

Example

x = -3:0.1:3;
y = exp(x);
plot(x,y)
grid on

Description

Plotting the real exponential function over the given interval. See Fig. 11.1.

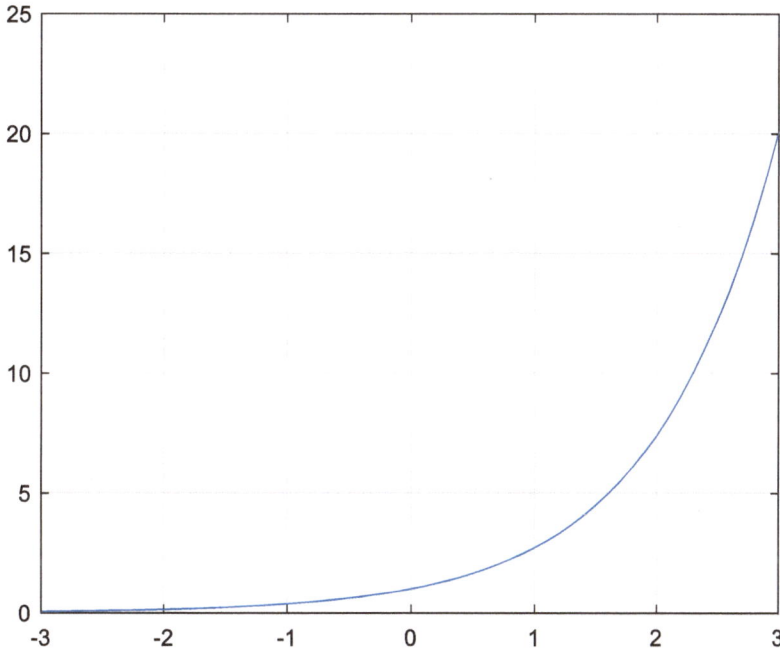

Fig. 11.1 The plot of real exponential function over the given interval

Exercise

Execute the codes below in your computer and write the result(s) in the assigned location(s).

x = [0 i*pi/2 -i*pi/2 i];
exp(x)

11.3   The Reference Function "sqrt": Square Root

- This reference function in the format sqrt(X) returns the square root of each element of the array X. The function's domain includes negative and complex numbers.
- For the negative or complex elements, sqrt(X) produces complex results.

Example

X = [-4 -1 0 1 4];
sqrt(X)

Output

ans =
    0.0000 + 2.0000i   0.0000 + 1.0000i   0.0000 + 0.0000i   1.0000 + 0.0000i   2.0000 + 0.0000i

Description

Computing the square root of each element of X.

Exercise

Execute the codes below in your computer and write the result(s) in the assigned location(s).

X = [-9 -i^2 i^2 9];
sqrt(X)

11.4   The Reference Function "nthroot": Real and Complex nth Root

- This reference function in the format nthroot(X,N) returns the real nth root of the elements of X. Both X and N must be real scalars or arrays of the same size. If an element in X is negative, then the corresponding element in N must be an odd integer.
- To compute the complex nth root of the elements of X, the command X^(1/N) for scalars and X.^(1./N) for matrices can be used.

Example

X = -27;
N = 3;
nthroot(-27, 3)

Output

ans =
    -3

Description

Computing the real root of a real quantity.

Example

X = -27;
N = 3;
X^(1/N)

Output

ans =
    1.5000 + 2.5981i

Description

Computing the complex root of a real quantity.

Example

X = [-2 -2 -2; 4 -3 -5]
N = [1 -1 3; 1/2 5 3]
nthroot(X,N)

Output

X =
    -2   -2   -2
     4   -3   -5

N =
    1.0000   -1.0000   3.0000
    0.5000   5.0000   3.0000

ans =
      -2.0000   -0.5000   -1.2599
      16.0000   -1.2457   -1.7100

Description

Computing the element-wise real root of a matrix including real quantities.

Example

X = [-2 -2 -2; 4 -3 -5]
N = [1 -1 3; 1/2 5 3]
X.^(1./N)

Output

X =
      -2   -2   -2
       4   -3   -5

N =
      1.0000   -1.0000   3.0000
      0.5000   5.0000   3.0000

ans =
      -2.0000 + 0.0000i   -0.5000 + 0.0000i   0.6300 + 1.0911i
      16.0000 + 0.0000i   1.0078 + 0.7322i   0.8550 + 1.4809i

Description

Computing the element-wise complex root of a matrix including real quantities.

Exercise

Execute the codes below in your computer and write the result(s) in the assigned location(s).

X = -8;
N = [5 3 -1];
a = nthroot(X,N)
b = X.^(1./N)

Output

a =

b =

11.5   The Reference Function "log": Natural Logarithm

- This reference function in the format log(X) returns the natural logarithm ln(x) of each element in array X. The function's domain includes negative and complex numbers.

Example

log(-1)

Output

ans =
  0.0000 + 3.1416i

Description

Computing the natural logarithm of an element.

Example

X = [-4 -1 0 1 4];
log(X)

Output

ans =
  1.3863 + 3.1416i   0.0000 + 3.1416i   -Inf + 0.0000i   0.0000 + 0.0000i   1.3863 + 0.0000i

Description

Computing the natural logarithm of each element of a vector.

Exercise

Execute the codes below in your computer and write the result(s) in the assigned location(s).

X = [-2.71828   2.71828^2 i];
log(X)

Output

ans =

11.6   The Reference Functions "log2": Logarithm with Base 2

- This reference function in the format log2(X) computes the base 2 logarithm of each element in array X. The function accepts both real and complex inputs.
- For real values of X in the interval (0, Inf), the function returns real values in the interval (-Inf, Inf).
- For complex and negative real values of X, the function returns complex values.

Example

X = [0 1 2 10 Inf];
log2(X)

Output

ans =
   -Inf   0   1.0000   3.3219   Inf

Description

Computing the base 2 logarithm of real elements of a vector.

Example

X = [-2i -i i 2i];
log2(X)

Output

ans =
   1.0000 − 2.2662i   0.0000 − 2.2662i   0.0000 + 2.2662i   1.0000 + 2.2662i

Description

Computing the base 2 logarithm of complex quantities of a vector.

Exercise

Execute the codes below in your computer and write the result(s) in the assigned location(s).

X = [4 8 16];
log2(X)

Output

ans =

11.7   The Reference Functions "log10": Logarithm with Base 10

- This reference function in the format log10(X) computes the base 10 logarithm (common logarithm) of each element in array X. The function accepts both real and complex inputs.
- For real values of X in the interval (0, Inf), the function returns real values in the interval (-Inf, Inf).
- For complex and negative real values of X, the function returns complex values.

Example

```
X = [0 1 2 10 Inf];
log10(X)
```

Output

```
ans =
   -Inf     0     0.3010     1.0000     Inf
```

Description

Computing the base 10 logarithm of real elements of a vector.

Example

```
X = [-10i -i i 10i];
log10(X)
```

Output

```
ans =
   1.0000 − 0.6822i     0.0000 − 0.6822i     0.0000 + 0.6822i     1.0000 + 0.6822i
```

Description

Computing the base 10 logarithm of complex quantities of a vector.

Exercise

Execute the codes below in your computer and write the result(s) in the assigned location(s).

```
X = [0.01 100];
log10(X)
```

Output

```
ans =
```

11.8   The Reference Function "factorial": Factorial Calculation

- This reference function in the format factorial(n) returns the product of all positive integers less than or equal to n, where n is a nonnegative integer value.
- If n is an array, then it contains the factorial of each value of n.

Example

n = [0 1 2 3 4 5 6];
factorial(n)

Output

ans =

   1   1   2   6   24   120   720

Description

Computing the factorial of each element of the vector.

Exercise

Execute the codes below in your computer and write the result(s) in the assigned location(s).

a = factorial(6);
b = factorial(5);
c = a / b

Output

c =

11.9   The Reference Function "sign": Sign Determination

- This reference function in the format sign(x) returns an array with the same size as x, where each element of that is 1 if the corresponding element of x is greater than 0, 0 if the corresponding element of x equals 0, and -1 if the corresponding element of x is less than 0.

Example

X = [-Inf -2 0 3 Inf];
sign(X)

Output

ans =
    -1   -1   0   1   1

Description

Determining the sign of each element of the vector.

Exercise

Execute the codes below in your computer and write the result(s) in the assigned location(s).

X = [-0.01 -i^2];
sign(X)

Output

ans =

11.10 The Reference Function "mod": Modulo

- This reference function in the format mod(a,m) returns the remainder after division of a by m, where a is the dividend and m is the divisor.

Example

mod(23,5)

Output

ans =
    3

Description

Computing 23 modulo 5.

Example

a = [-5 -4 -1 7 9];
m = 3;
mod(a,m)

Output

ans =
   1   2   2   1   0

Description

Applying the function on a vector including the positive and negative quantities.

Exercise

Execute the codes below in your computer and write the result(s) in the assigned location(s).

```
a = [-5 -4 -1 7 9];
m = 5;
mod(a,m)
```

Output

ans =

11.11   The Reference Function "ceil": Rounding Toward Greater Nearest Integer

- This reference function in the format ceil(X) rounds each element of X to the nearest integer greater than or equal to that element.

Example

```
X = [-1.9 -0.2 3.4 5.6 7];
ceil(X)
```

Output

ans =
  -1   0   4   6   7

Description

Applying the function on a vector with real quantities.

Example

```
X = 2.4+3.6i;
ceil(X)
```

ans =
   3.0000 + 4.0000i

Applying the function on a complex quantity.

Execute the codes below in your computer and write the result(s) in the assigned location(s).

X = [-0.99 0.01 1.01];
ceil(X)

ans =

11.12   The Reference Function "floor": Rounding Toward Smaller Nearest Integer

- This reference function in the format floor(X) rounds each element of X to the nearest integer smaller than or equal to that element.

X = [-1.9 -0.2 3.4 5.6 7];
floor(X)

ans =
   -2   -1   3   5   7

Applying the function on a vector with real quantities.

X = 2.4+3.6i;
floor(X)

Output

ans =
    2.0000 + 3.0000i

Description

Applying the function on a complex quantity.

Exercise

Execute the codes below in your computer and write the result(s) in the assigned location(s).

X = [-0.99 0.01 1.01];
floor(X)

Output

ans =

11.13 The Reference Function "round": Rounding to Nearest Integer

• This reference function in the format round(X) rounds each element of X to the nearest integer.

Example

X = [-1.4 -1.6 3.4 3.6 9];
round(X)

Output

ans =
    -1   -2   3   4   9

Description

Applying the function on a vector with real quantities.

Example

X = 2.4+3.6i;
round(X)

Output

ans =

  2.0000 + 4.0000i

Description

Applying the function on a complex quantity.

Exercise

Execute the codes below in your computer and write the result(s) in the assigned location(s).

X = [-0.49 -0.51 0.49 0.51];
round(X)

Output

ans =

References

1. MATLAB 2023a.
2. Rahmani-Andebili, M. (2024). *Mathematics of engineering and science – Practice problems, methods, and solutions*. Springer Nature.
3. Rahmani-Andebili, M. (2022). *Differential equations – Practice problems, methods, and solutions*. Springer Nature.
4. Rahmani-Andebili, M. (2023). *Calculus III – Practice problems, methods, and solutions*. Springer Nature.
5. Rahmani-Andebili, M. (2023). *Calculus II – Practice problems, methods, and solutions*. Springer Nature.
6. Rahmani-Andebili, M. (2023). *Calculus I – Practice problems, methods, and solutions* (2nd ed.). Springer Nature.
7. Rahmani-Andebili, M. (2021). *Calculus – Practice problems, methods, and solutions*. Springer Nature.
8. Rahmani-Andebili, M. (2024). *Precalculus – Practice problems, methods, and solutions* (2nd ed.). Springer Nature.
9. Rahmani-Andebili, M. (2021). *Precalculus – Practice problems, methods, and solutions*. Springer Nature.

Abstract

In this chapter, the reference functions concerned with the plotting features in MATLAB are presented and described. In this regard, several examples and exercises for each section of the chapter are presented. The exercises that include writing the codes, executing them, and achieving the results need to be done by students to master programming skills. In this book, the codes, outputs, and descriptions are in blue, black, and green colors, respectively. To program in MATLAB, a script file can be created and saved with an appropriate name (e.g. untitled01) in the preferred directory of a computer. The program can be run by clicking on the "Run" available on the top toolbar of the script in MATLAB or calling the script by typing its name in Command Window or in the other scripts.

12.1   The Reference Function "Plot": Plotting

- This reference function in the format plot(X,Y) creates a 2-D line plot of the data in Y versus the corresponding values in X. The vectors X and Y must be of the same length [1–9].
- This reference function in the format plot(X,Y1,X,Y2) creates multiple line plots of the data in Y1 and Y2 versus the corresponding values in X. The vectors X, Y1, and Y2 must be of the same length.
- This reference function in the format plot(Y) creates multiple line plots from a matrix, where each column of the matrix Y is considered as a specific line.

Example

```
x = 0:pi/100:2*pi;
y = sin(x);
plot(x,y)
```

Description

Creating a line plot of the data. The vector x is linearly spaced between 0 and 2π with the increment of $\pi/100$. The vector y is the sine values of the vector x. See Fig. 12.1.

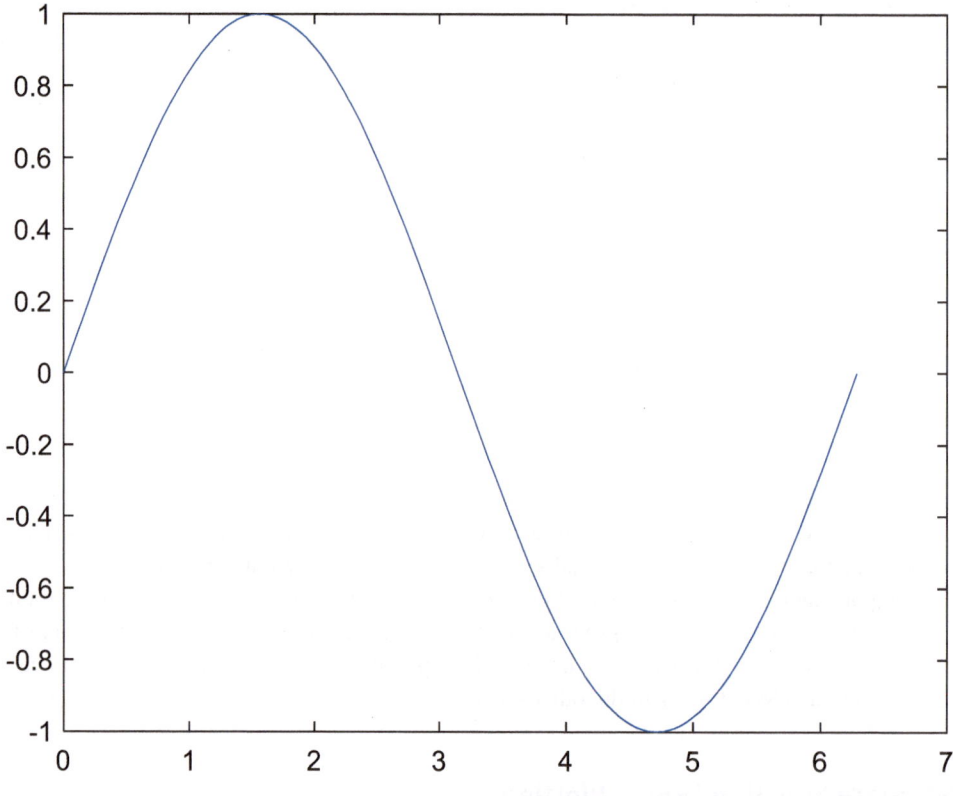

Fig. 12.1 A line plot of the data

x = 0:pi/100:2*pi;
y1 = sin(x);
y2 = cos(x);
plot(x,y1,x,y2)

Creating multiple line plots of the data. The vectors y1 and y2 are the sine and cosine values of the vector x. See Fig. 12.2.

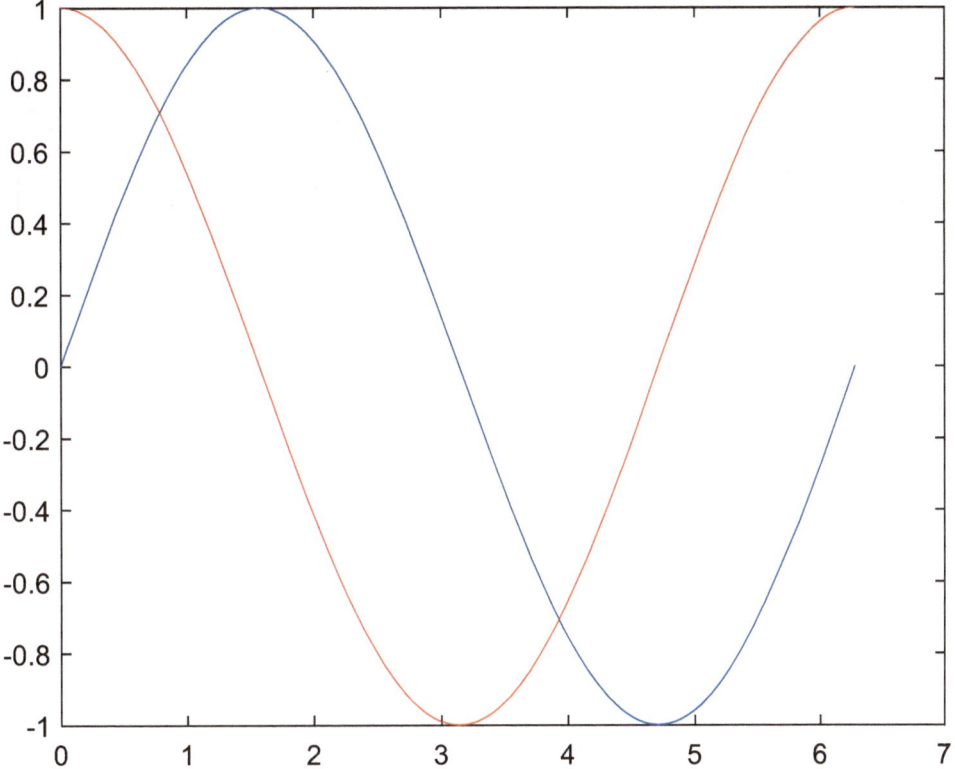

Fig. 12.2 Multiple line plots of the data

```
y = [16      2      3     13
      5     11     10      8
      9      7      6     12
      4     14     15      1];
plot(y)
```

Creating multiple line plots from a matrix. Herein, each matrix column is considered as a specific line. See Fig. 12.3.

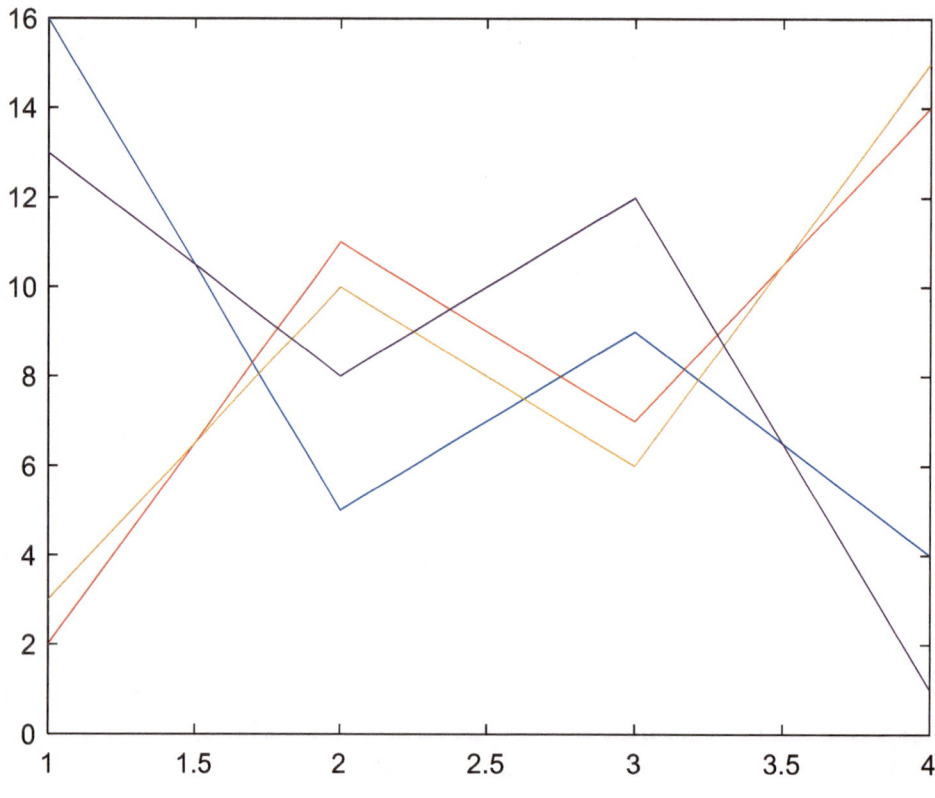

Fig. 12.3 Multiple line plots from a matrix

Exercise

(a) *Write the codes to plot the graph of tangent function for the interval (−pi/2, pi/2).*

(b) *Execute the codes in your computer, then draw the result(s) on the axes.*

12.2   The Plotting Property: Line Style Selection

- This property in the format plot(X,Y1,'*LineStyle1*',X,Y2,'*LineStyle2*') creates multiple line plots of the data with the specified line styles. Figure 12.4 shows the line styles available in MATLAB.

Line Style	Description	Resulting Line
" - "	Solid line	————
" - - "	Dashed line	- — — — —
" : "	Dotted line	....
" - . "	Dash-dotted line	—.—.—.—..

Fig. 12.4 The list of line styles available in MATLAB

Example

```
x = 0:pi/100:2*pi;
y1 = sin(x);
y2 = sin(x-0.25);
y3 = sin(x-0.5);
y4 = sin(x-0.75);
plot(x,y1,'-',x,y2,'--',x,y3,'-.',x,y4,':')
```

Description

Specifying line styles. See Fig. 12.5.

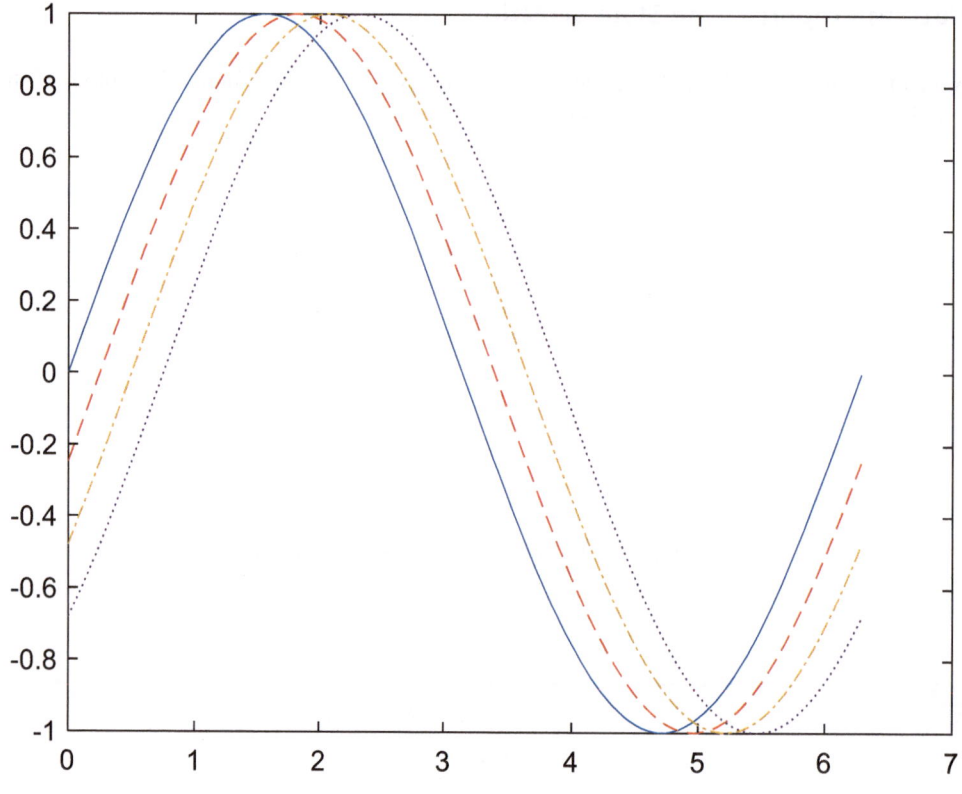

Fig. 12.5   Plotting with different line styles

Exercise

(a)   *Write the codes to plot the graph of cotangent function for the interval (0, pi) using a dashed line.*

(b)   *Execute the codes in your computer, then draw the result(s) on the axes.*

12.3   The Plotting Property: Line Marker Selection

- This property in the format plot(X,Y1,'*LineMarker1*',X,Y2,'*LineMarker2*') creates multiple line plots of the data with the specified line markers. Figure 12.6 shows the line markers available in MATLAB.

Marker	Description	Resulting Marker
"o"	Circle	○
"+"	Plus sign	+
"*"	Asterisk	✳
"."	Point	•
"x"	Cross	×
"_"	Horizontal line	—
"|"	Vertical line	|
"square"	Square	▢
"diamond"	Diamond	◇
"^"	Upward-pointing triangle	△
"v"	Downward-pointing triangle	▽
">"	Right-pointing triangle	▷
"<"	Left-pointing triangle	◁
"pentagram"	Pentagram	☆
"hexagram"	Hexagram	✡

Fig. 12.6 The list of line markers available in MATLAB

Example

```
x = 0:pi/10:2*pi;
y1 = sin(x);
y2 = sin(x-0.25);
plot(x,y1,'--*',x,y2,'--o')
```

Description

Specifying line markers. See Fig. 12.7.

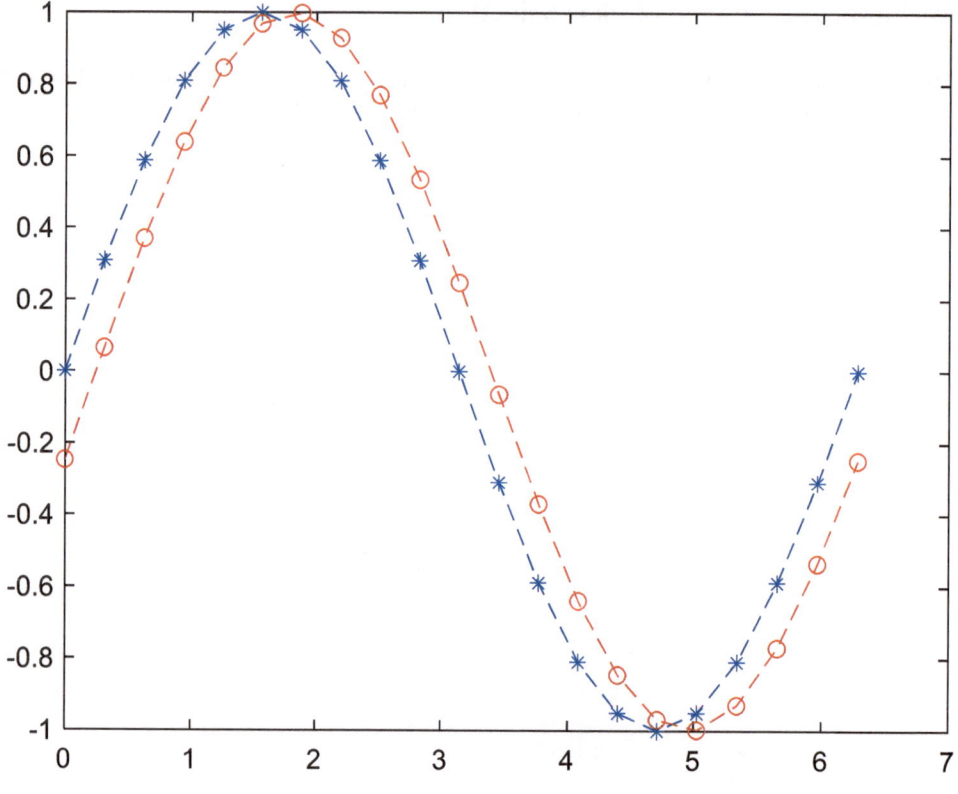

Fig. 12.7 Plotting with different line markers

Exercise

(a) *Write the codes to plot the graph of tangent function for the interval (−pi/2, pi/2) using the diamond line marker.*
(b) *Execute the codes in your computer, then draw the result(s) on the axes.*

12.4 The Plotting Property: Line Color Selection

• This property in the format plot(X,Y1,'*LineColor1*',X,Y2,'*LineColor2*') creates multiple line plots of the data with the specified line colors. Figure 12.8 shows the line colors available in MATLAB.

Color Name	Short Name	Appearance
"red"	"r"	
"green"	"g"	
"blue"	"b"	
"cyan"	"c"	
"magenta"	"m"	
"yellow"	"y"	
"black"	"k"	
"white"	"w"	

Fig. 12.8 The list of line colors available in MATLAB

Example

```
x = 0:pi/100:2*pi;
y1 = sin(x);
y2 = sin(x-0.25);
plot(x,y1,'g',x,y2,'k')
```

Description

Specifying line colors. See Fig. 12.9.

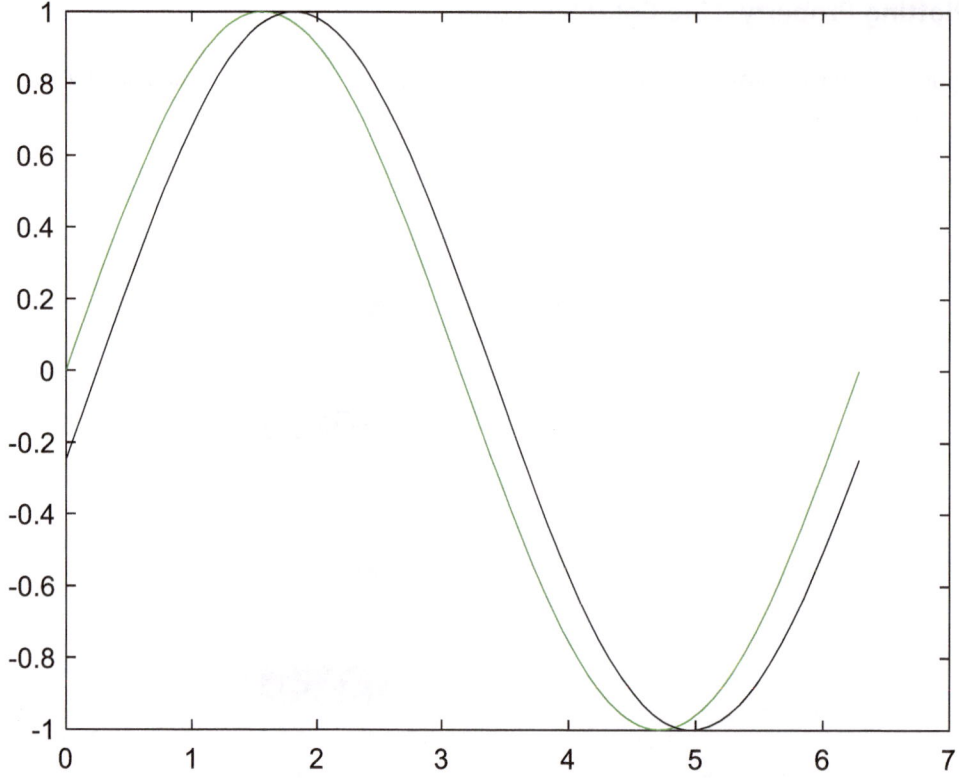

Fig. 12.9 Plotting with different line colors

Exercise

(a)   *Write the codes to plot the graph of cotangent function for the interval (0, pi) using magenta color.*

(b)   *Execute the codes in your computer, then draw the result(s) on the axes.*

12.5   The Plotting Property: Line Width Selection

- This property in the format plot(X,Y1,X,Y2,'LineWidth',*Number*) creates multiple line plots of the data with the specified line width.

```
x = 0:pi/100:2*pi;
y1 = sin(x);
y2 = cos(x);
plot(x,y1,x,y2,'LineWidth',2)
```

Specifying line width. See Fig. 12.10.

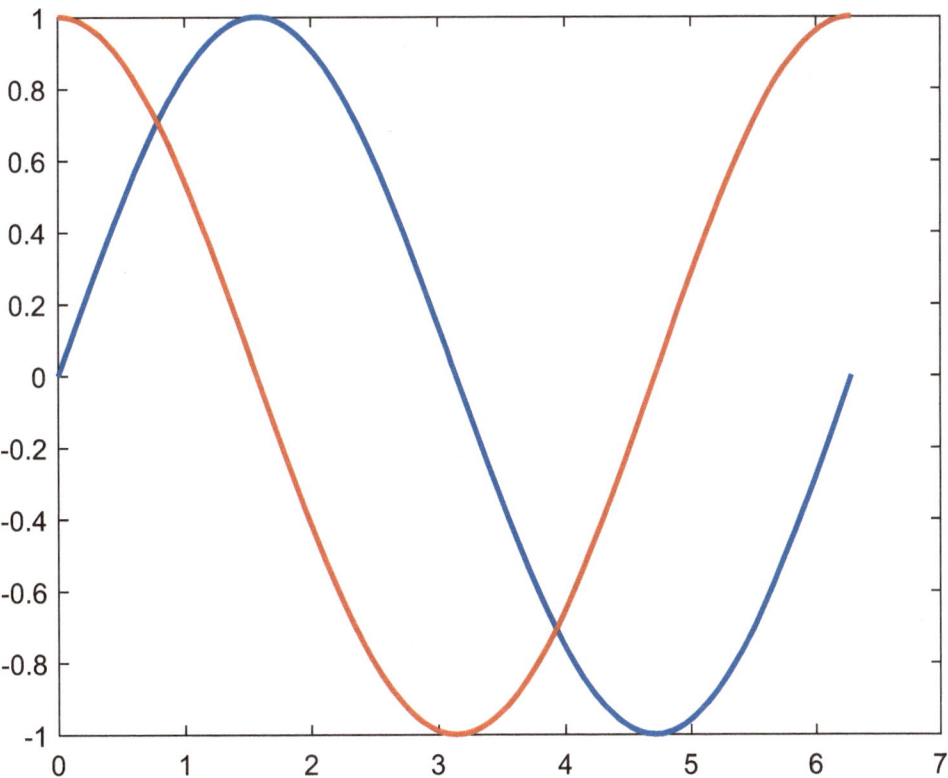

Fig. 12.10 Plotting with different line width

(a)   *Write the codes to plot the graph of tangent function for the interval (−pi/2, pi/2) using the line width 3.*
(b)   *Execute the codes in your computer, then draw the result(s) on the axes.*

12.6   The Reference Functions "title," "xlabel," "legend": Title, Axis Labels, and Legend Addition

- This reference function in the format title('*YourTitle*') adds a title to the graph.
- This reference function in the format xlabel('*YourXLabel*'), adds a label to the x-axis of the graph. This command can be used for y-axis and z-axis as well.
- This reference function in the format legend("*Legend1*", "*Legend2*") adds a legend to the graph.

Example

```
x = 0:pi/100:2*pi;
y1 = sin(x);
y2 = cos(x);
plot(x,y1,x,y2)
title('2-D Line Plot')
xlabel('x')
ylabel('sin(x) and cos(x)')
legend("sin(x)", "cos(x)")
```

Description

Adding title, axis labels, and legend. See Fig. 12.11.

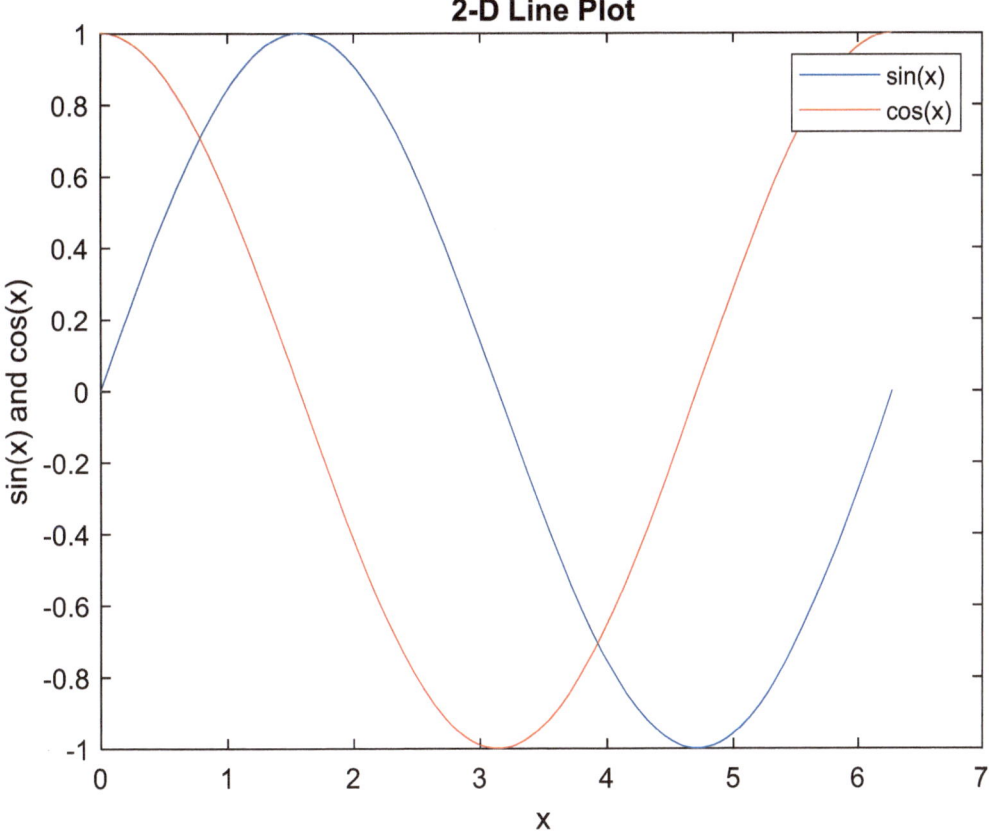

Fig. 12.11   Adding title, axis labels, and legend to the plot

Exercise

(a) *Write the codes to plot the graph of tangent and cotangent functions for the interval (0, pi/2). Add an appropriate title, axis labels, and legends.*

(b) *Execute the codes in your computer, then draw the result(s) on the axes.*

12.7   The Reference Function "xticks": Ticks Addition

- This reference function in the format xticks([*TicksVector*]) sets x-axis tick values, which are the locations along the x-axis where the tick marks appear. Specify ticks as a vector of increasing values. This command can be used for y-axis and z-axis as well.

Example

x = 0:pi/100:2*pi;
y1 = sin(x);
plot(x,y1)
xticks([0 2 4 6])
yticks([-1 -0.5 0 0.5 1])

Description

Adding x-axis and y-axis tick values on the graph. See Fig. 12.12.

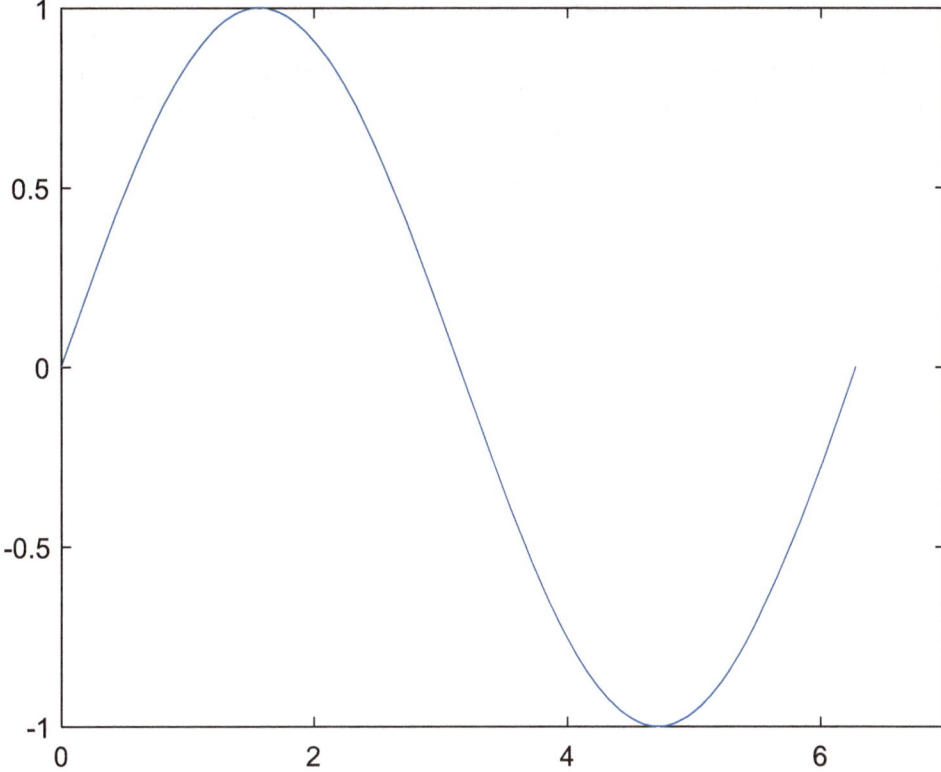

Fig. 12.12   Adding x-axis and y-axis tick values to the plot

Exercise

(a)   *Write the codes to plot the graph of tangent function for the interval (−pi/2, pi/2). Add ticks at −pi/2, −pi/3, −pi/6, 0, pi/6, pi/3, pi/2.*

(b)   *Execute the codes in your computer, then draw the result(s) on the axes.*

12.8   The Reference Function "xticklabels": Tick Labels Addition

- This reference function in the format xticklabels({*Labels*}) sets the x-axis tick labels for the current axes. Specify labels as a string array or a cell array of character vectors (e.g., 'x=x0','x=x1'). Tick labels must be set after tick values. This command can be used for y-axis and z-axis as well.

Example

```
x = 0:pi/100:2*pi;
y1 = sin(x);
plot(x,y1)
xticks([0 2 4 6])
xticklabels({'x = 0','x = 2','x = 4','x = 6'})
yticks([-1 -0.5 0 0.5 1])
yticklabels({'y = -1','y = -0.5','y = 0','y = 0.5','y = 1'})
```

Description

Adding x-axis and y-axis tick labels on the graph. See Fig. 12.13.

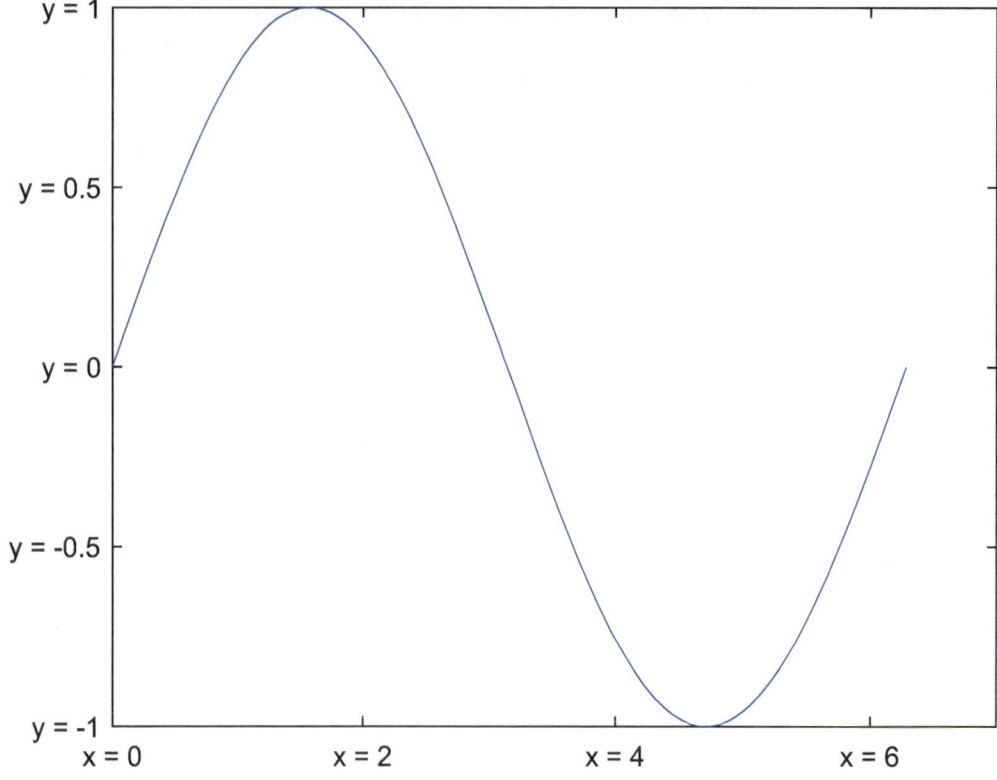

Fig. 12.13 Adding x-axis and y-axis tick labels to the plot

(a) *Write the codes to plot the graph of cotangent function for the interval (0, pi). Add tick labels at 0, pi/6, pi/3, pi/2, 2*pi/3, 5*pi/6, pi.*

(b) *Execute the codes in your computer, then draw the result(s) on the axes.*

12.9 The Reference Function "xtickangle": Tick Angle Selection

• This reference function in the format xtickangle(*angle*) rotates the x-axis tick labels for the current axes to the specified angle in degrees, where 0 is horizontal. Specify a positive value for counterclockwise rotation or a negative value for clockwise rotation. This command can be used for y-axis and z-axis as well.

```
x = 0:pi/100:2*pi;
y = sin(x);
plot(x,y)
xticks([0 2 4 6])
xticklabels({'x = 0','x = 2','x = 4','x = 6'})
xtickangle(45)
```

Rotating the x-axis tick labels about 45-degrees from the horizontal plane. See Fig. 12.14.

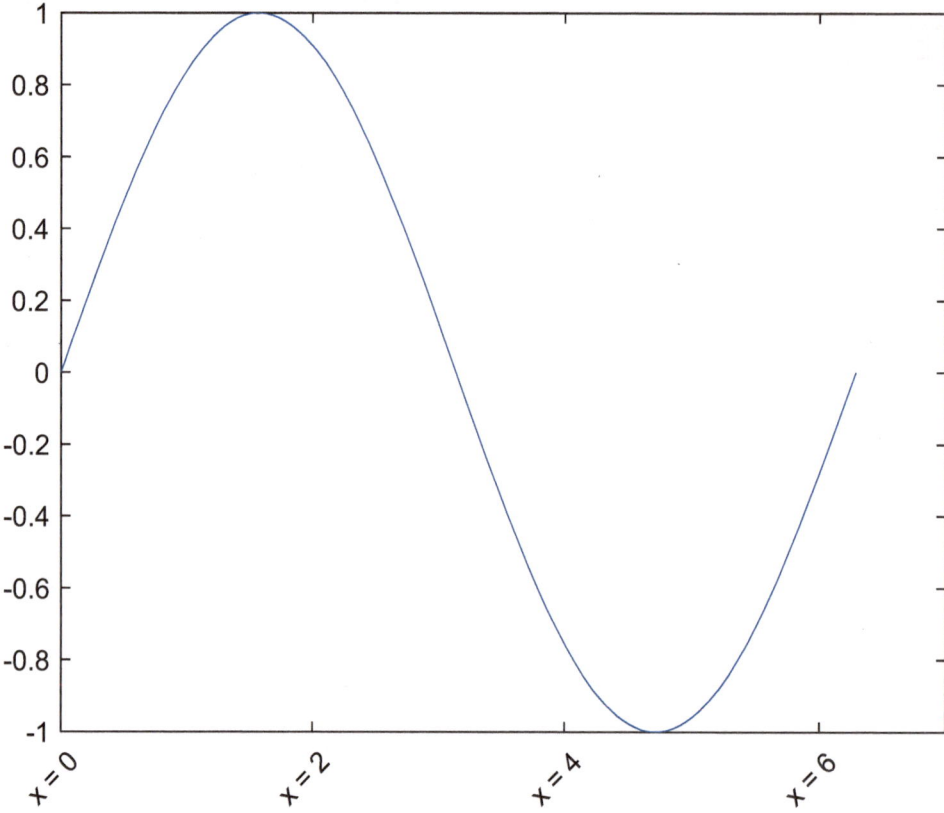

Fig. 12.14 The plot with rotated x-axis tick labels

12.10 The Reference Functions "xlim," "ylim," and "axis": Axis Limits Addition

- This reference function in the format xlim([xmin xmax]) sets the x-axis limits. This command can be used for y-axis and z-axis as well.
- This reference function in the format axis([xmin xmax ymin ymax zmin zmax]) specifies the limits for the current axes.

Example

x = 0:pi/100:2*pi;
y = sin(x);
plot(x,y)
xlim([0 5])
ylim([-1 0.9])

Description

Setting the limits on the axes. See Fig. 12.15.

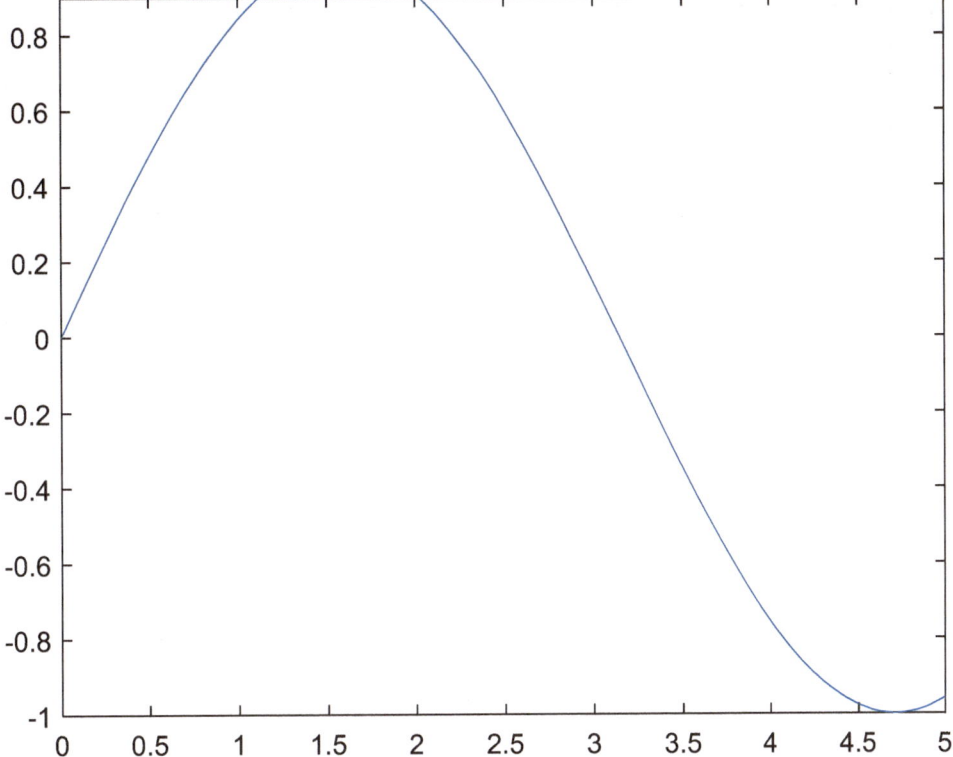

Fig. 12.15 The plot with the limits set on the axes using the reference functions "xlim" and "ylim"

```
x = 0:pi/100:2*pi;
y = sin(x);
plot(x,y)
axis([0 3*pi -1.5 1.5])
```

Setting the limits on the axes. See Fig. 12.16.

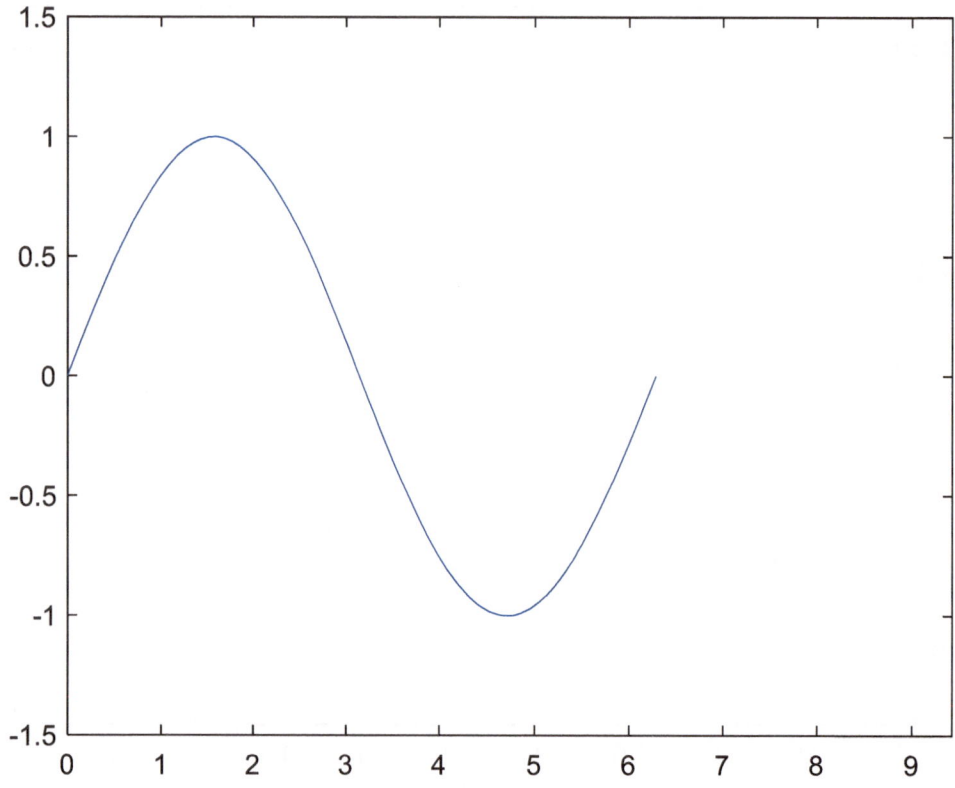

Fig. 12.16 The plot with the limits set on the axes using the reference function "axis"

12.11 The Reference Function "text": Text Addition

- This reference function in the format text(x,y,'text') adds a text description to one or more data points in the current axes using the text specified. To add text to one point, specify x and y as scalars. To add text to multiple points, specify x and y as vectors with equal length.

Example

```
x = 0:pi/20:2*pi;
y = sin(x);
plot(x,y)
text(pi,0,'\leftarrow sin(\pi)')
```

Description

Adding a text to the specified point. See Fig. 12.17.

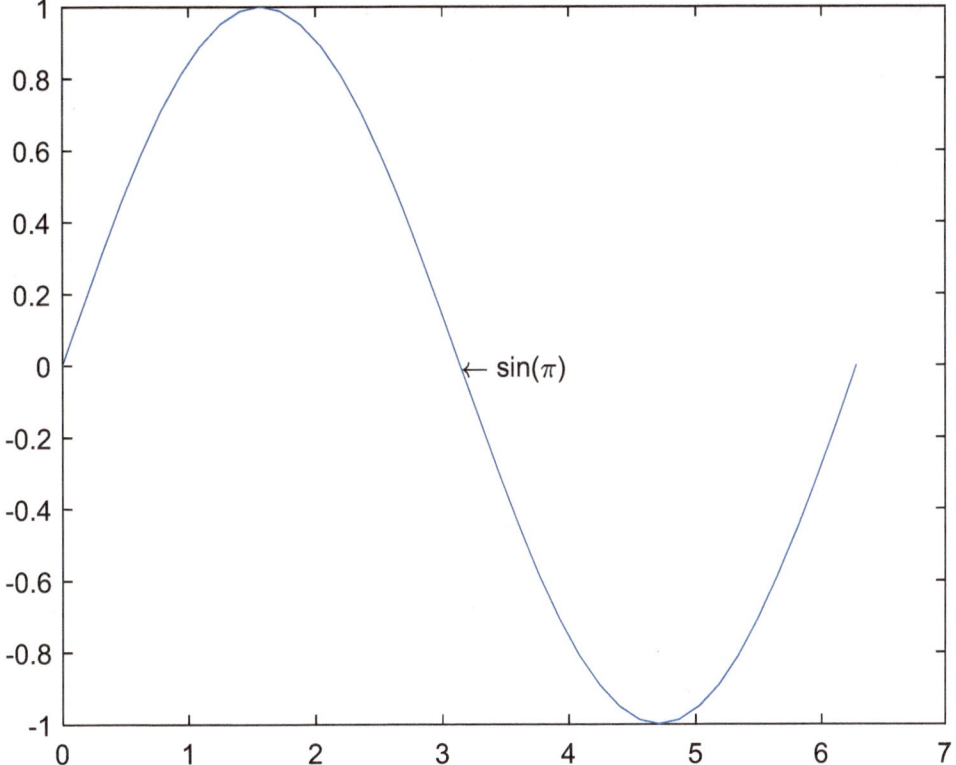

Fig. 12.17   Adding a text to the specified point of the graph

Example

```
x = -5:0.1:5;
y = x.^3-12*x;
plot(x,y)
xt = [-2 2];
yt = [16 -16];
str = 'dy/dx = 0';
text(xt,yt,str)
```

Description

Adding multiple texts to the specified points. See Fig. 12.18.

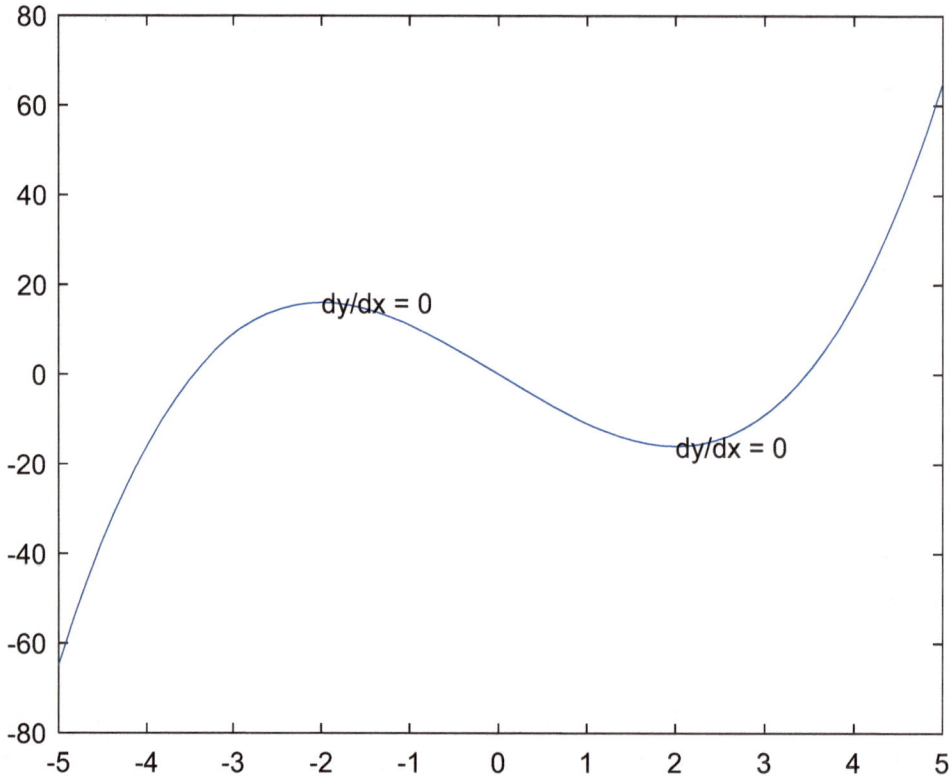

Fig. 12.18 Adding multiple texts to the specified points of the graph

```
x = -5:0.1:5;
y = x.^3-12*x;
plot(x,y)
str = {'This is a local','maximum ponit'};
text(-2,16,str)
```

Adding a text, including more than one line, to the specified point. See Fig. 12.19.

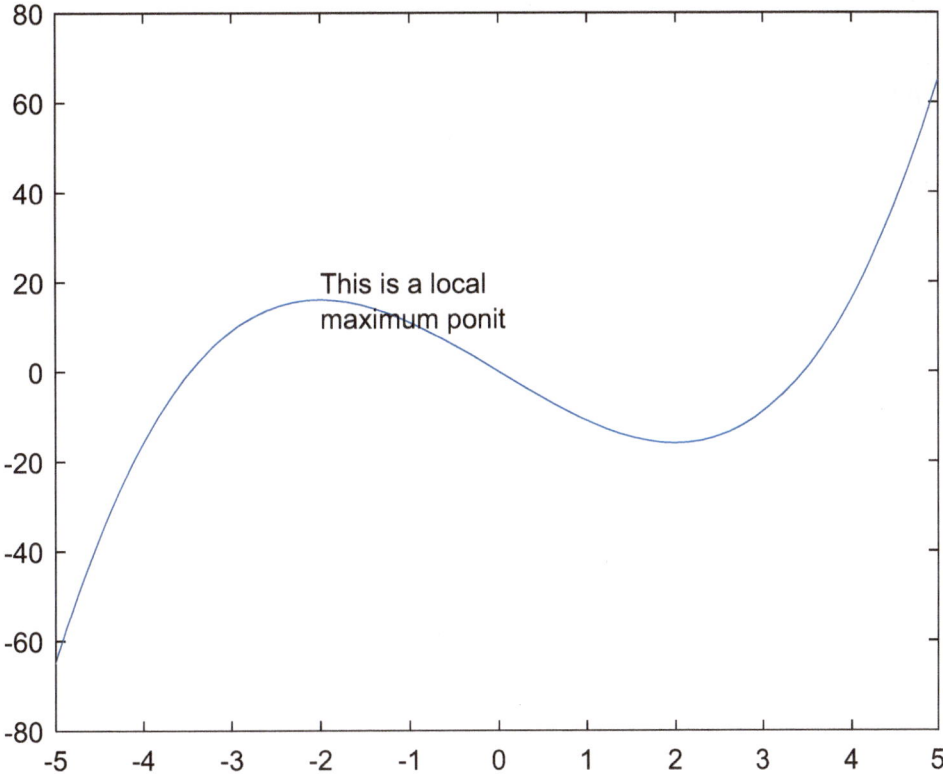

Fig. 12.19 Adding a text, including more than one line, to the specified point of the graph

```
x = -5:0.1:5;
y = x.^3-12*x;
plot(x,y)
text(-2,16,'local max','Color','red','FontSize',14)
```

Description

Adding a text with desirable color and size to the specified point. See Fig. 12.20.

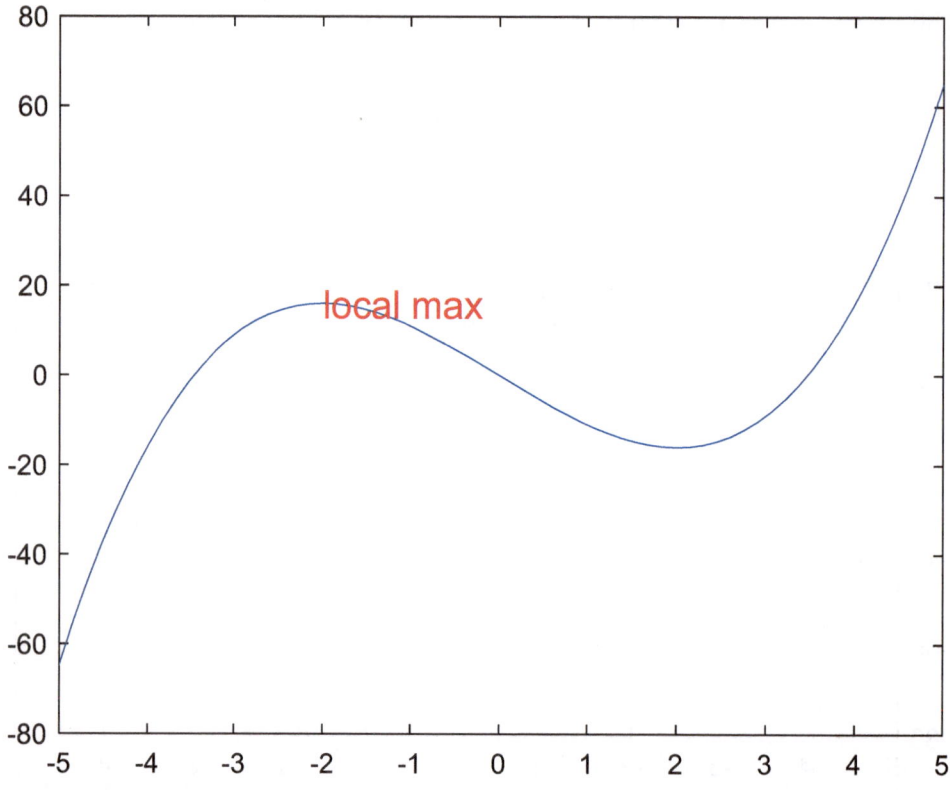

Fig. 12.20 Adding a text with desirable color and size to the specified point of the graph

Exercise

(a)   *Write the codes to plot the graph of tangent function for the interval (−pi/2, pi/2). Add a text using a left arrow to show the point (pi/4,1).*

(b)   *Execute the codes in your computer, then draw the result(s) on the axes.*

12.12 The Reference Function "grid on," "grid off": Grid Lines Addition

- The reference functions "grid on" and "grid off" display and hide the major grid lines for the current axes, respectively. Major grid lines extend from each tick mark.
- The reference function "grid minor" displays the minor grid lines for the current axes.

Example

```
x = 0:pi/100:2*pi;
y = sin(x);
plot(x,y)
grid on
```

Description

Adding major grid lines to the graph. See Fig. 12.21.

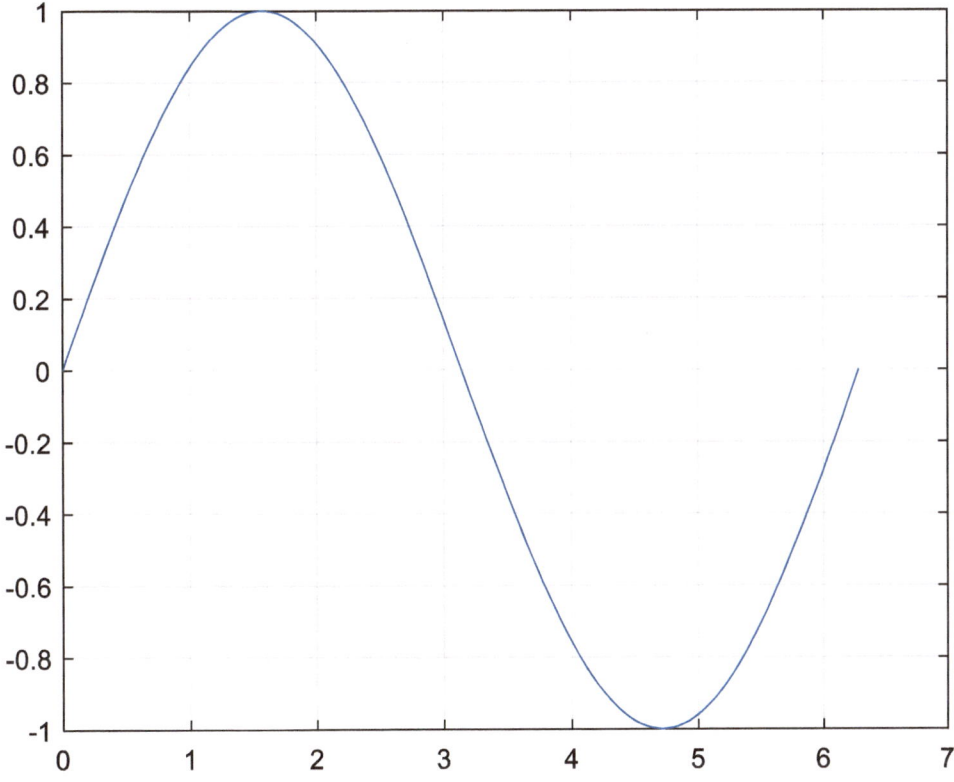

Fig. 12.21 Adding major grid lines to the graph

Example

```
x = 0:pi/100:2*pi;
y = sin(x);
plot(x,y)
grid minor
```

Description

Adding minor grid lines to the graph. See Fig. 12.22.

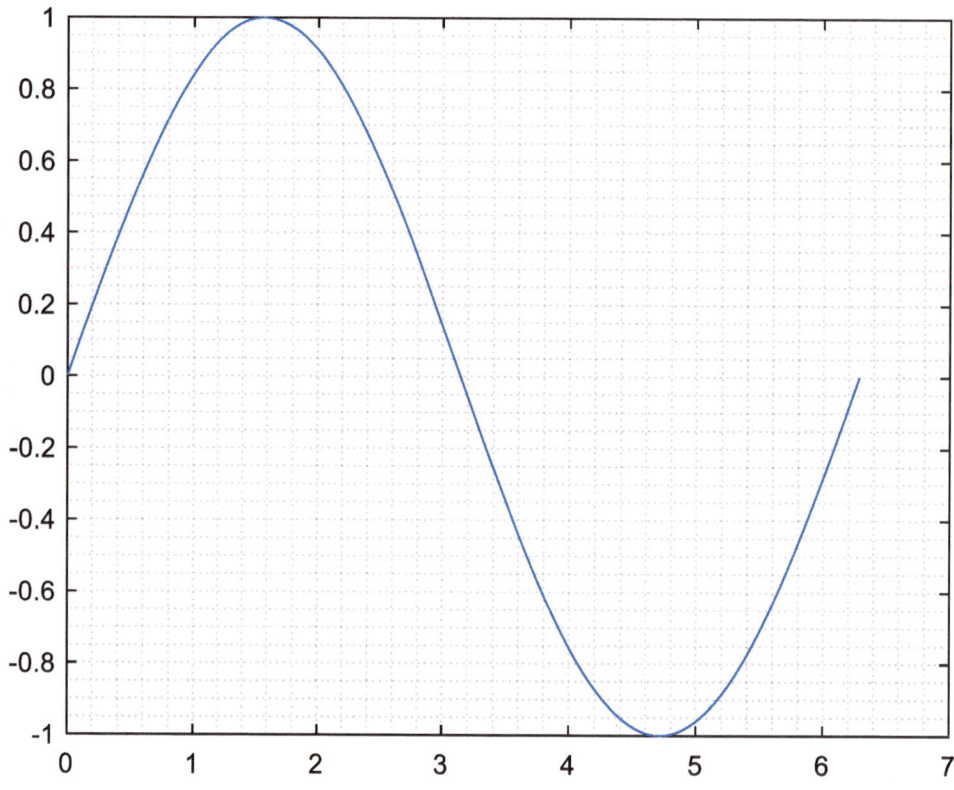

Fig. 12.22 Adding minor grid lines to the graph

12.13 The Reference Functions "hold on," "hold off": Plots Holding

- The reference function "hold on" retains plots in the current axes so that new plots added to the axes do not delete existing plots. New plots use the next colors and line styles based on the ColorOrder and LineStyleOrder properties of the axes.
- The reference function "hold off" sets the hold state to off so that new plots added to the axes clear existing plots and reset all axes properties. The next plot added to the axes uses the first color and line style based on the ColorOrder and LineStyleOrder properties of the axes.

Example

```
x = 0:pi/100:2*pi;
y1 = sin(x);
plot(x,y1)
hold on
y2 = cos(x);
plot(x,y2)
hold off
```

Description

Using the functions to add a second-line plot without deleting the existing line plot. See Fig. 12.23.

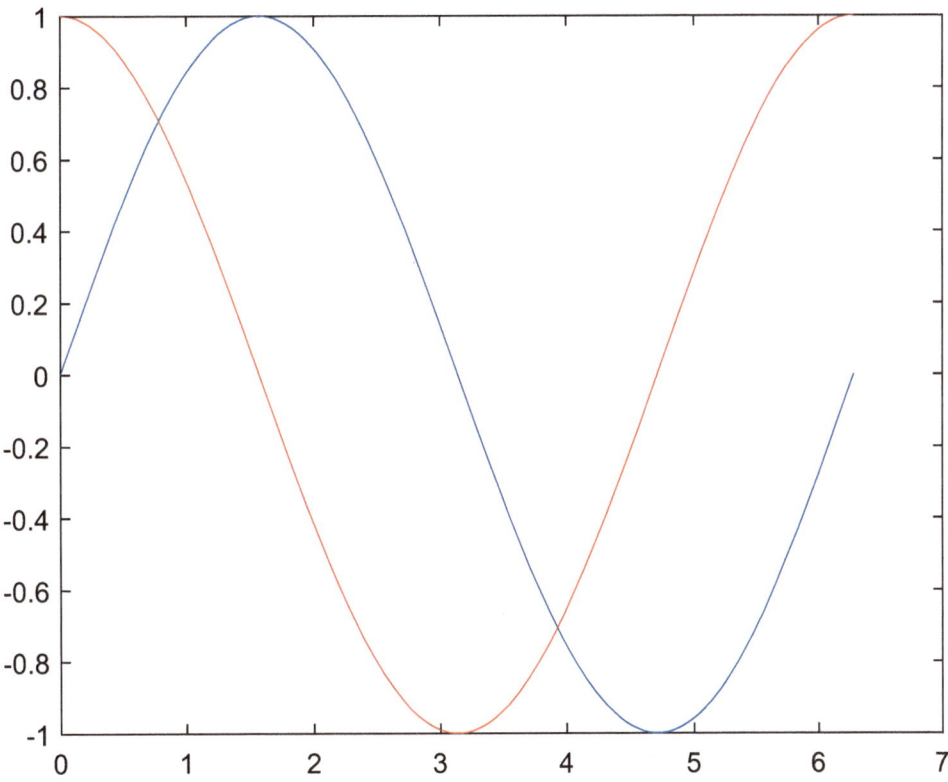

Fig. 12.23   Adding a second-line plot while holding the existing line plot

Exercise

(a) *Write the codes to plot the graph of tangent and cotangent functions for the interval (0, pi/2) using the "hold on" reference function.*

(b) *Execute the codes in your computer, then draw the result(s) on the axes.*

12.14 The Reference Function "subplot": Stacked Plots Creation

- The reference function in the format subplot(m,n,p) divides the current figure into an m-by-n grid and creates axes in the position specified by p. The first subplot is the first column of the first row, the second subplot is the second column of the first row, and so on.

Example

```
x = 0:pi/100:2*pi;

subplot(2,2,1);
y1 = sin(1*x);
plot(x,y1)
title('Subplot 1: sin(x)')

subplot(2,2,2);
y2 = sin(2*x);
plot(x,y2)
title('Subplot 2: sin(2x)')

subplot(2,2,3);
y3 = sin(3*x);
plot(x,y3)
title('Subplot 3: sin(3x)')

subplot(2,2,4);
y4 = sin(4*x);
plot(x,y4)
title('Subplot 4: sin(4x)')
```

Description

Creating a figure with four stacked subplots. See Fig. 12.24.

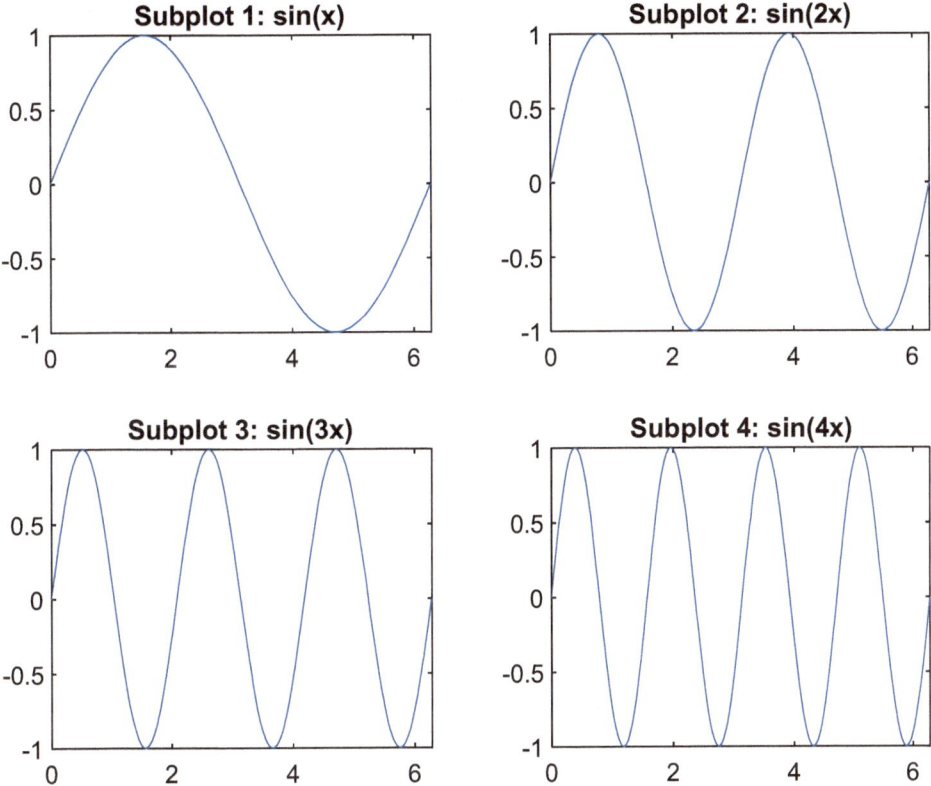

Fig. 12.24   Creation of a figure with four stacked subplots using the reference function "subplot"

Exercise

(a)   *Write the codes to plot the graph of tangent function for the interval (−pi/2, pi/2) and the graph of cotangent function for the interval (0, pi) using the "subplot" reference function.*

(b)   *Execute the codes in your computer, then draw the result(s) on the axes.*

12.15 The Reference Function "tiledlayout": Stacked Plots Creation

• The reference function in the format tiledlayout(m,n) creates a tiled chart layout for displaying multiple plots in the current figure. The layout has a fixed m-by-n tile arrangement that can display up to m*n plots.

Example

```
x = 0:pi/100:2*pi;
tiledlayout(2,2);

nexttile
y1 = sin(1*x);
plot(x,y1)
title('Subplot 1: sin(x)')

nexttile
y2 = sin(2*x);
plot(x,y2)
```

```
title('Subplot 2: sin(2x)')

nexttile
y3 = sin(3*x);
plot(x,y3)
title('Subplot 3: sin(3x)')

nexttile
y4 = sin(4*x);
plot(x,y4)
title('Subplot 4: sin(4x)')
```

Description

Creating a figure with four stacked subplots. See Fig. 12.25.

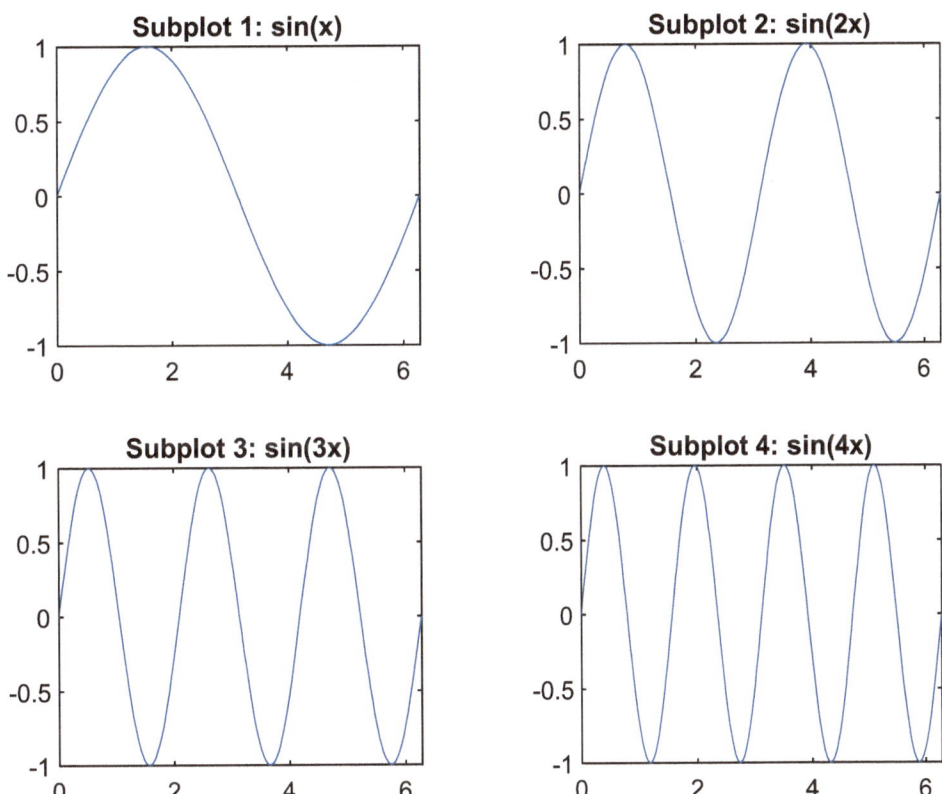

Fig. 12.25 Creation of a figure with four stacked subplots using the reference function "tiledlayout"

Exercise

(a) *Write the codes to plot the graph of tangent function for the interval (−pi/2, pi/2) and the graph of cotangent function for the interval (0, pi) using the "tiledlayout" reference function.*

(b) *Execute the codes in your computer, then draw the result(s) on the axes.*

12.16 The Reference Function "yyaxis": Two Y-Axes Plots Creation

- The reference function in the format "yyaxis left" activates the side of the current axes associated with the left y-axis. Subsequent graphics commands target the left side. If the current axes do not include two y-axes, then this command adds a second y-axis. If there are no axes, then this command first creates them.
- The reference function in the format "yyaxis right" activates the side of the current axes associated with the right y-axis. Subsequent graphics commands target the right side.

Example

```
x = 0:0.01:10;
y = sin(3*x);
yyaxis left
plot(x,y)
z = sin(3*x).*exp(0.5*x);
yyaxis right
plot(x,z)
ylim([-150 150])
```

Creating axes with a y-axis on both the left and right sides. See Fig. 12.26.

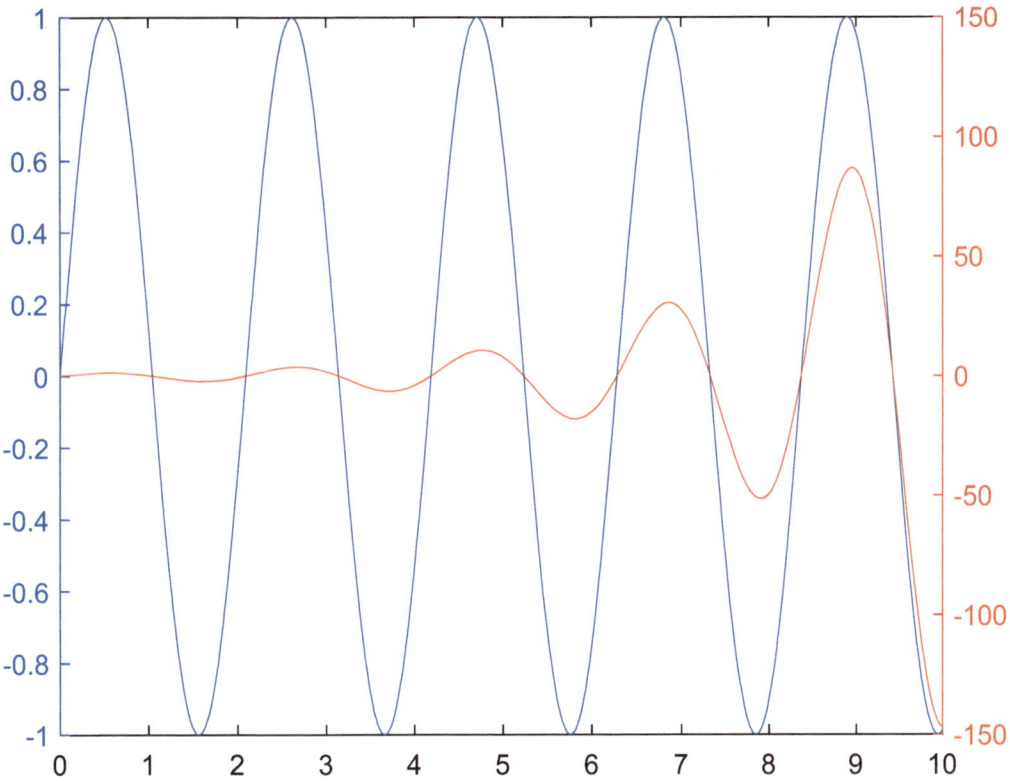

Fig. 12.26 Creation of axes with a y-axis on both the left and right sides

Exercise

(a)   *Write the codes to plot the graph of cosine and tangent functions for the interval (−pi/2, pi/2) using two y-axes.*
(b)   *Execute the codes in your computer, then draw the result(s) on the axes.*

12.17   The Reference Function "linspace": Linearly Spaced Vector Generation

- This reference function in the format linspace(x1,x2) generates a row vector of 100 evenly spaced points between x1 and x2.
- This reference function in the format linspace(x1,x2,n) generates n points, where the spacing between the points is (x2−x1)/(n−1).

Example

linspace(-4,4,9)

Output

ans =

  -4   -3   -2   -1   0   1   2   3   4

Description

Generating a linearly spaced vector with nine elements.

x = linspace(0,2*pi);
y = sin(x);
plot(x,y)

Plotting a sine function using the reference function "linspace." See Fig. 12.27.

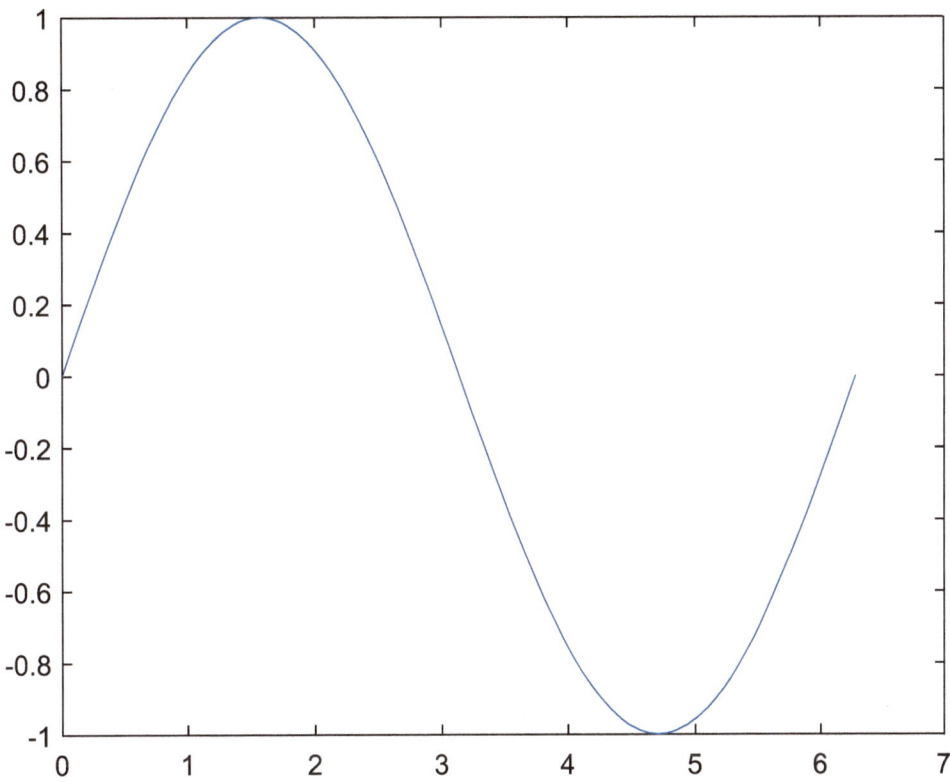

Fig. 12.27 Plotting a sine function using the reference function "linspace"

Exercise

(a)   *Write the codes to plot the graph of cosine function for the interval (−pi, pi) using the first format of the reference function.*

(b)   *Execute the codes in your computer, then draw the result(s) on the axes.*

12.18 The Reference Function "logspace": Logarithmically Spaced Vector Generation

- This reference function in the format y = logspace(a,b) generates a row vector of 50 logarithmically spaced points between decades 10^a and 10^b.
- This reference function in the format logspace(a,b,n) generates n points between decades 10^a and 10^b.

Example

logspace(-2,2,5)

Output

ans =
   0.0100   0.1000   1.0000   10.0000   100.0000

Description

Generating a logarithmically spaced vector with five elements.

12.19 The Reference Function "figure": Figure Window Creation

- The reference function in the format "figure" creates a new figure window using default property values.

- The reference function in the format *Name*.Position returns a vector with four elements as the position and dimensions of the figure. The elements, respectively, indicate the number of pixels to the right as the horizontal position, the number of pixels above the bottom left corner of the primary display as the vertical position, the number of pixels as the width of the figure, and the number of pixels as the height of the figure.
- The reference function in the formats figure('Name','*Your Preferred Name*') and figure('Name','*Your Preferred Name'*,'NumberTitle','off') assigns a name to a figure.
- The reference function in the format Name.Position(3:4) = [a b] modifies the size of the current figure in which "a" and "b" are the number of pixels of the width and the number of pixels of the height of the figure, respectively.

Example

Name = figure;

Description

Creating a new figure window. See Fig. 12.28.

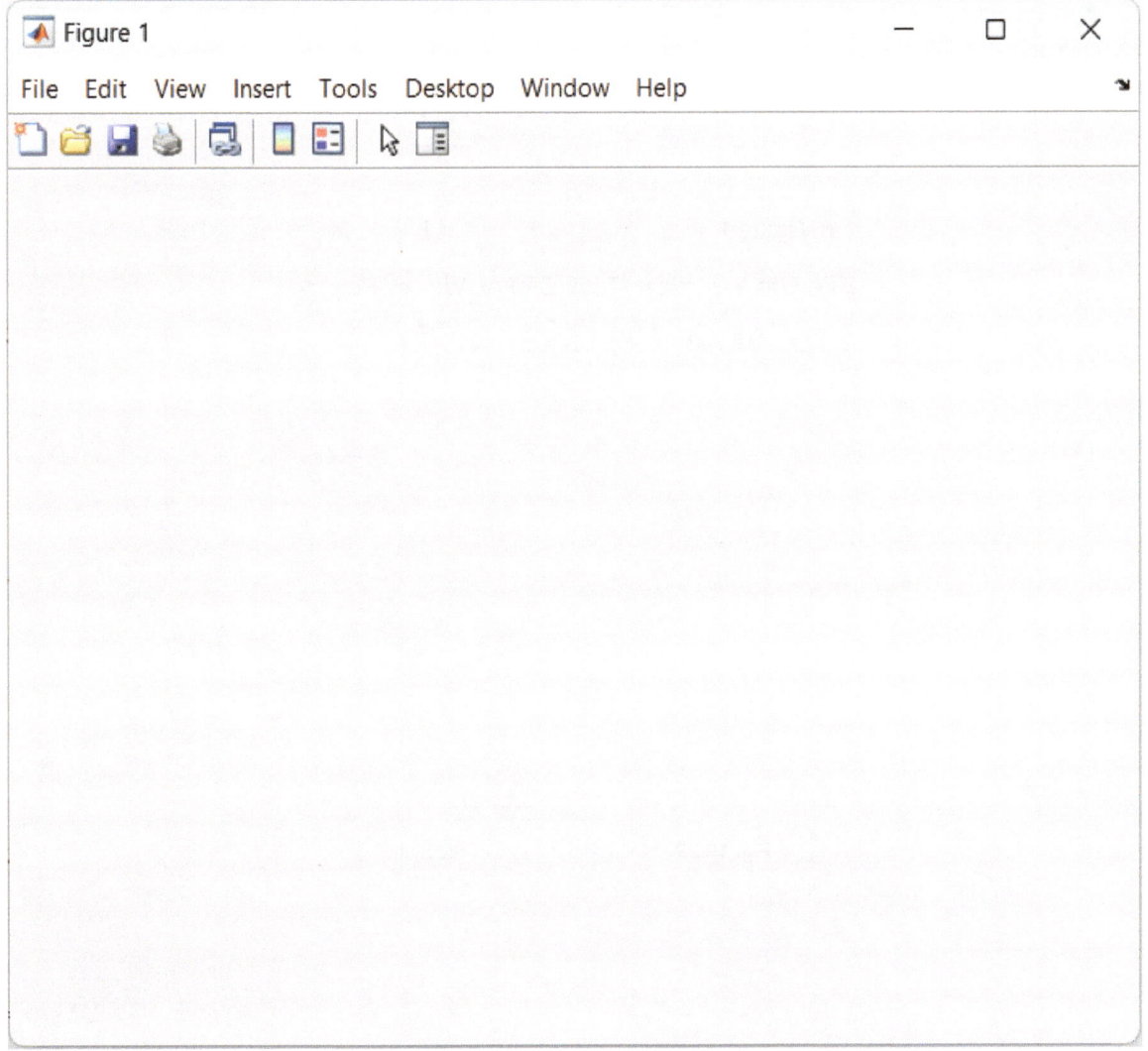

Fig. 12.28 Creation of a new figure window

Example

Name.Position

Output

ans =
   488   242   560   420

Description

Getting the location, width, and height of the figure.

Example

Name.Position(3:4) = [280 210];

Description

Halving the figure width and height by adjusting the third and fourth elements of the position vector. See Fig. 12.29.

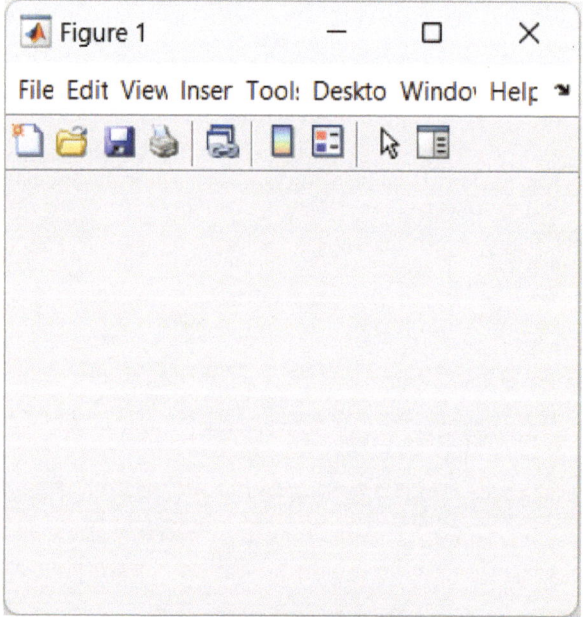

Fig. 12.29 Modifying the size of the figure

Example

figure('Name','Name of This Figure is . . .');

Description

Specifying a name to the figure. See Fig. 12.30.

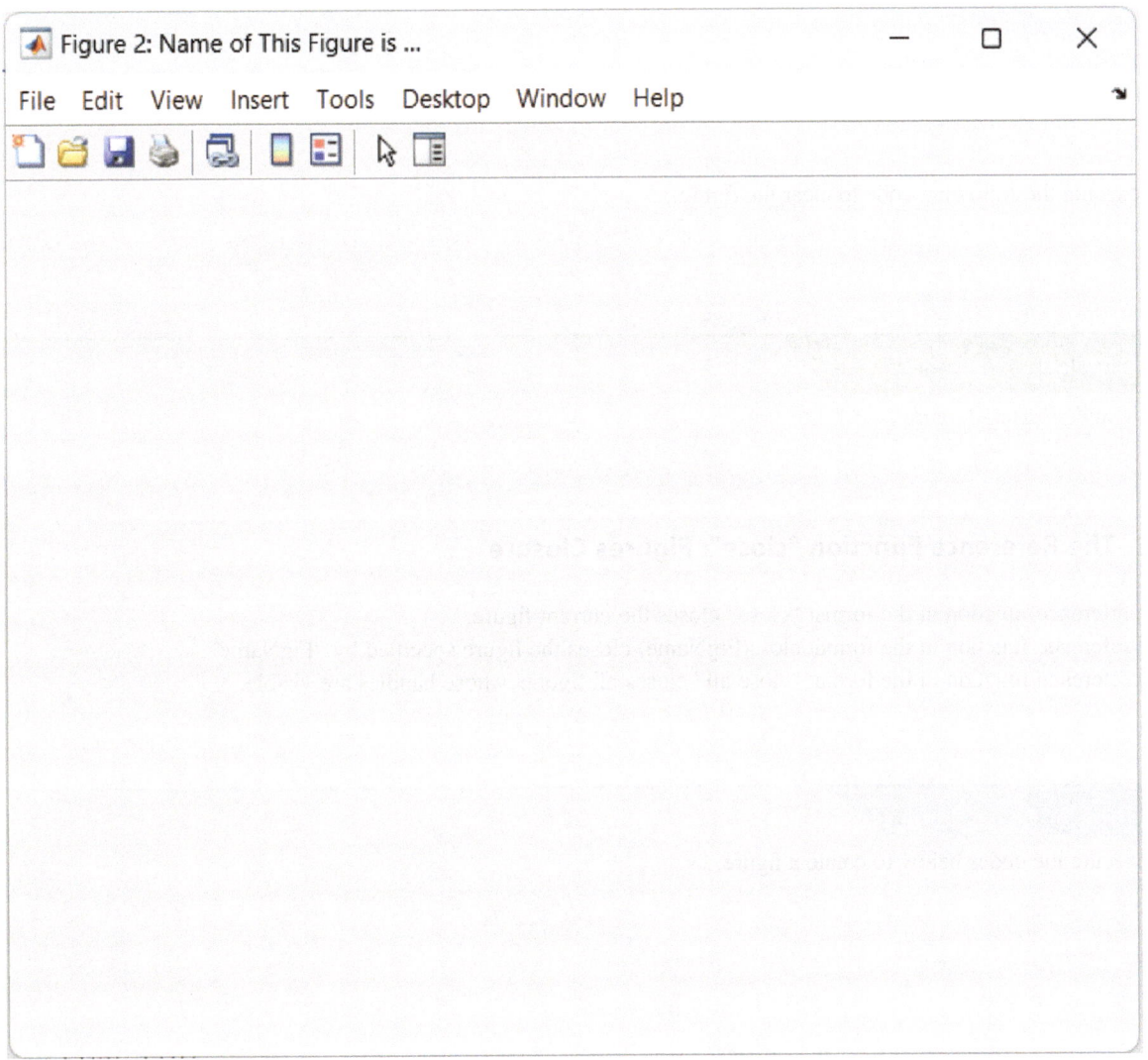

Fig. 12.30   Specification of a name to the figure

Exercise

Execute the codes below in your computer and write the result(s) in the assigned location(s).

figure('Name','Name of This Figure is . . .','NumberTitle','off');

Question: What difference do you see between the result of these codes and the one of the last example?

12.20 The Reference Function "clf": Figures Clearance

- The reference function in the format "clf" clears the current figure.
- The reference function in the format "clf(FigName)" clears the figure specified by "FigName".

Example

First, execute the codes below to create a figure.

```
x = 0:pi/100:2*pi;
y = sin(x);
plot(x,y)
```

Then, execute the following code to clear the figure.

```
clf
```

Description

Creating and clearing a figure.

12.21 The Reference Function "close": Figures Closure

- The reference function in the format "close" closes the current figure.
- The reference function in the format close(FigName) closes the figure specified by "FigName".
- The reference function in the format "close all" closes all figures whose handles are visible.

Example

First, execute the codes below to create a figure.

```
x = 0:pi/100:2*pi;
y = sin(x);
plot(x,y)
```

Then, execute the following code to close the figure.

```
close
```

Description

Creating and closing a figure.

References

1. MATLAB 2023a.
2. Rahmani-Andebili, M. (2024). *Mathematics of engineering and science – Practice problems, methods, and solutions*. Springer Nature.
3. Rahmani-Andebili, M. (2022). *Differential equations – Practice problems, methods, and solutions*. Springer Nature.
4. Rahmani-Andebili, M. (2023). *Calculus III – Practice problems, methods, and solutions*. Springer Nature.
5. Rahmani-Andebili, M. (2023). *Calculus II – Practice problems, methods, and solutions*. Springer Nature.
6. Rahmani-Andebili, M. (2023). *Calculus I – Practice problems, methods, and solutions* (2nd ed.). Springer Nature.
7. Rahmani-Andebili, M. (2021). *Calculus – Practice problems, methods, and solutions*. Springer Nature.
8. Rahmani-Andebili, M. (2024). *Precalculus – Practice problems, methods, and solutions* (2nd ed.). Springer Nature.
9. Rahmani-Andebili, M. (2021). *Precalculus – Practice problems, methods, and solutions*. Springer Nature.

Abstract

In this chapter, the 3-D plot in MATLAB is presented and described. In this regard, several examples and exercises for each section of the chapter are presented. The exercises that include writing the codes, executing them, and achieving the results need to be done by students to master programming skills. In this book, the codes, outputs, and descriptions are in blue, black, and green colors, respectively. To program in MATLAB, a script file can be created and saved with an appropriate name (e.g. untitled01) in the preferred directory of a computer. The program can be run by clicking on the "Run" available on the top toolbar of the script in MATLAB or calling the script by typing its name in Command Window or in the other scripts.

13.1 3-D Plot

In the following, the description of 3-D plot is presented and exemplified [1–9].

- This plot in the format plot3(X,Y,Z) plots coordinates in 3-D space.
- To plot a set of coordinates connected by line segments, specify X, Y, and Z as vectors of the same length.
- To plot multiple sets of coordinates on the same set of axes, specify at least one of X, Y, or Z as a matrix and the others as vectors.

Example

```
t = 0:pi/50:10*pi;
x = sin(t);
y = cos(t);
plot3(x,y,t)
```

Description

Plotting a 3-D function. See Fig. 13.1.

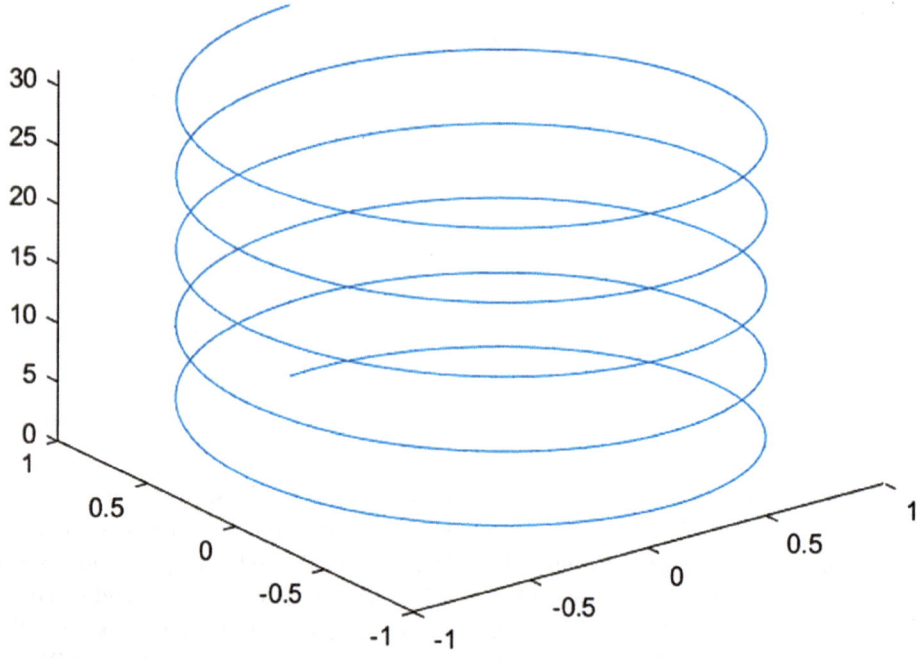

Fig. 13.1 Plotting a 3-D function

Exercise

(a)   *Write the codes to plot the previous graph while rotating in the opposite direction.*
(b)   *Execute the codes in your computer, then draw the result(s) on the axes.*

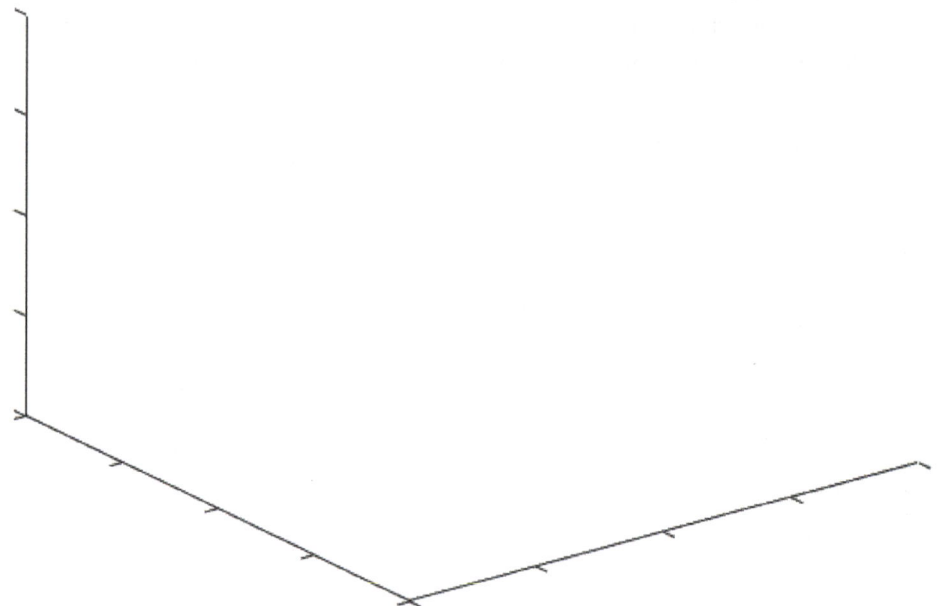

Example

```
t = 0:pi/500:pi;
xt1 = sin(t).*cos(10*t);
yt1 = sin(t).*sin(10*t);
zt1 = cos(t);
xt2 = sin(t).*cos(12*t);
yt2 = sin(t).*sin(12*t);
zt2 = cos(t);
plot3(xt1,yt1,zt1,xt2,yt2,zt2)
```

Description

Plotting multiple 3-D functions. See Fig. 13.2.

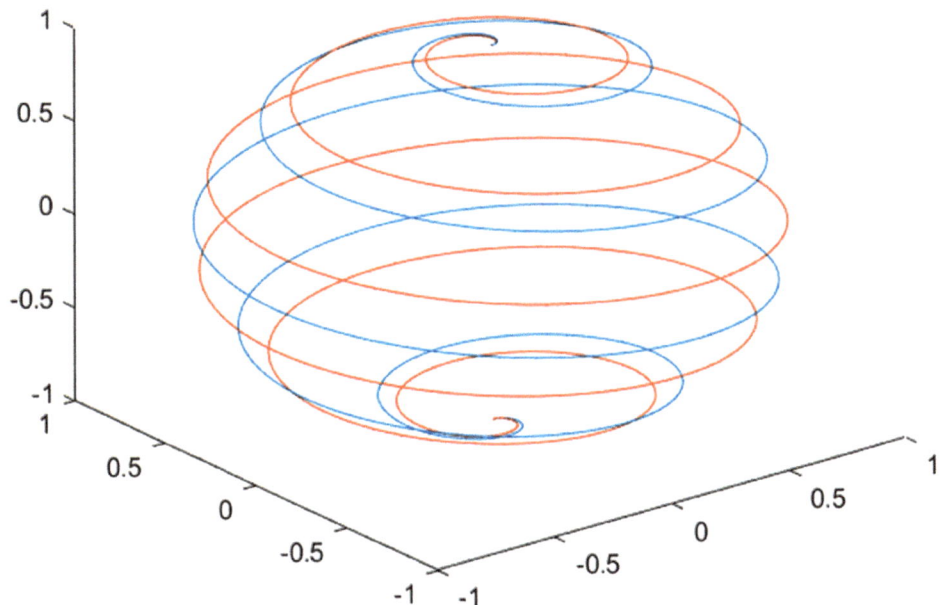

Fig. 13.2 Plotting multiple 3-D functions

Demonstration

Execute the codes in your computer. Then, draw the result(s) on the axes.

```
t = 0:pi/500:pi;
xt1 = sin(t).*sin(10*t);
yt1 = sin(t).*cos(10*t);
zt1 = cos(t);
xt2 = sin(t).*sin(12*t);
yt2 = sin(t).*cos(12*t);
zt2 = cos(t);
plot3(xt1,yt1,zt1,xt2,yt2,zt2)
```

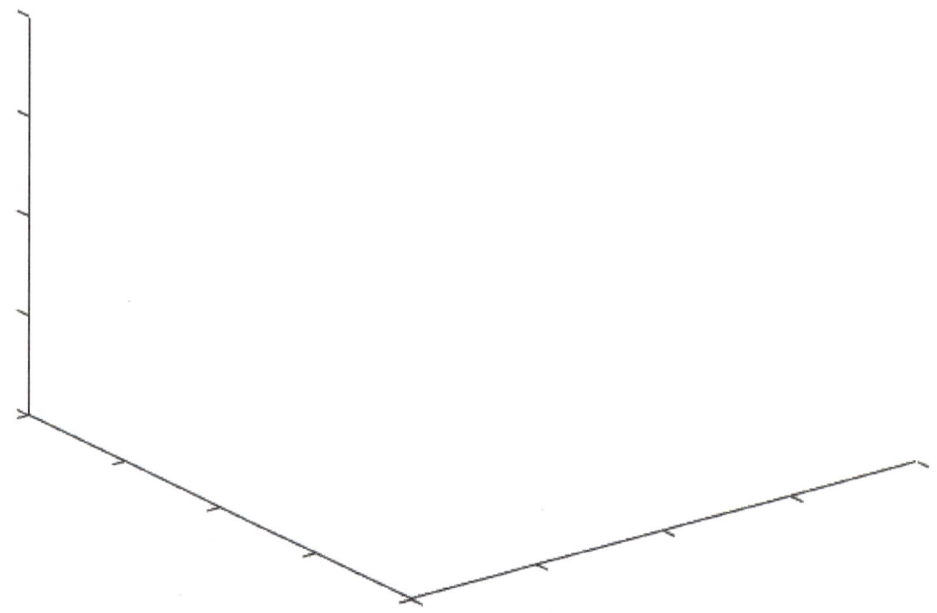

Example

t = 0:pi/500:pi;
X = [sin(t).*cos(10*t); sin(t).*cos(12*t); sin(t).*cos(20*t)];
Y = [sin(t).*sin(10*t); sin(t).*sin(12*t); sin(t).*sin(20*t)];
Z = cos(t);
plot3(X,Y,Z)

Description

Plotting multiple 3-D functions using matrices. See Fig. 13.3.

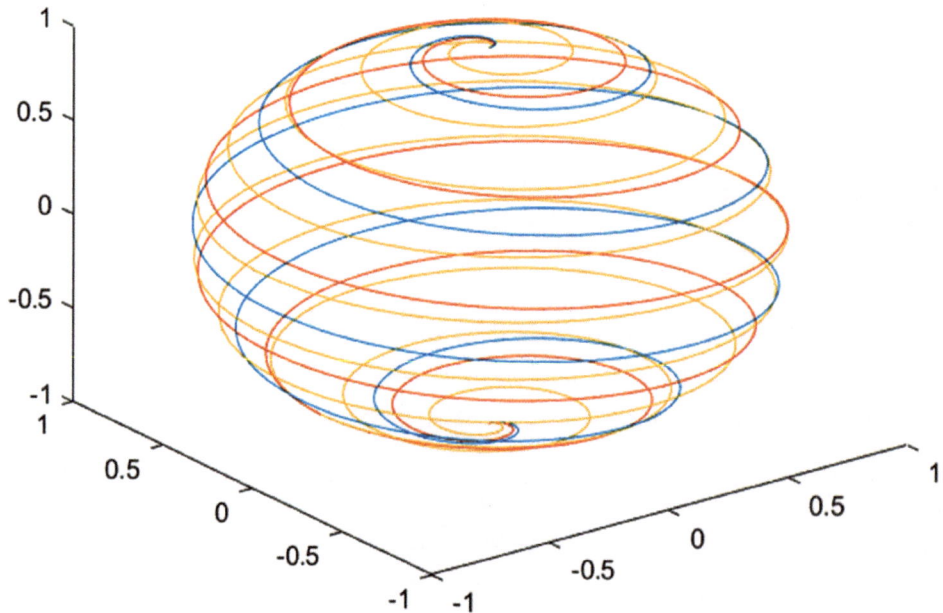

Fig. 13.3 Plotting multiple 3-D functions using matrices

```
t = 0:pi/500:pi;
X = [sin(t).*cos(10*t); sin(t).*cos(12*t); sin(t).*cos(20*t)];
Y = [sin(t).*sin(10*t); sin(t).*sin(12*t); sin(t).*sin(20*t)];
Z = cos(t);
plot3(X,Y,Z)
xlabel('x(t)')
ylabel('y(t)')
zlabel('z(t)')
title('3-D Plots')
grid on
```

Adding x, y, z labels, title, and grid to the graph. See Fig. 13.4.

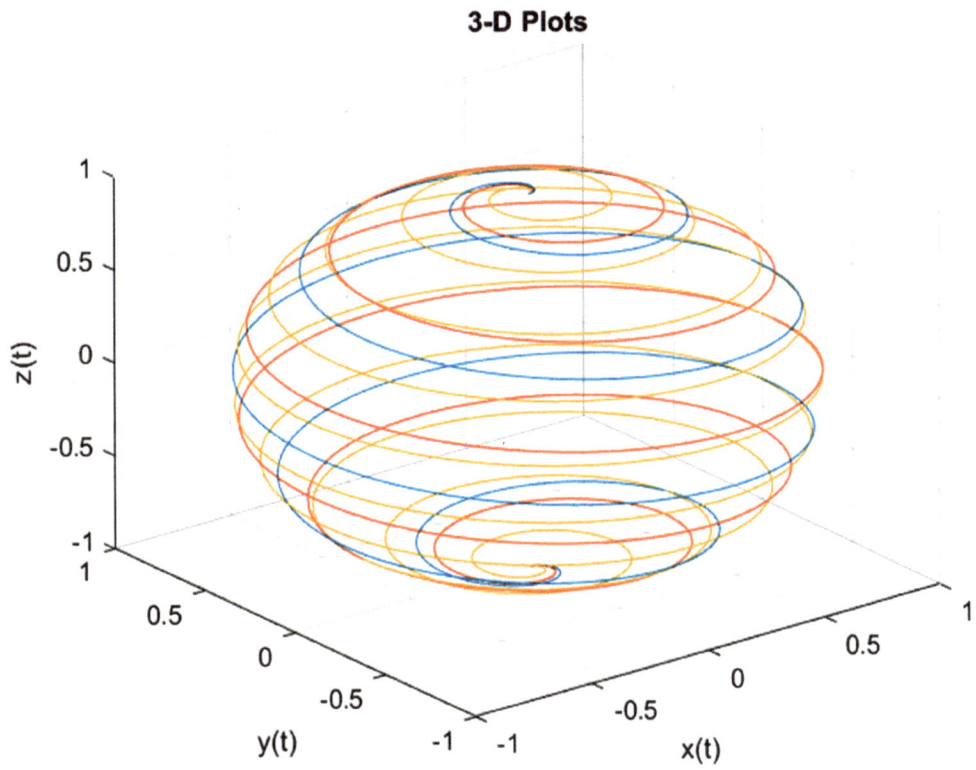

Fig. 13.4 Adding x, y, z labels, title, and grids to the graph

```
t = 0:pi/500:40*pi;
xt = (3 + cos(sqrt(32)*t)).*cos(t);
yt = sin(sqrt(32) * t);
zt = (3 + cos(sqrt(32)*t)).*sin(t);
plot3(xt,yt,zt,'m')
axis equal
```

Description

Plotting the graph with a specific color (See Fig. 13.5). The list of line colors available in MATLAB is shown in Fig. 13.6.

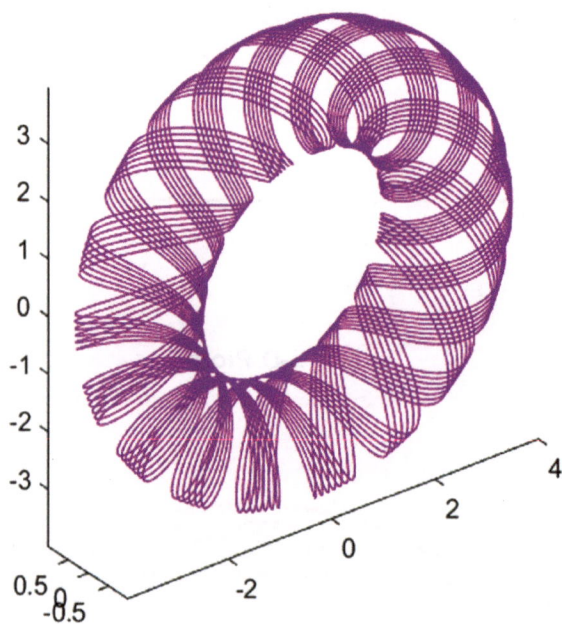

Fig. 13.5 Plotting the graph with a specific color

Color Name	Short Name	Appearance
"red"	"r"	
"green"	"g"	
"blue"	"b"	
"cyan"	"c"	
"magenta"	"m"	
"yellow"	"y"	
"black"	"k"	
"white"	"w"	

Fig. 13.6 The list of line colors available in MATLAB

```
t = 0:pi/20:10*pi;
xt = sin(t);
yt = cos(t);
plot3(xt,yt,t,'o')
```

Plotting the graph with a specific marker (See Fig. 13.7). The list of line markers available in MATLAB is shown in Fig. 13.8.

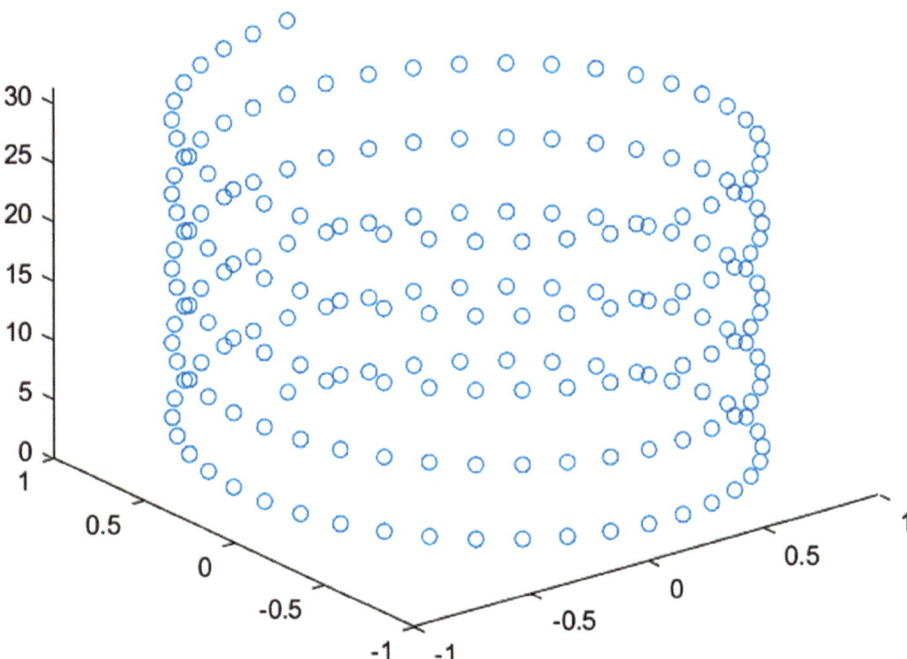

Fig. 13.7 Plotting the graph with a specific marker

Marker	Description	Resulting Marker
"o"	Circle	○
"+"	Plus sign	+
"*"	Asterisk	✳
"."	Point	•
"x"	Cross	×
"_"	Horizontal line	—
"|"	Vertical line	|
"square"	Square	□
"diamond"	Diamond	◇
"^"	Upward-pointing triangle	△
"v"	Downward-pointing triangle	▽
">"	Right-pointing triangle	▷
"<"	Left-pointing triangle	◁
"pentagram"	Pentagram	☆
"hexagram"	Hexagram	✡

Fig. 13.8 The list of line markers available in MATLAB

Example

t = 0:pi/20:10*pi;
xt1 = sin(t);
yt1 = cos(t);
xt2 = sin(2*t);
yt2 = cos(2*t);
plot3(xt1,yt1,t,xt2,yt2,t,'--')

Description

Plotting the graph with specific line styles (See Fig. 13.9). The list of line styles available in MATLAB is shown in Fig. 13.10.

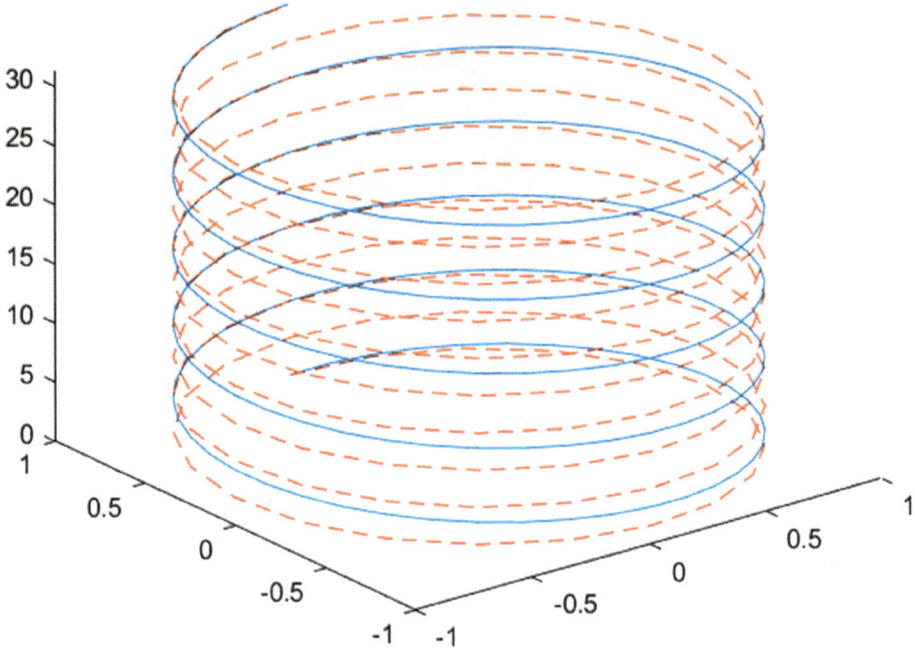

Fig. 13.9 Plotting the graph with specific line styles

Line Style	Description	Resulting Line
"‐"	Solid line	————
"‐‐"	Dashed line	‐ — — — —
":"	Dotted line	················
"‐."	Dash-dotted line	—.—.—.—..

Fig. 13.10 The list of line styles available in MATLAB

Example

t = linspace(-10,10,1000);
xt = exp(-t./10).*sin(5*t);
yt = exp(-t./10).*cos(5*t);
plot3(xt,yt,t,'LineWidth',2);

Description

Plotting the graph with a specific line width. See Fig. 13.11.

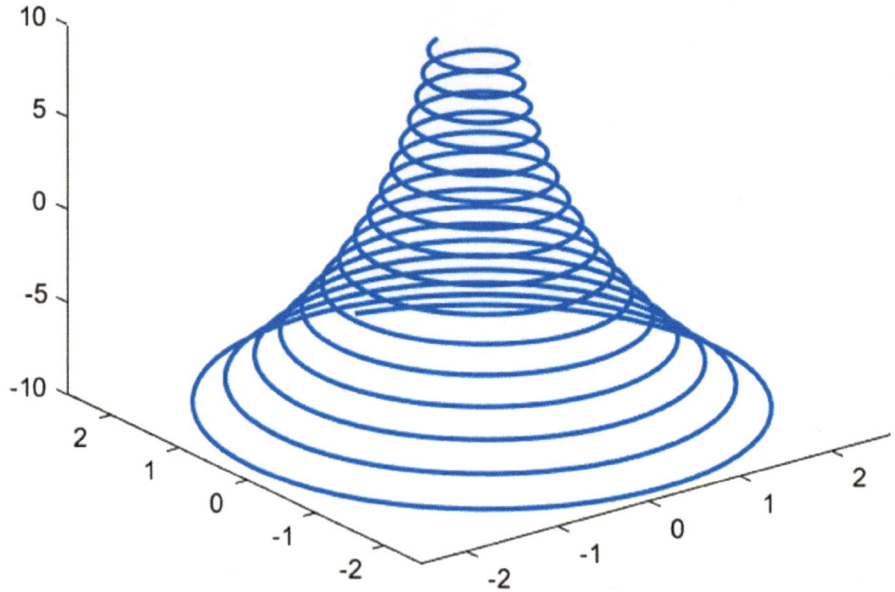

Fig. 13.11 Plotting the graph with a specific line width

Exercise

Execute the codes in your computer. Then, draw the result(s) on the axes.

```
t = linspace(-10,10,1000);
xt = exp(-t./10).*sin(5*t);
yt = exp(-t./10).*cos(5*t);
plot3(xt,yt,t);
hold on
t = linspace(-10,10,1000);
xt = exp(t./10).*sin(5*t);
yt = exp(t./10).*cos(5*t);
plot3(xt,yt,t);
```

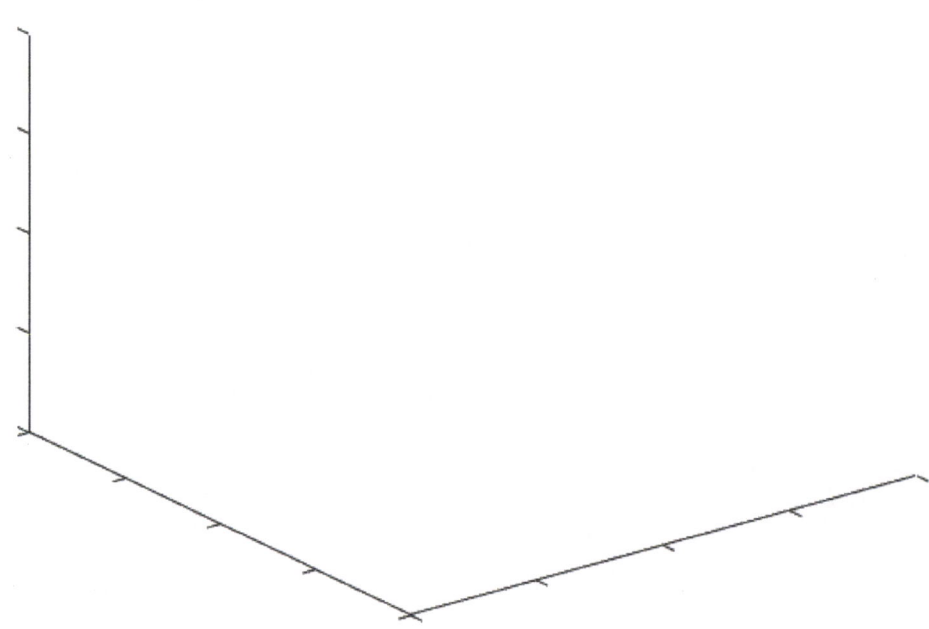

References

1. MATLAB 2023a.
2. Rahmani-Andebili, M. (2024). *Mathematics of engineering and science – Practice problems, methods, and solutions*. Springer Nature.
3. Rahmani-Andebili, M. (2022). *Differential equations – Practice problems, methods, and solutions*. Springer Nature.
4. Rahmani-Andebili, M. (2023). *Calculus III – Practice problems, methods, and solutions*. Springer Nature.
5. Rahmani-Andebili, M. (2023). *Calculus II – Practice problems, methods, and solutions*. Springer Nature.
6. Rahmani-Andebili, M. (2023). *Calculus I – Practice problems, methods, and solutions* (2nd ed.). Springer Nature.
7. Rahmani-Andebili, M. (2021). *Calculus – Practice problems, methods, and solutions*. Springer Nature.
8. Rahmani-Andebili, M. (2024). *Precalculus – Practice problems, methods, and solutions* (2nd ed.). Springer Nature.
9. Rahmani-Andebili, M. (2021). *Precalculus – Practice problems, methods, and solutions*. Springer Nature.

Abstract

In this chapter, the stairstep graph, which is one of the plot types in MATLAB, is presented and described. In this regard, several examples and exercises for each section of the chapter are presented. The exercises that include writing the codes, executing them, and achieving the results need to be done by students to master programming skills. In this book, the codes, outputs, and descriptions are in blue, black, and green colors, respectively. To program in MATLAB, a script file can be created and saved with an appropriate name (e.g., untitled01) in the preferred directory of a computer. The program can be run by clicking on the "Run" available on the top toolbar of the script in MATLAB or by calling the script by typing its name in Command Window or in the other scripts.

14.1 Stairstep Graph

In the following, the description of stairstep graph plot is presented and exemplified [1].

- This plot in the format stairs(Y) draws a stairstep graph of the elements in Y. If Y is a vector, then it draws one line. If Y is a matrix, then it draws one line per matrix column.
- This plot in the format stairs(X,Y) plots the elements in Y at the locations specified by X. The inputs X and Y must be vectors or matrices of the same size. Additionally, X can be a row or column vector and Y must be a matrix with length (X) rows.

Example

```
X = linspace(0,4*pi,40);
Y = sin(X);
stairs(Y)
```

Description

Creating a stairstep plot of a function. See Fig. 14.1.

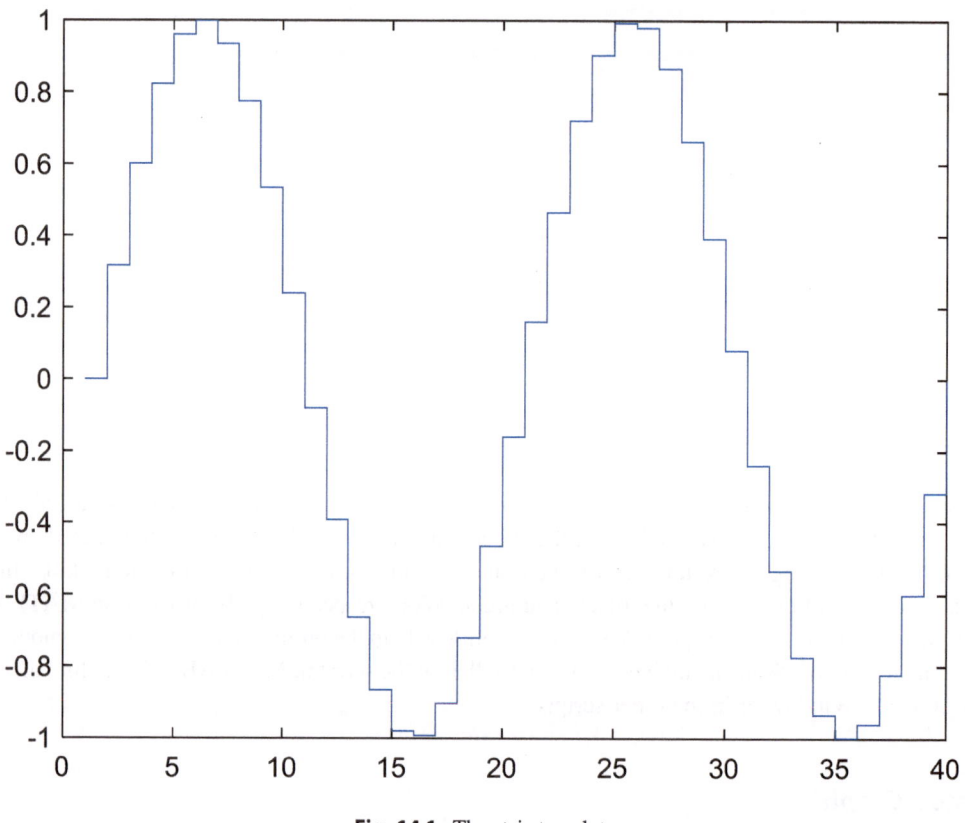

Fig. 14.1 The stairstep plot

Exercise

(a) *Write the codes to plot the graph of the cosine function for the interval (0,4*pi).*
(b) *Execute the codes in your computer, then draw the result(s) on the axes.*

Example

X = linspace(0,4*pi,50)';
Y = [0.5*cos(X), 2*cos(X)];
stairs(Y)

Description

Creating multiple stairstep plots. See Fig. 14.2.

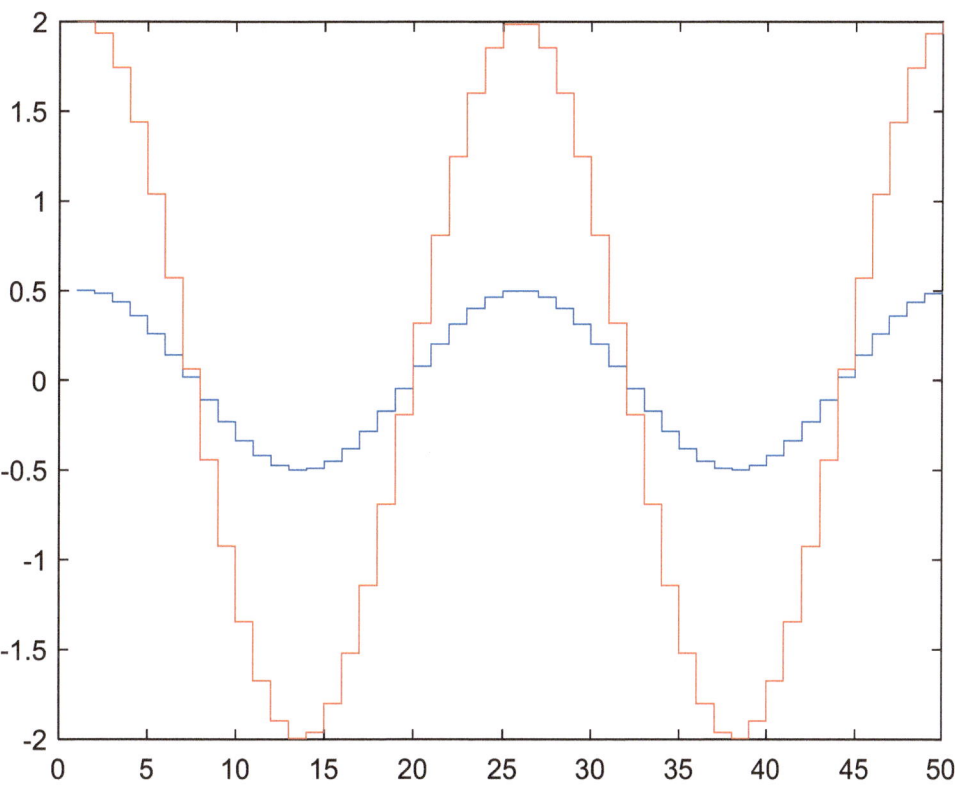

Fig. 14.2   The multiple stairstep plots

Example

X = linspace(0,4*pi,20);
Y = sin(X);
stairs(Y, '-.om')

Description

Plotting the graph by using specific line style, marker, and color (See Fig. 14.3). The lists of line styles, markers, and colors available in MATLAB are shown in Figs. 14.4, 14.5 and 14.6, respectively.

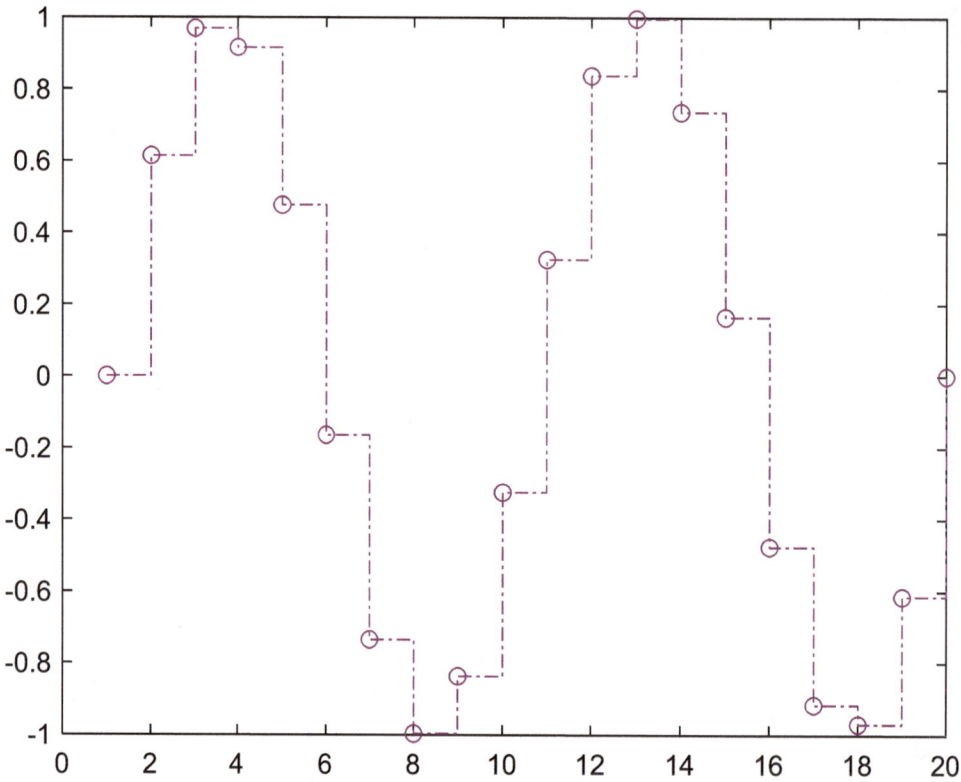

Fig. 14.3 Plotting the graph by using specific line style, marker, and color

Line Style	Description	Resulting Line
" - "	Solid line	——————
" - - "	Dashed line	- — — — —
" : "	Dotted line	....
" - . "	Dash-dotted line	—.—.—.—..

Fig. 14.4 The list of line styles available in MATLAB

Marker	Description	Resulting Marker
"o"	Circle	○
"+"	Plus sign	+
"*"	Asterisk	✳
"."	Point	•
"x"	Cross	×
"_"	Horizontal line	—
"|"	Vertical line	|
"square"	Square	□
"diamond"	Diamond	◇
"^"	Upward-pointing triangle	△
"v"	Downward-pointing triangle	▽
">"	Right-pointing triangle	▷
"<"	Left-pointing triangle	◁
"pentagram"	Pentagram	☆
"hexagram"	Hexagram	✡

Fig. 14.5 The list of line markers available in MATLAB

Color Name	Short Name	Appearance
"red"	"r"	
"green"	"g"	
"blue"	"b"	
"cyan"	"c"	
"magenta"	"m"	
"yellow"	"y"	
"black"	"k"	
"white"	"w"	

Fig. 14.6 The list of line colors available in MATLAB

Exercise

(a) *Write the codes to plot the graph of the cosine function for the interval (0,4*pi) using dotted line, asterisk marker, and yellow color.*

(b) *Execute the codes in your computer, then draw the result(s) on the axes.*

Example

X = linspace(0,4*pi,20);
Y = sin(X);
stairs(Y,'LineWidth',2)

Description

Plotting the graph by using a specific line width. See Fig. 14.7.

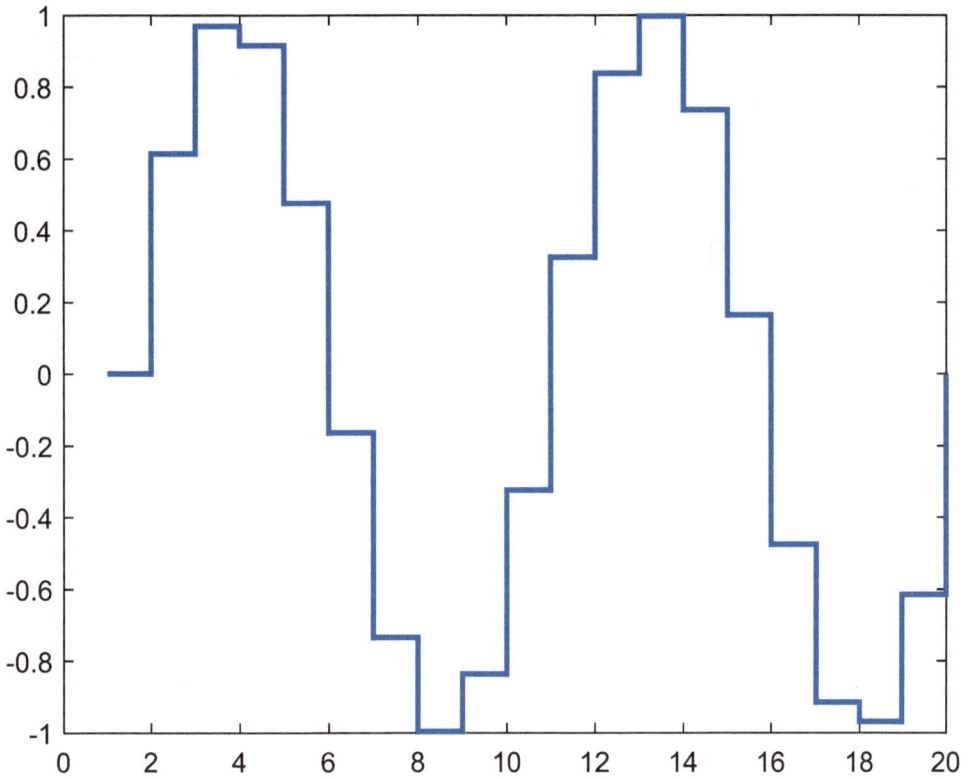

Fig. 14.7 Plotting the graph by using a specific line width

```
X = linspace(0,4*pi,50)';
Y = [0.5*cos(X), 2*cos(X)];
stairs(Y)
title('Stairstep Graph')
xlabel('x')
ylabel('0.5cos(x) and 2cos(x)')
legend("0.5cos(x)", "2cos(x)")
grid on
```

Adding x and y labels, title, legend, and grids to the graph. See Fig. 14.8.

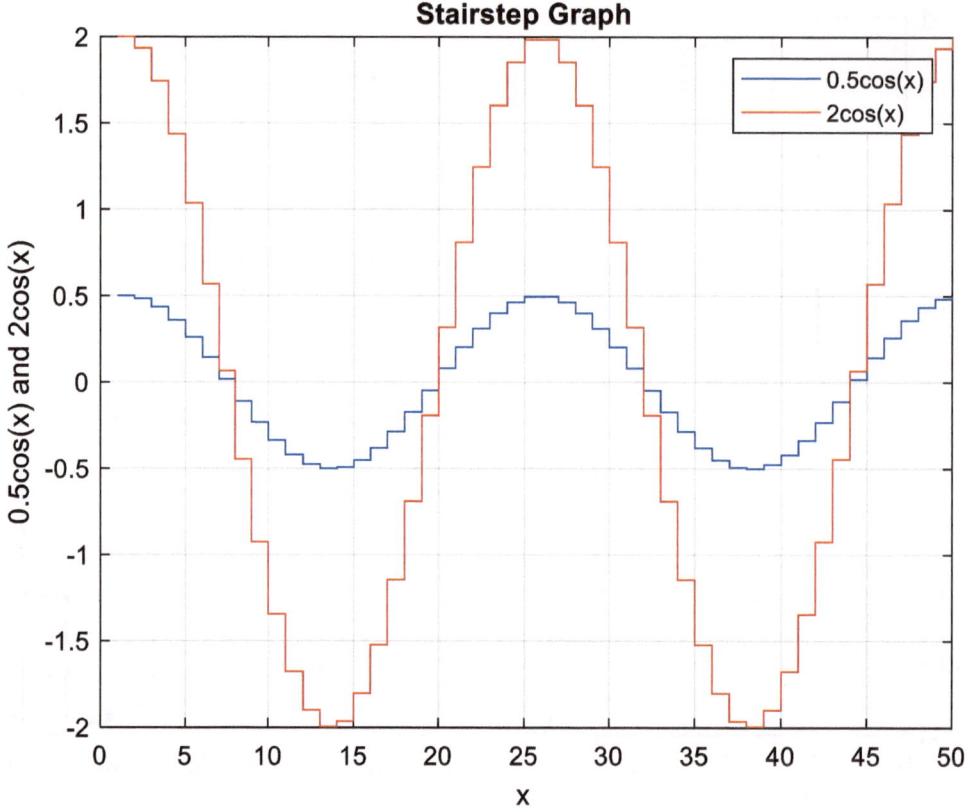

Fig. 14.8 Adding x and y labels, title, legend, and grids to the graph

Reference

1. MATLAB 2023a.

Abstract

In this chapter, the line plot with error bars in MATLAB is presented and described. In this regard, several examples and exercises for each section of the chapter are presented. The exercises that include writing the codes, executing them, and achieving the results need to be done by students to master programming skills. In this book, the codes, outputs, and descriptions are in blue, black, and green colors, respectively. To program in MATLAB, a script file can be created and saved with an appropriate name (e.g. untitled01) in the preferred directory of a computer. The program can be run by clicking on the "Run" available on the top toolbar of the script in MATLAB or calling the script by typing its name in Command Window or in the other scripts.

15.1 Line Plot with Error Bars

In the following, the description of line plot with error bars is presented and exemplified [1].

- This plot in the format errorbar(y,err) creates a line plot of the data in y and draws a vertical error bar at each data point. The values in err determine the lengths of each error bar above and below the data points. Thus, the total error bar lengths are double the err values.
- This plot in the format errorbar(x,y,err) plots y versus x and draws a vertical error bar at each data point.
- This plot in the format errorbar(x,y,neg,pos) draws a vertical error bar at each data point, where neg determines the length below the data point and pos determines the length above the data point, respectively.
- This plot in the format errorbar(,ornt) sets the orientation of the error bars. Specify ornt as "horizontal" for horizontal error bars or "both" for both horizontal and vertical error bars. The default for ornt is "vertical," which draws vertical error bars.

Example

```
x = 1:10:100;
y = [10 16 22 20 31 35 40 35 45 40];
err = 6*ones(size(y));
errorbar(x,y,err)
```

Description

Displaying vertical error bars (that are equal in length) on the line plot. See Fig. 15.1.

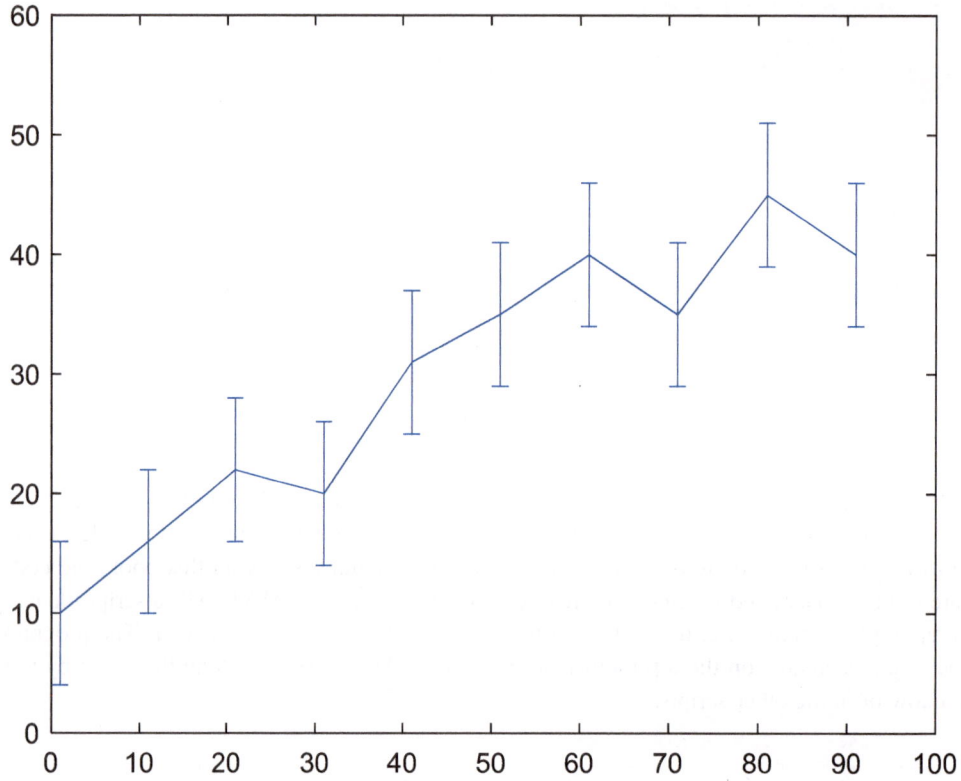

Fig. 15.1   Displaying vertical error bars (with equal length) on the line plot

Exercise

Execute the codes in your computer. Then, draw the result(s) on the axes.

x = linspace(0,10,15);
y = sin(x/2);
err = 0.3*ones(size(y));
errorbar(x,y,err,"-s")

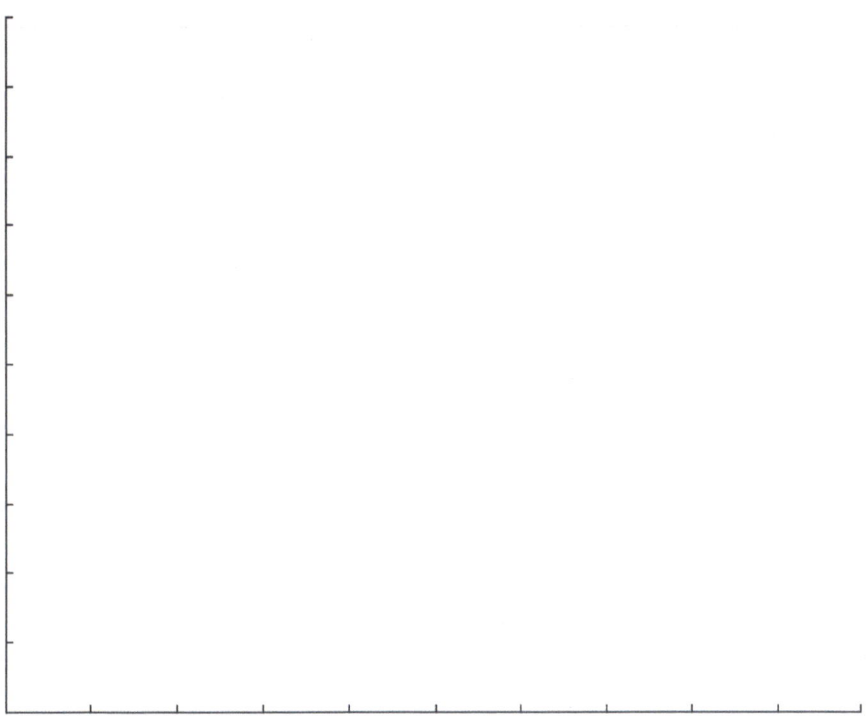

Example

x = 1:10:100;
y = [10 16 22 20 31 35 40 35 45 40];
err = [1 3 2 4 3 3 7 3 1 3];
errorbar(x,y,err,'--r')

Description

Displaying different vertical error bars at each data point using a specific line style and color (See Fig. 15.2). The lists of line colors and style available in MATLAB are illustrated in Figs. 15.3 and 15.4, respectively.

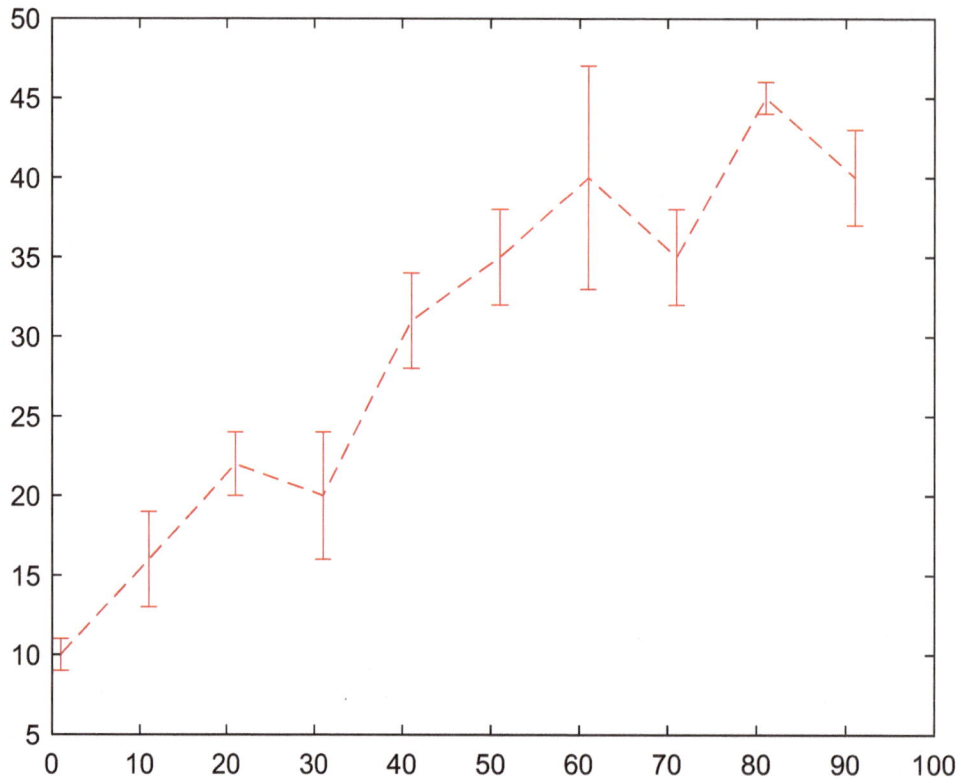

Fig. 15.2 Displaying different vertical error bars at each data point using a specific line style and color

Color Name	Short Name	Appearance
"red"	"r"	
"green"	"g"	
"blue"	"b"	
"cyan"	"c"	
"magenta"	"m"	
"yellow"	"y"	
"black"	"k"	
"white"	"w"	

Fig. 15.3 The list of line colors available in MATLAB

Line Style	Description	Resulting Line
" - "	Solid line	———————
" -- "	Dashed line	- — — — —
" : "	Dotted line	·················
" - . "	Dash-dotted line	_ . _ . _ . _ ..

Fig. 15.4 The list of line styles available in MATLAB

Exercise

Execute the codes in your computer. Then, draw the result(s) on the axes.

```
x = linspace(0,3,10);
y = exp(x);
err = [1 0.5 1 2 2 1.5 1 0.9 0.5 0.3];
errorbar(x,y,err)
```

Example

```
x = 1:10:100;
y = [10 16 22 20 31 35 40 35 45 40];
err = [1 3 2 4 3 3 1 3 1 3];
errorbar(x,y,err,'horizontal')
```

Description

Displaying different horizontal error bars at each data point. See Fig. 15.5.

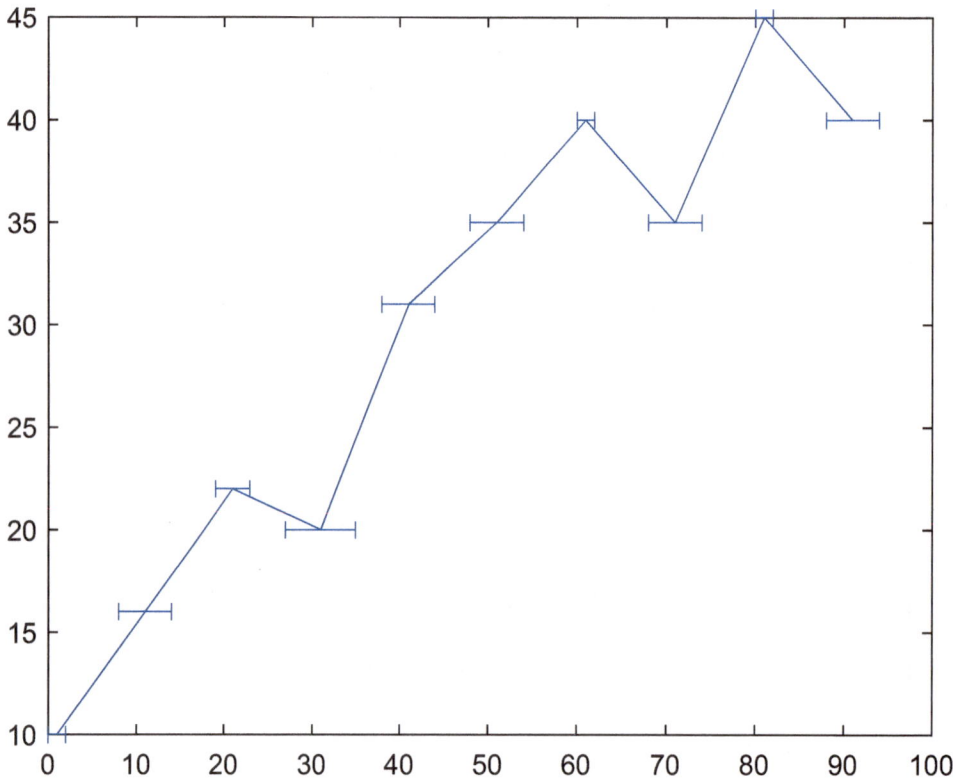

Fig. 15.5 Displaying different horizontal error bars at each data point

Example

x = 1:10:100;
y = [10 16 22 20 31 35 40 35 45 40];
err = [1 3 2 4 3 3 1 3 1 3];
errorbar(x,y,err,'both')

Description

Displaying different vertical and horizontal error bars at each data point. See Fig. 15.6.

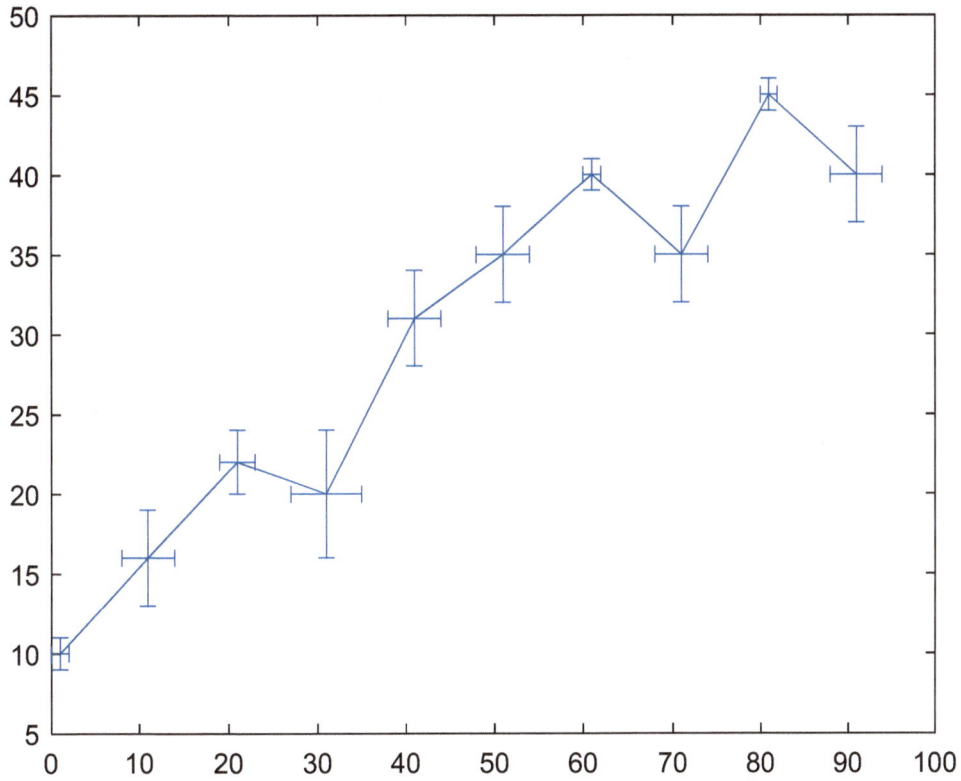

Fig. 15.6 Displaying different vertical and horizontal error bars at each data point

Execute the codes in your computer. Then, draw the result(s) on the axes.

```
x = linspace(0,15,10);
y = 2*x;
err = [1 0.5 1 2 2 1.5 1 0.9 0.5 0.3];
errorbar(x,y,err,'both')
```

Example

x = 1:10:100;
y = [10 16 22 20 31 35 40 35 45 40];
err = [1 3 2 4 3 3 1 3 1 3];
errorbar(x,y,err,"both","o")

Description

Plotting different vertical and horizontal error bars at each data point without a line plot using a specific marker (See Fig. 15.7). The list of line markers available in MATLAB is shown in Fig. 15.8.

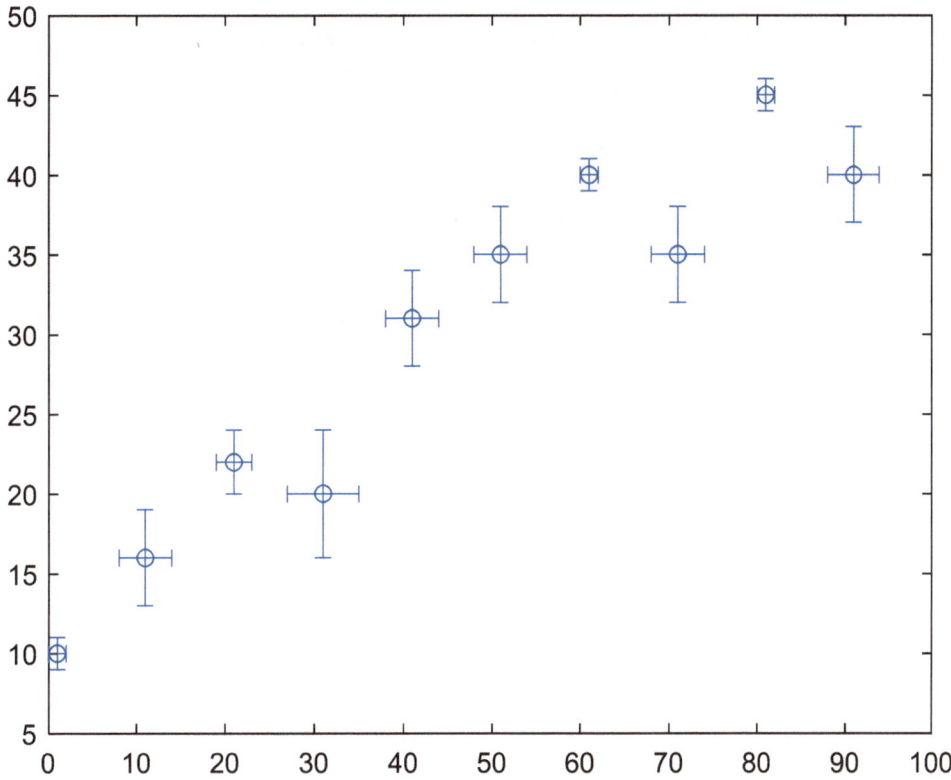

Fig. 15.7 Plotting different vertical and horizontal error bars at each data point without a line plot using a specific marker

Marker	Description	Resulting Marker
"o"	Circle	○
"+"	Plus sign	+
"*"	Asterisk	✳
"."	Point	•
"x"	Cross	×
"_"	Horizontal line	—
"|"	Vertical line	|
"square"	Square	◻
"diamond"	Diamond	◇
"^"	Upward-pointing triangle	△
"v"	Downward-pointing triangle	▽
">"	Right-pointing triangle	▷
"<"	Left-pointing triangle	◁
"pentagram"	Pentagram	☆
"hexagram"	Hexagram	✡

Fig. 15.8 The list of line markers available in MATLAB

Example

x = 1:10:100;
y = [10 16 22 20 31 35 40 35 45 40];
err = [1 3 2 4 3 3 1 3 1 3];
errorbar(x,y,err,"both","LineStyle","none")

Description

Plotting different vertical and horizontal error bars at each data point without a line plot and line marker. See Fig. 15.9.

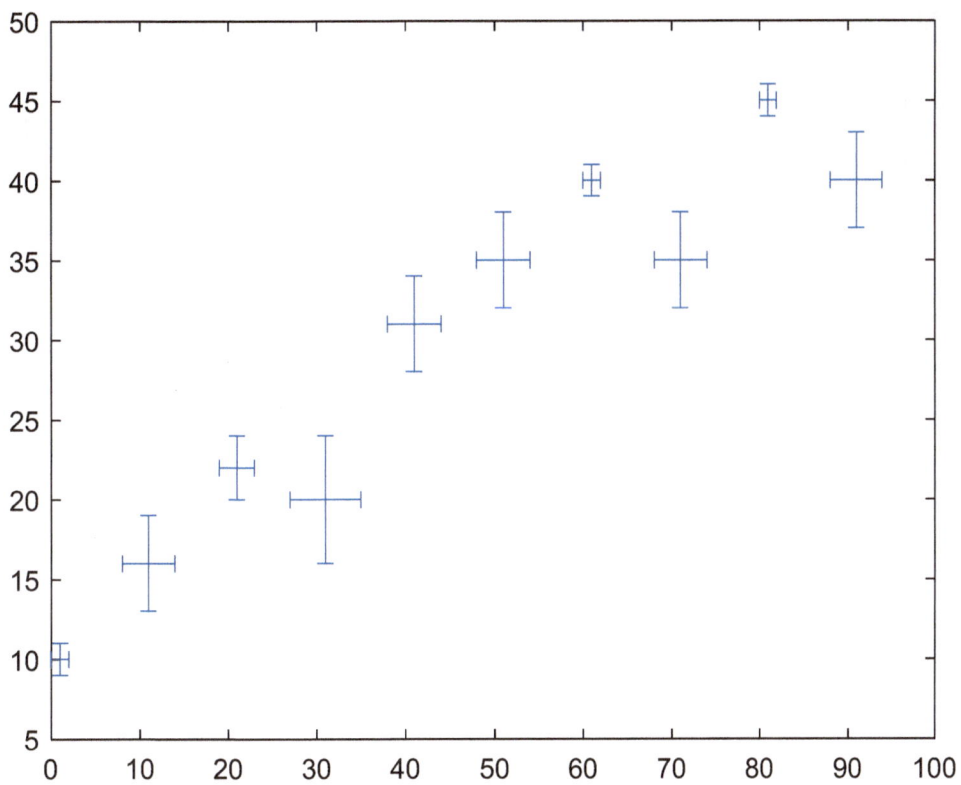

Fig. 15.9 Plotting different vertical and horizontal error bars at each data point without a line plot and line marker

x = 1:10:100;
y = [20 30 45 40 60 65 80 75 95 90];
yneg = [1 3 5 3 5 3 6 4 3 3];
ypos = [2 5 3 5 2 5 2 2 5 5];
xneg = [1 3 5 3 5 3 6 4 3 3];
xpos = [2 5 3 5 2 5 2 2 5 5];
errorbar(x,y,yneg,ypos,xneg,xpos,'o')

Description

Plotting different vertical and horizontal error bars at each data point without a line plot while controlling the lower and upper lengths of the vertical error bars and the left and right lengths of the horizontal error bars. See Fig. 15.10.

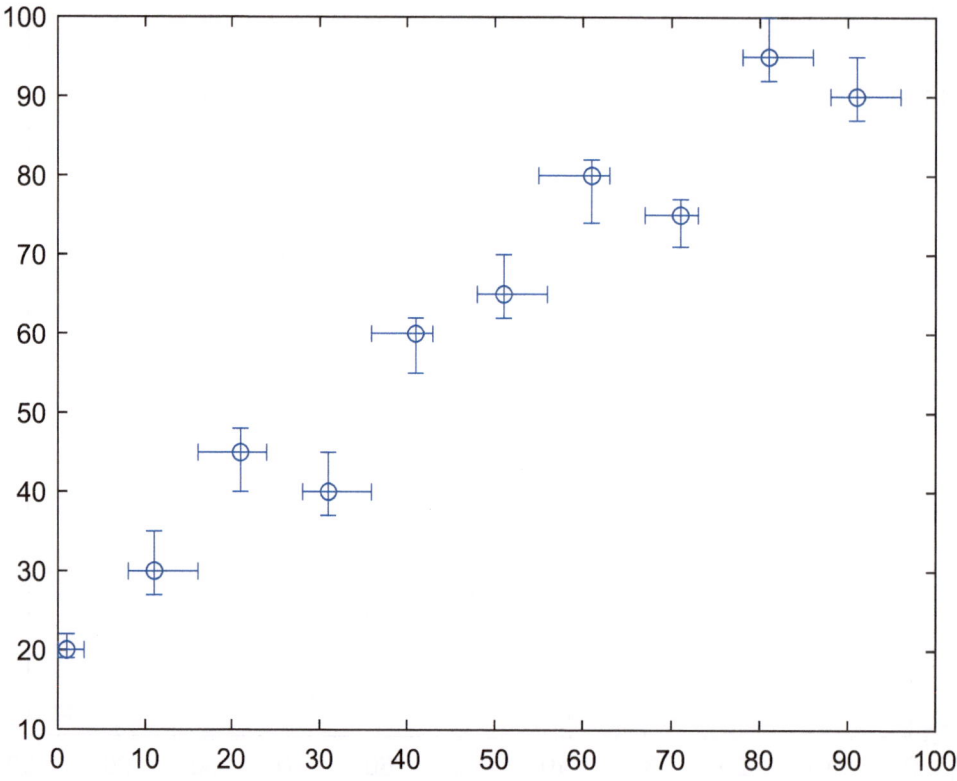

Fig. 15.10 Plotting different vertical and horizontal error bars at each data point while controlling the lower/upper and left/right lengths of the vertical and horizontal error bars

---

Reference

1. MATLAB 2023a.

Abstract

In this chapter, the stacked plot in MATLAB is presented and described. In this regard, several examples and exercises for each section of the chapter are presented. The exercises that include writing the codes, executing them, and achieving the results need to be done by students to master programming skills. In this book, the codes, outputs, and descriptions are in blue, black, and green colors, respectively. To program in MATLAB, a script file can be created and saved with an appropriate name (e.g., untitled01) in the preferred directory of a computer. The program can be run by clicking on the "Run" available on the top toolbar of the script in MATLAB or calling the script by typing its name in Command Window or in the other scripts.

16.1   Stacked Plot

In the following, the description of stacked plot is presented and exemplified [1].

- This plot in the format stackedplot(X,Y) plots the columns of Y versus the vector X up to a maximum of 25 columns.
- This plot in the format stackedplot(Y) plots the columns of Y versus their row number. The x-axis scale ranges from 1 to the number of rows in Y.

Example

```
X = [0:4:20];
Y = [42      38      46
      31      55      49
      53      66      81
      62      57      15
      64      36      43
      60      98      72];
stackedplot(X,Y)
```

Description

Plotting three stacked graphs. See Fig. 16.1.

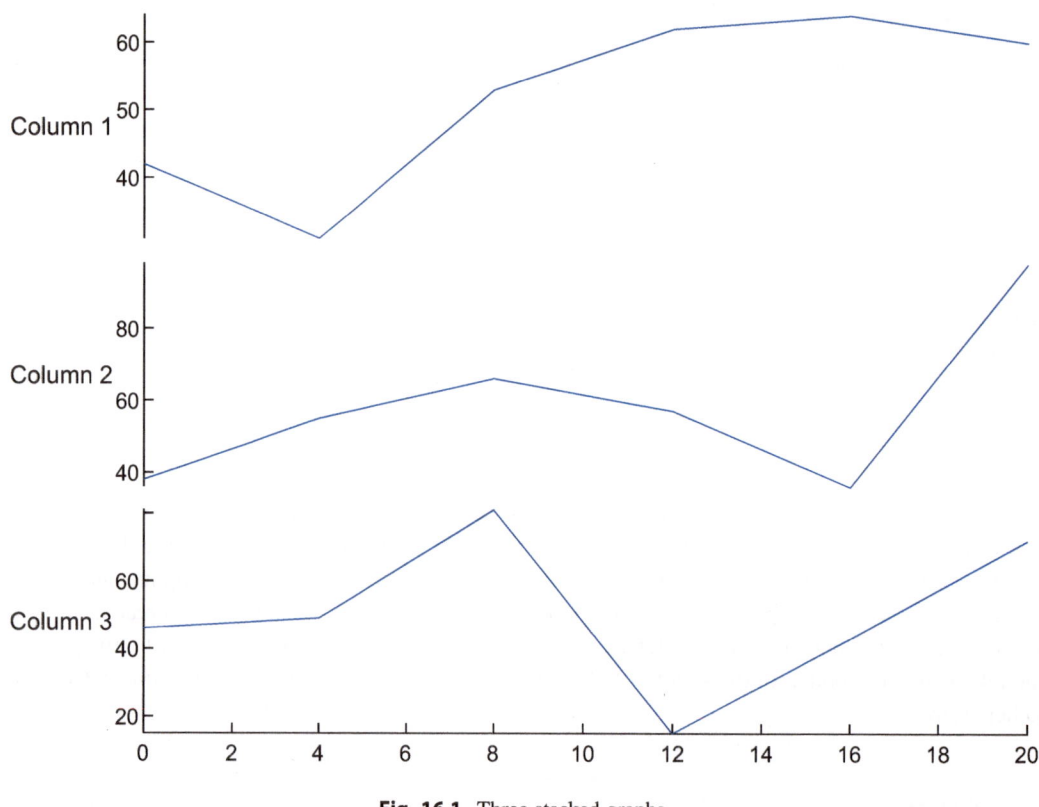

Fig. 16.1   Three stacked graphs

Description

Choosing a specific line style, marker, color, and width for the stacked plots (See Fig. 16.2). The lists of line styles, markers, and colors available in MATLAB are illustrated in Figs. 16.3, 16.4 and 16.5, respectively.

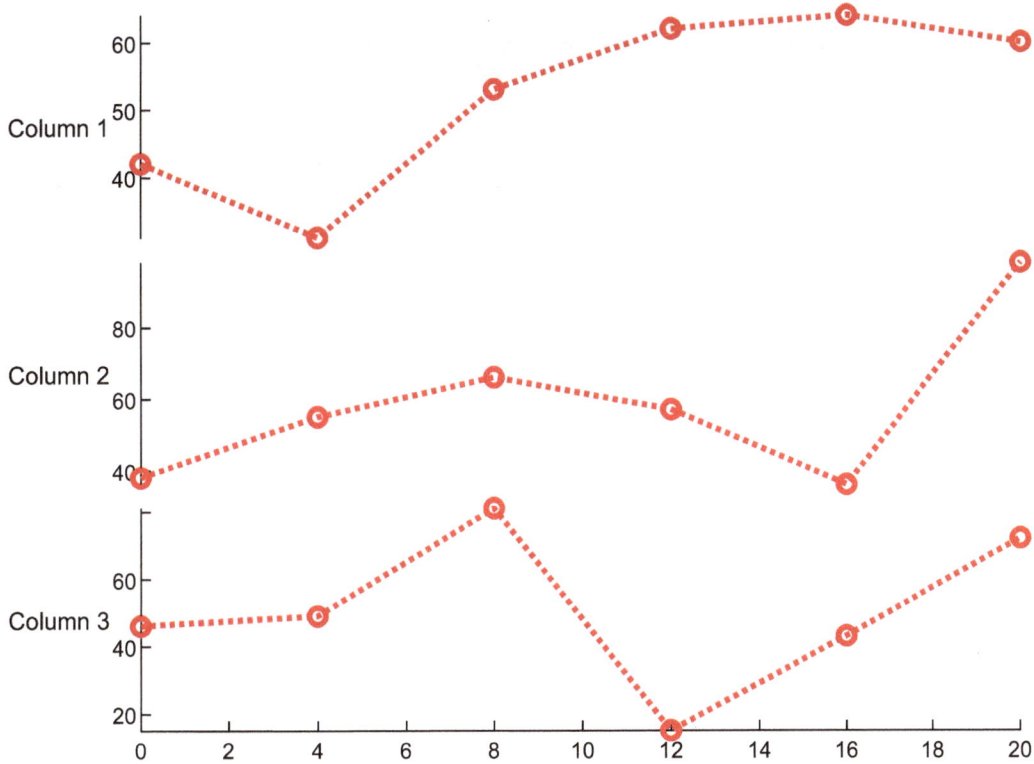

Fig. 16.2 The stacked plots by using a specific line style, marker, color, and width

Line Style	Description	Resulting Line
" - "	Solid line	——————
" - - "	Dashed line	- — — — —
" : "	Dotted line	··············
" - . "	Dash-dotted line	— · — · — · — ··

Fig. 16.3 The list of line styles available in MATLAB

Marker	Description	Resulting Marker
"o"	Circle	○
"+"	Plus sign	+
"*"	Asterisk	✳
"."	Point	•
"x"	Cross	×
"_"	Horizontal line	—
"|"	Vertical line	|
"square"	Square	□
"diamond"	Diamond	◇
"^"	Upward-pointing triangle	△
"v"	Downward-pointing triangle	▽
">"	Right-pointing triangle	▷
"<"	Left-pointing triangle	◁
"pentagram"	Pentagram	☆
"hexagram"	Hexagram	✡

Fig. 16.4 The list of line markers available in MATLAB

Color Name	Short Name	Appearance
"red"	"r"	▇
"green"	"g"	▇
"blue"	"b"	▇
"cyan"	"c"	▇
"magenta"	"m"	▇
"yellow"	"y"	▇
"black"	"k"	▇
"white"	"w"	▢

Fig. 16.5 The list of line colors available in MATLAB

Reference

1. MATLAB 2023a.

Abstract

In this chapter, the log-log scale plot in MATLAB is presented and described. In this regard, several examples and exercises for each section of the chapter are presented. The exercises that include writing the codes, executing them, and achieving the results need to be done by students to master programming skills. In this book, the codes, outputs, and descriptions are in blue, black, and green colors, respectively. To program in MATLAB, a script file can be created and saved with an appropriate name (e.g. untitled01) in the preferred directory of a computer. The program can be run by clicking on the "Run" available on the top toolbar of the script in MATLAB or calling the script by typing its name in Command Window or in the other scripts.

17.1 Log-Log Scale Plot

In the following, the description of log-log scale plot is presented and exemplified [1–9].

- This plot in the format loglog(X,Y) plots x- and y-coordinates using a base-10 logarithmic scale on the x-axis and the y-axis.
- To plot a set of coordinates connected by line segments, specify X and Y as vectors of the same length.
- To plot multiple sets of coordinates on the same set of axes, specify at least one of X or Y as a matrix.

Example

```
x = logspace(-1,2);
y = 2.^x;
loglog(x,y)
grid on
```

Description

Plotting the function in the log-log scale. See Fig. 17.1.

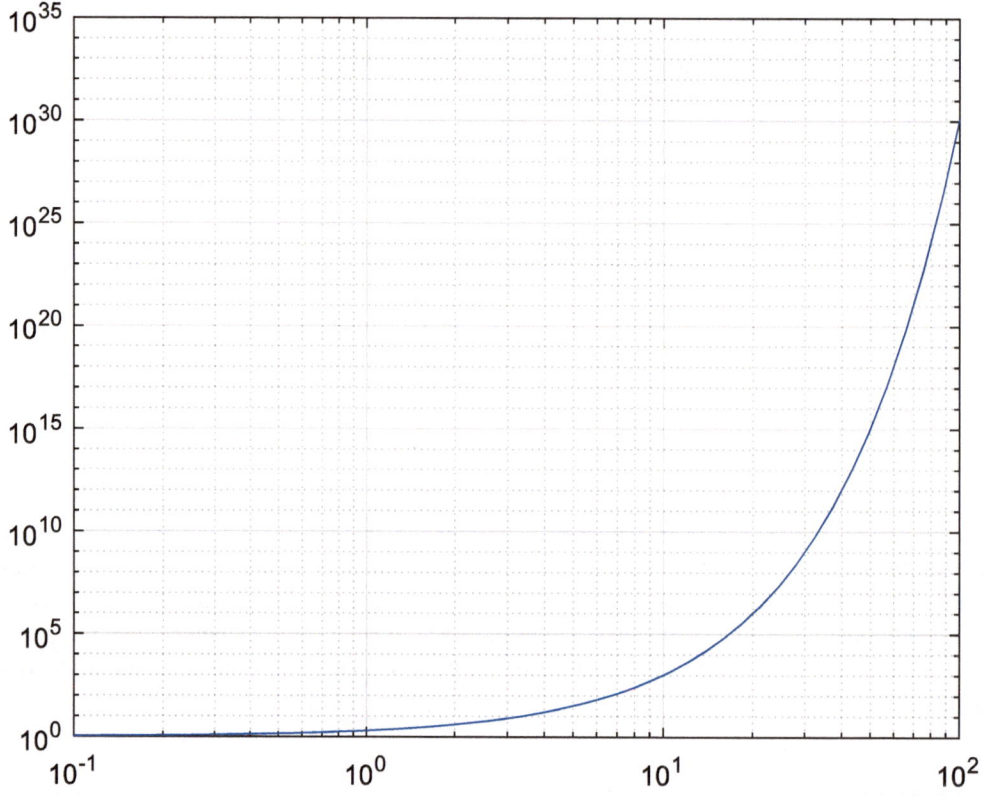

Fig. 17.1 Plotting the function in the log-log scale

Exercise

(a)   *Write the codes to plot the graph of the function y=0.5.^x for the logarithmic interval (-1,2).*
(b)   *Execute the codes in your computer, then draw the result(s) on the axes.*

```
x = logspace(-1,2);
y1 = 10.^x;
y2 = 1./10.^x;
loglog(x,y1,x,y2)
grid on
```

Alternative codes using a matrix:

```
x = logspace(-1,2);
y1 = 10.^x;
y2 = 1./10.^x;
y = [y1;y2];
loglog(x,y)
grid on
```

Plotting the multiple functions in the log-log scale. See Fig. 17.2.

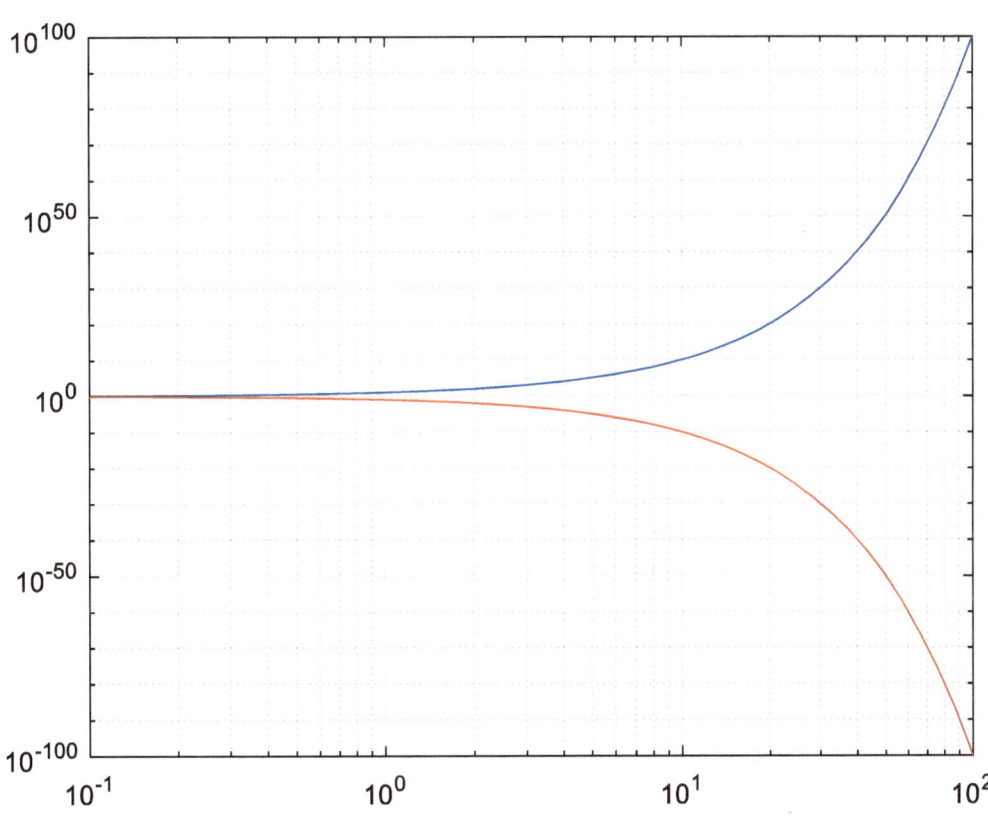

Fig. 17.2 Plotting the multiple functions in the log-log scale

x = logspace(-1,2,10000);
y = 5 + 3*sin(x);
loglog(x,y)

Plotting the function in the log-log scale. See Fig. 17.3.

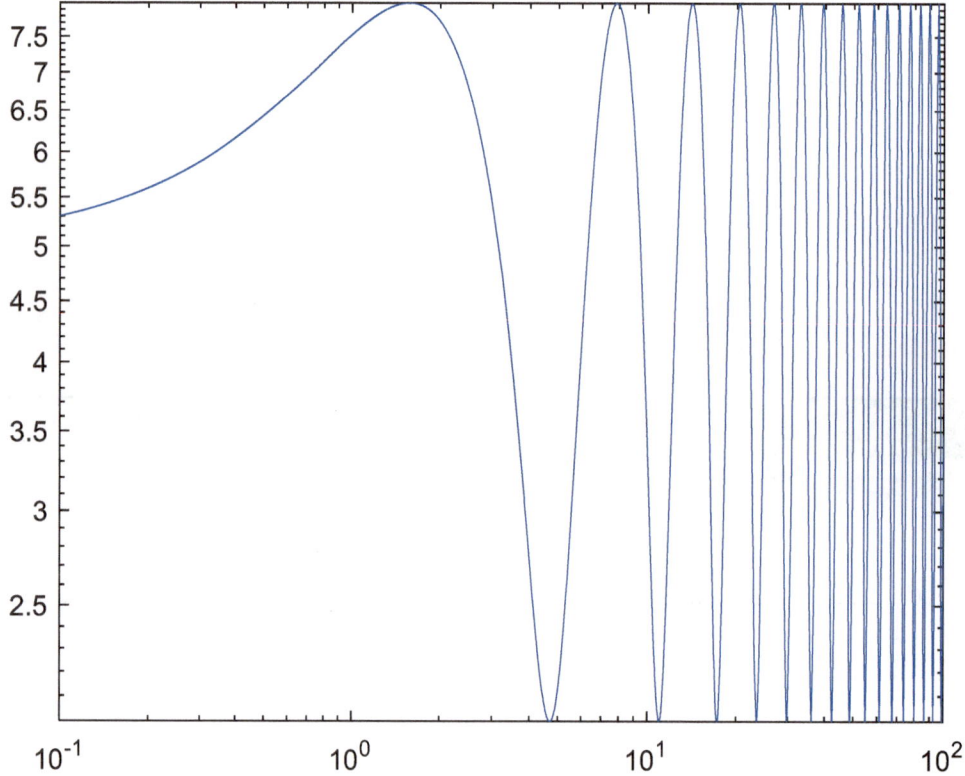

Fig. 17.3 Plotting the function in the log-log scale

x = logspace(-1,2,10000);
y = 5 + 3*sin(x);
loglog(x,y)
xlabel('x')
ylabel('5 + 3 sin(x)')
title('2-D Log-Log Plot')

Adding x and y labels and title to the graph. See Fig. 17.4.

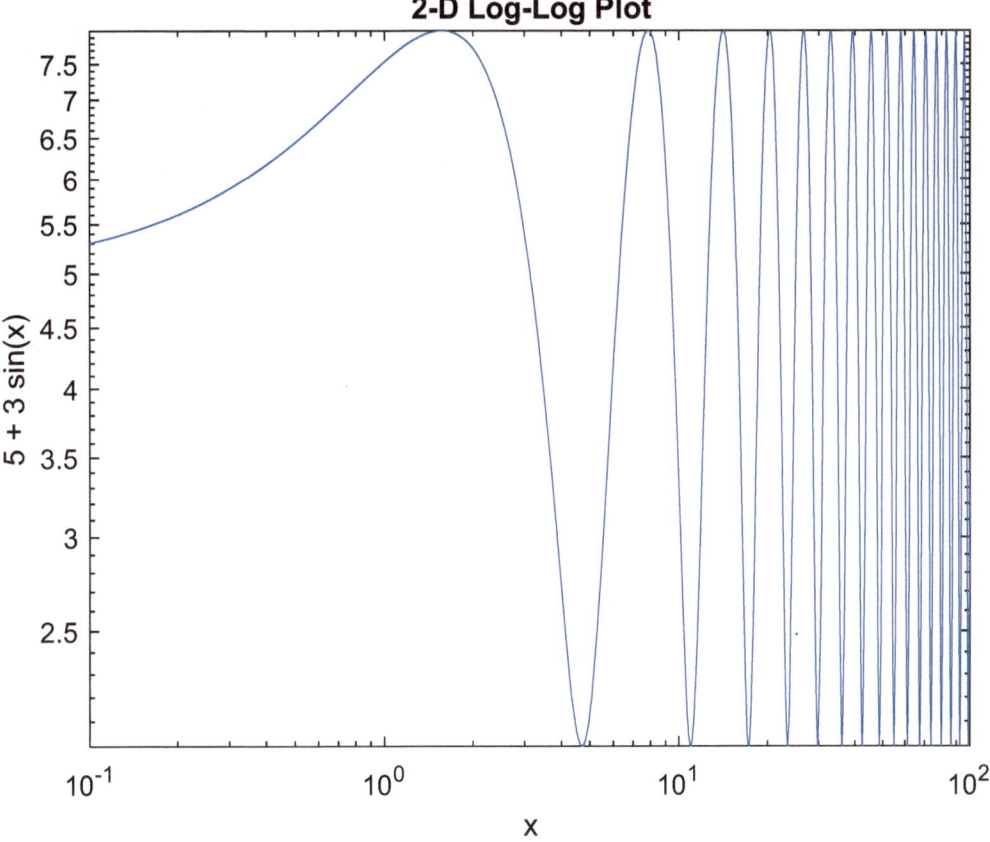

Fig. 17.4 Adding x and y labels and title to the graph

```
x = logspace(-1,2,10000);
y = 5 + 3*sin(x);
loglog(x,y)
xlabel('x')
ylabel('5 + 3 sin(x)')
title('2-D Log-Log Plot')
yticks([3 4 5 6 7])
```

Adding tick values to the graph. See Fig. 17.5.

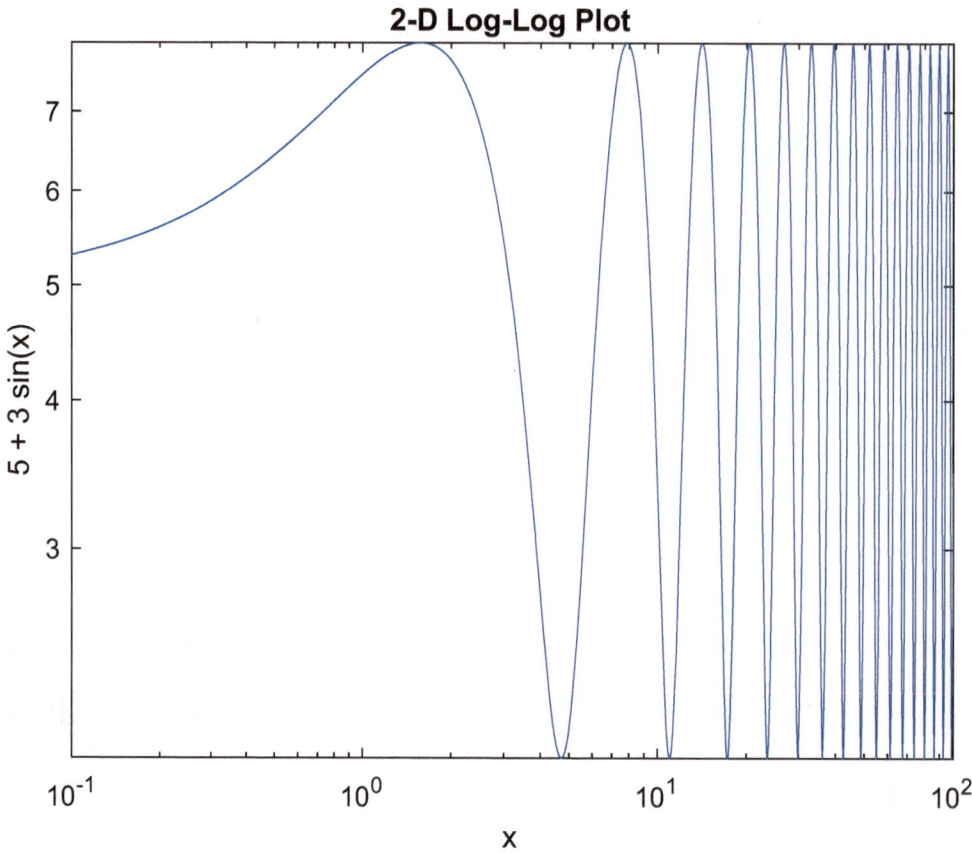

Fig. 17.5 Adding tick values to the graph

```
x = logspace(-1,2,10000);
y = 5 + 3*sin(x);
loglog(x,y)
xlabel('x')
ylabel('5 + 3 sin(x)')
title('2-D Log-Log Plot')
yticks([3 4 5 6 7])
yticklabels({'y = 3','y = 4','y = 5','y = 6','y = 7'})
```

Adding tick labels to the graph. See Fig. 17.6.

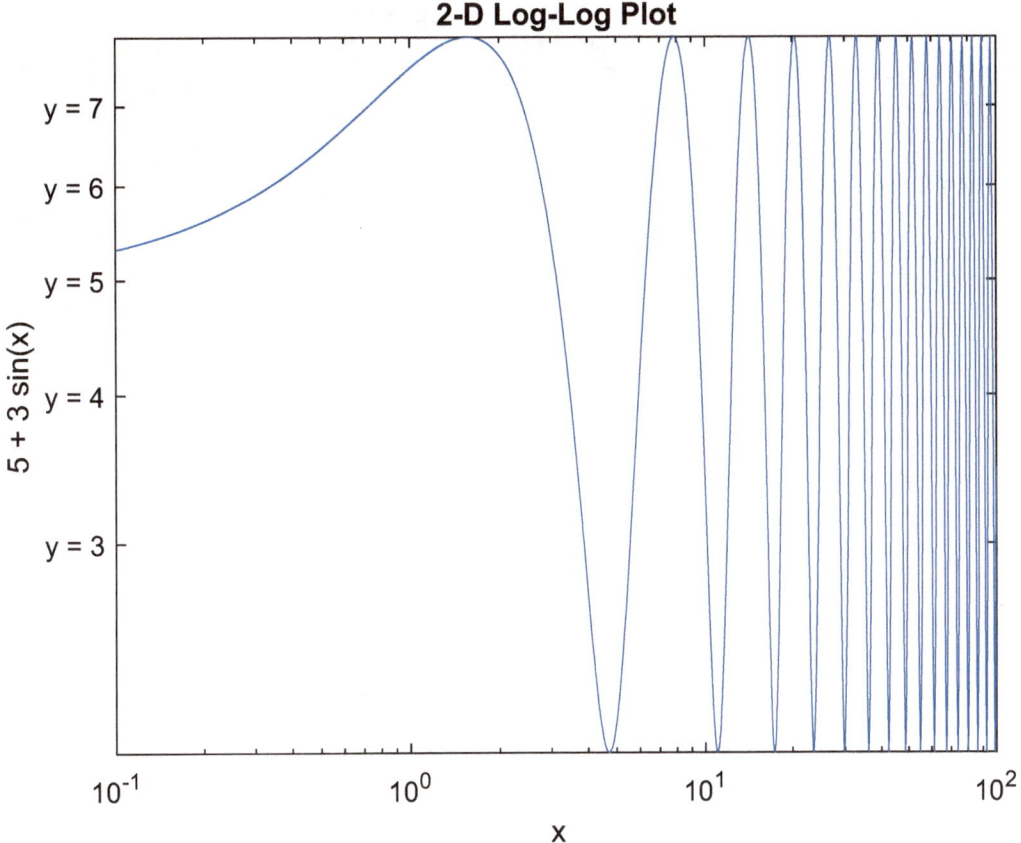

Fig. 17.6 Adding tick labels to the graph

```
x = logspace(-1,2,1000);
y = 5 + 3*sin(x);
loglog(x,y,'--xm')
xlabel('x')
ylabel('5 + 3 sin(x)')
title('2-D Log-Log Plot')
yticks([3 4 5 6 7])
yticklabels({'y = 3','y = 4','y = 5','y = 6','y = 7'})
```

Plotting the graph using a specific line style, color, and marker (See Fig. 17.7). The lists of line styles, markers, and colors available in MATLAB are illustrated in Figs. 17.8, 17.9 and 17.10, respectively.

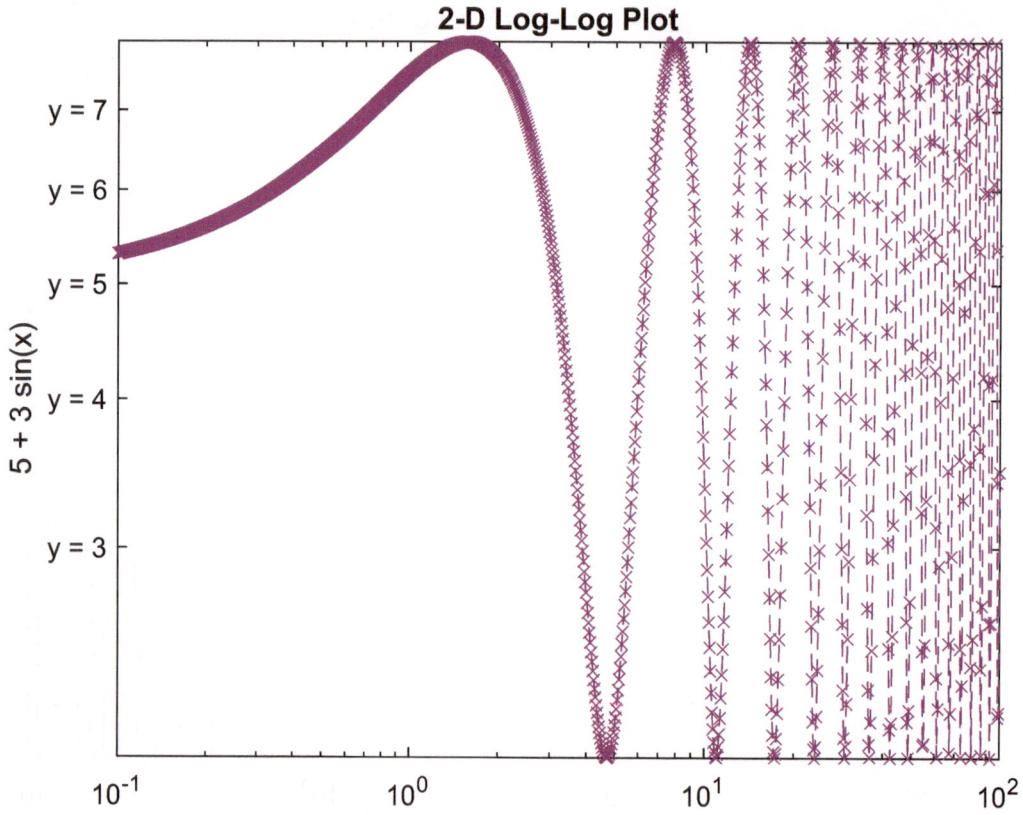

Fig. 17.7 Plotting the graph using a specific line style, color, and marker

Line Style	Description	Resulting Line
"-"	Solid line	——————
"--"	Dashed line	- — — — —
":"	Dotted line	················
"-."	Dash-dotted line	—·—·—·—··

Fig. 17.8 The list of line styles available in MATLAB

Color Name	Short Name	Appearance
"red"	"r"	
"green"	"g"	
"blue"	"b"	
"cyan"	"c"	
"magenta"	"m"	
"yellow"	"y"	
"black"	"k"	
"white"	"w"	

Fig. 17.9 The list of line colors available in MATLAB

Marker	Description	Resulting Marker
"o"	Circle	○
"+"	Plus sign	+
"*"	Asterisk	✳
"."	Point	•
"x"	Cross	×
"_"	Horizontal line	—
"|"	Vertical line	|
"square"	Square	□
"diamond"	Diamond	◇
"^"	Upward-pointing triangle	△
"v"	Downward-pointing triangle	▽
">"	Right-pointing triangle	▷
"<"	Left-pointing triangle	◁
"pentagram"	Pentagram	☆
"hexagram"	Hexagram	✡

Fig. 17.10 The list of line markers available in MATLAB

```
x = logspace(-1,2,10000);
y1 = 5 + 3*sin(x/4);
y2 = 5 - 3*sin(x/4);
loglog(x,y1,x,y2,'--')
legend('Signal 1','Signal 2','Location','northwest')
```

Description

Adding a legend to the specific place (northwest) of the graph. See Fig. 17.11.

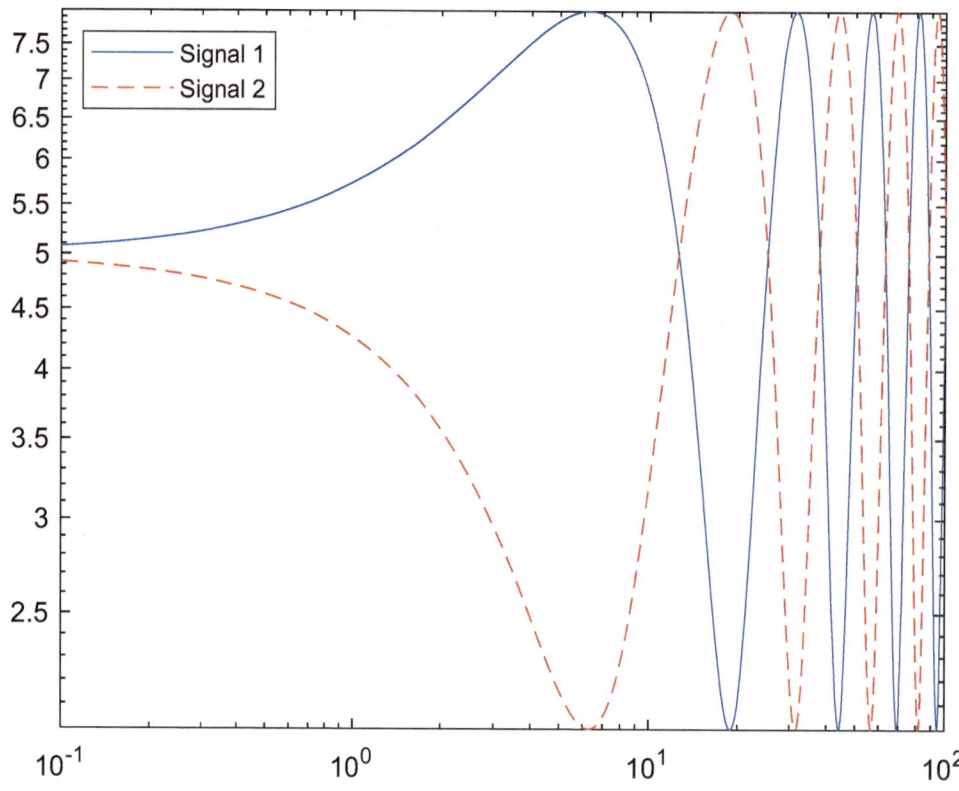

Fig. 17.11   Adding a legend to the specific place (northwest) of the graph

Exercise

(a)   *Write the codes to plot the graph of the functions y=0.9.^x and y=1.1.^x for the logarithmic interval (-1,1). Add appropriate x and y labels and legend. Also, select the dashed line style for one of them.*

(b)   *Execute the codes in your computer, then draw the result(s) on the axes.*

References

1. MATLAB 2023a.
2. Rahmani-Andebili, M. (2024). *Mathematics of engineering and science – Practice problems, methods, and solutions*. Springer Nature.
3. Rahmani-Andebili, M. (2022). *Differential equations – Practice problems, methods, and solutions*. Springer Nature.
4. Rahmani-Andebili, M. (2023). *Calculus III – Practice problems, methods, and solutions*. Springer Nature.
5. Rahmani-Andebili, M. (2023). *Calculus II – Practice problems, methods, and solutions*. Springer Nature.
6. Rahmani-Andebili, M. (2023). *Calculus I – Practice problems, methods, and solutions* (2nd ed.). Springer Nature.
7. Rahmani-Andebili, M. (2021). *Calculus – Practice problems, methods, and solutions*. Springer Nature.
8. Rahmani-Andebili, M. (2024). *Precalculus – Practice problems, methods, and solutions* (2nd ed.). Springer Nature.
9. Rahmani-Andebili, M. (2021). *Precalculus – Practice problems, methods, and solutions*. Springer Nature.

Plot Types in MATLAB: Semilog Scale Plot (for X-Axis)

18

Abstract

In this chapter, the semilog scale plot (for x-axis) in MATLAB is presented and described. In this regard, several examples and exercises for each section of the chapter are presented. The exercises that include writing the codes, executing them, and achieving the results need to be done by students to master programming skills. In this book, the codes, outputs, and descriptions are in blue, black, and green colors, respectively. To program in MATLAB, a script file can be created and saved with an appropriate name (e.g., untitled01) in the preferred directory of a computer. The program can be run by clicking on the "Run" available on the top toolbar of the script in MATLAB or by calling the script by typing its name in Command Window or in the other scripts.

18.1 Semilog Scale Plot (for X-Axis)

In the following, the description of semilog scale plot is presented and exemplified [1–9].

- This plot in the format semilogx(X,Y) plots x- and y-coordinates using a base-10 logarithmic scale on the x-axis and a linear scale on the y-axis.
- To plot a set of coordinates connected by line segments, specify X and Y as vectors of the same length.
- To plot multiple sets of coordinates on the same set of axes, specify at least one of X or Y as a matrix.

Example

```
x = logspace(-1,2);
y = x;
semilogx(x,y)
grid on
```

Description

Plotting the function in the semilog scale (x-axis). See Fig. 18.1.

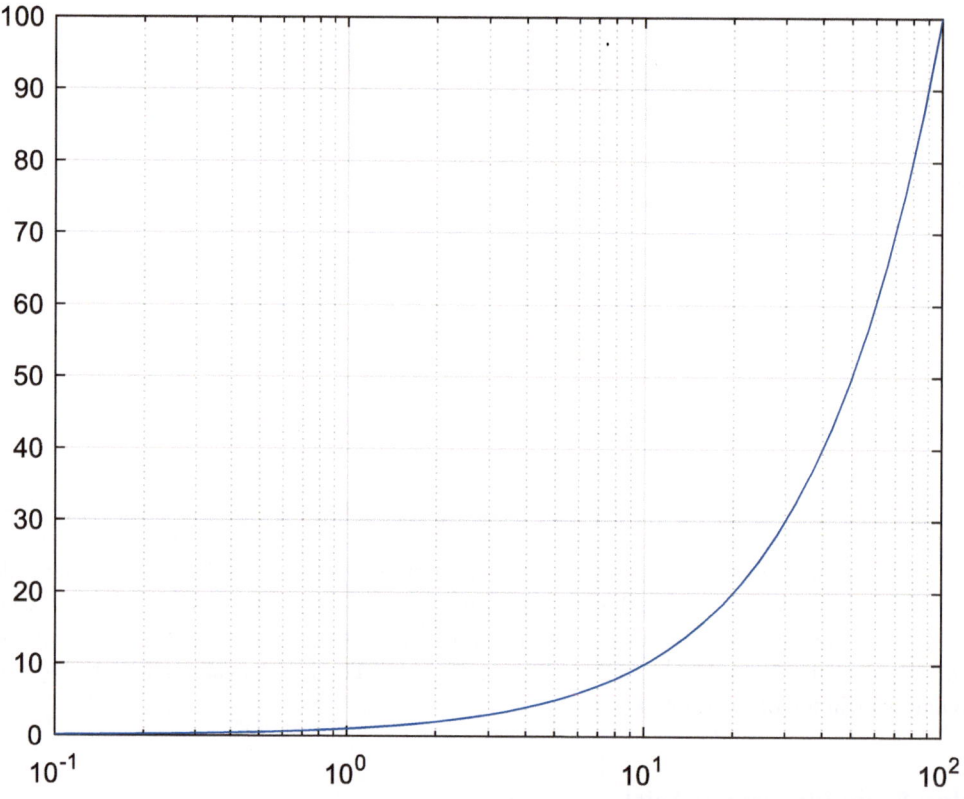

Fig. 18.1 Plotting the function in the semilog scale (x-axis)

Exercise

(a) *Write the codes to plot the graph of the function y=0.5x for the logarithmic interval (-1,2).*

(b) *Execute the codes in your computer, then draw the result(s) on the axes.*

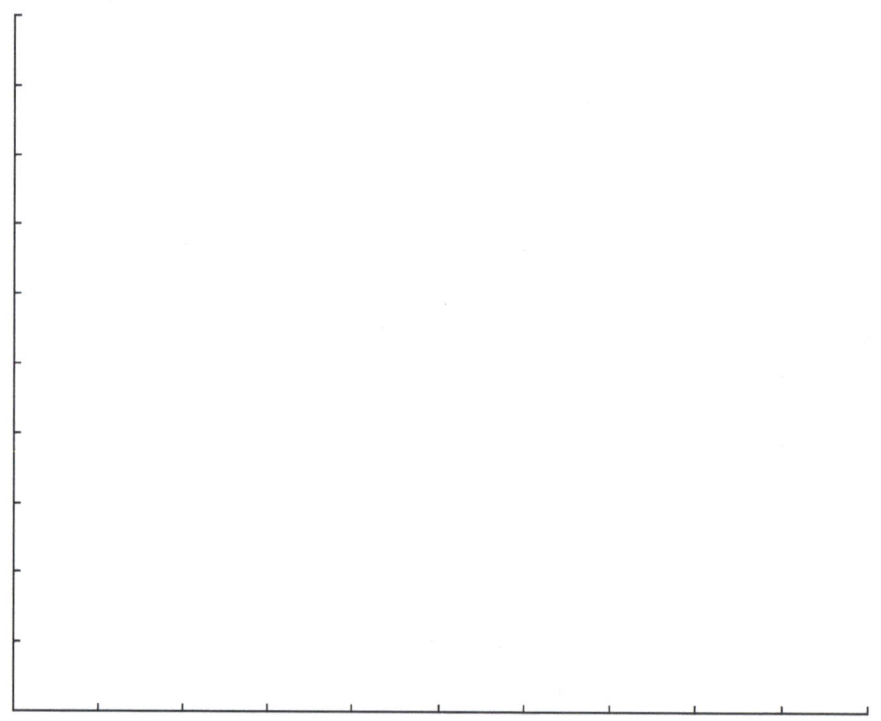

Example

```
x = logspace(-1,2);
y1 = x;
y2 = -x;
semilogx(x,y1,x,y2)
grid on
```

Description

Alternative codes using a matrix:

Example

```
x = logspace(-1,2);
y1 = x;
y2 = -x;
y = [y1;y2];
semilogx(x,y)
grid on
```

Description

Plotting the multiple functions in the semilog scale (x-axes). See Fig. 18.2.

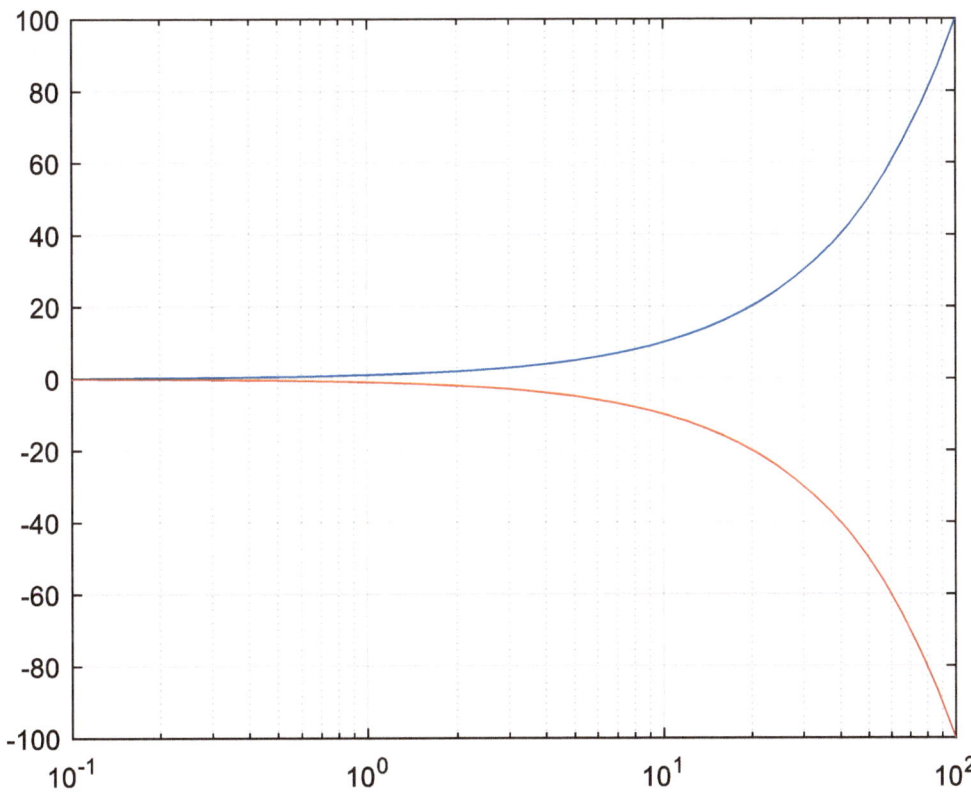

Fig. 18.2 Plotting the multiple functions in the semilog scale (x-axes)

Example

```
f = logspace(1,5,100);
v = linspace(-50,50,100);
gain = (1-exp(5*(2.5*v.^2)./7500))/14;
semilogx(f,gain)
grid on
```

Description

Plotting the function in the semilog scale (x-axes). See Fig. 18.3.

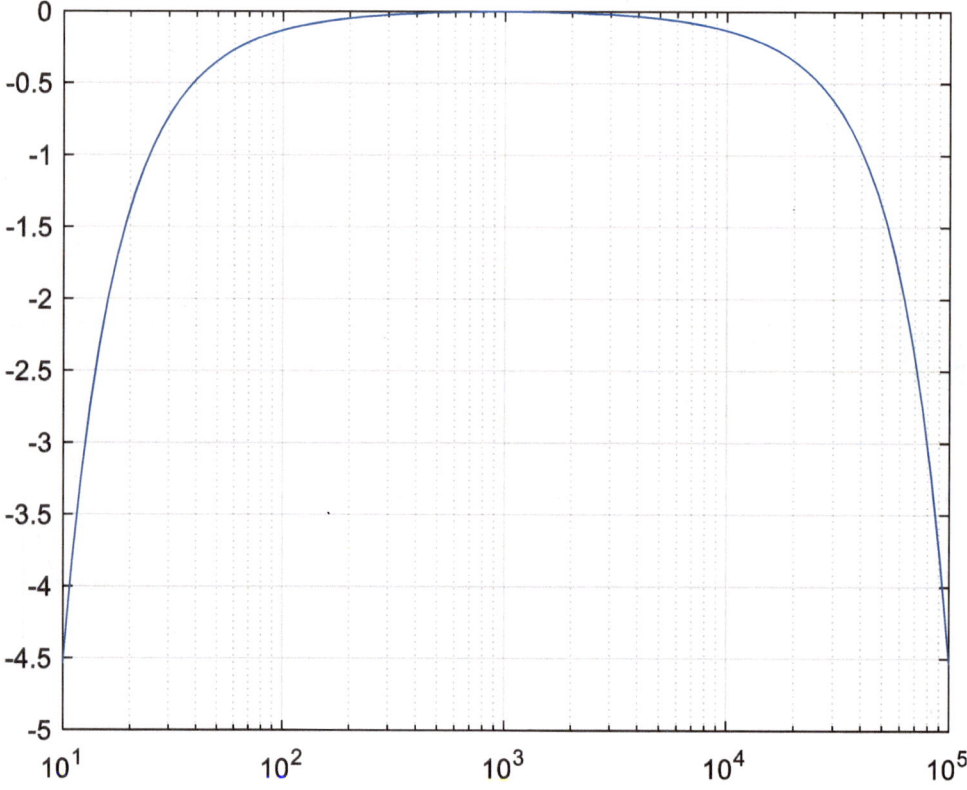

Fig. 18.3 Plotting the function in the semilog scale (x-axes).

Example

```
f = logspace(1,5,100);
v = linspace(-50,50,100);
gain = (1-exp(5*(2.5*v.^2)./7500))/14;
semilogx(f,gain)
grid on
xlabel ('Frequency (Hz)')
ylabel('Power Gain (dB)')
title('2-D Semilog Plot')
```

Adding x and y labels and title to the graph. See Fig. 18.4.

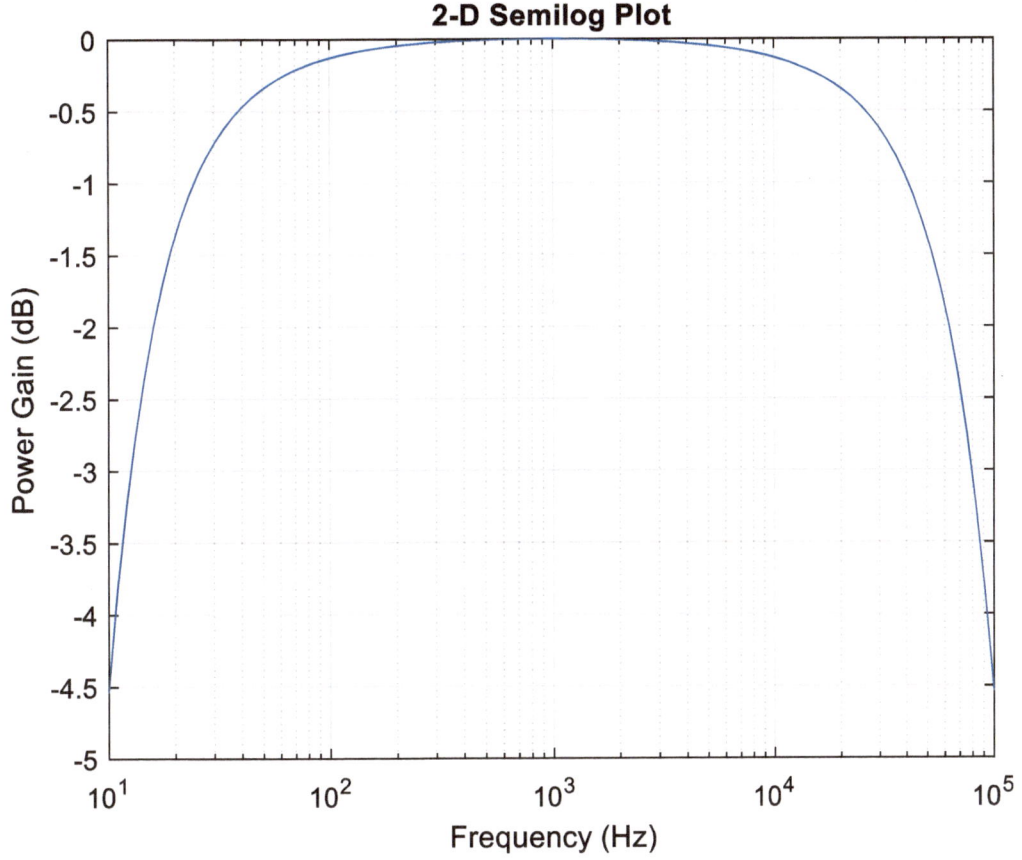

Fig. 18.4   Adding x and y labels and title to the graph

Example

```
f = logspace(1,5,100);
v = linspace(-50,50,100);
gain = (1-exp(5*(2.5*v.^2)./7500))/14;
semilogx(f,gain,'--om')
grid on
xlabel ('Frequency (Hz)')
ylabel('Power Gain (dB)')
title('2-D Semilog Plot')
```

Description

Plotting the graph using a specific line style, color, and marker (See Fig. 18.5). The lists of line styles, markers, and colors available in MATLAB are illustrated in Figs. 18.6, 18.7 and 18.8, respectively.

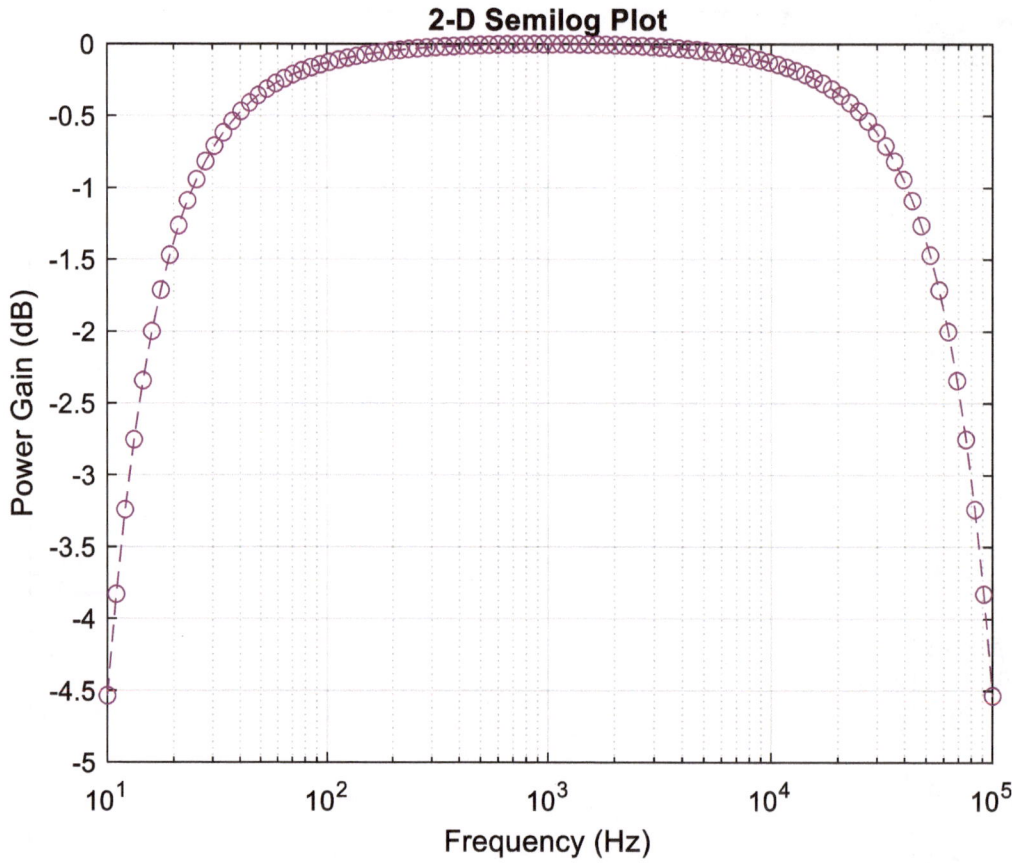

Fig. 18.5 Plotting the graph using a specific line style, color, and marker

Line Style	Description	Resulting Line
" - "	Solid line	——————
" - - "	Dashed line	- — — — —
" : "	Dotted line	··············
" - . "	Dash-dotted line	—·—·—·—··

Fig. 18.6 The list of line styles available in MATLAB

Marker	Description	Resulting Marker
"o"	Circle	○
"+"	Plus sign	+
"*"	Asterisk	✳
"."	Point	•
"x"	Cross	✕
"_"	Horizontal line	—
"|"	Vertical line	|
"square"	Square	▢
"diamond"	Diamond	◇
"^"	Upward-pointing triangle	△
"v"	Downward-pointing triangle	▽
">"	Right-pointing triangle	▷
"<"	Left-pointing triangle	◁
"pentagram"	Pentagram	☆
"hexagram"	Hexagram	✡

Fig. 18.7 The list of line markers available in MATLAB

Color Name	Short Name	Appearance
"red"	"r"	
"green"	"g"	
"blue"	"b"	
"cyan"	"c"	
"magenta"	"m"	
"yellow"	"y"	
"black"	"k"	
"white"	"w"	

Fig. 18.8 The list of line colors available in MATLAB

Example

```
x = logspace(1,4,100);
v = linspace(-50,50,100);
y1 = 100*exp(-1*((v+5).^2)./200);
y2 = 100*exp(-1*(v.^2)./200);
semilogx(x,y1,x,y2,'--','LineWidth',2)
legend('Measured','Estimated')
grid on
```

Description

Plotting using a specific line width and adding a legend to the specific place (northwest) of the graph. See Fig. 18.9.

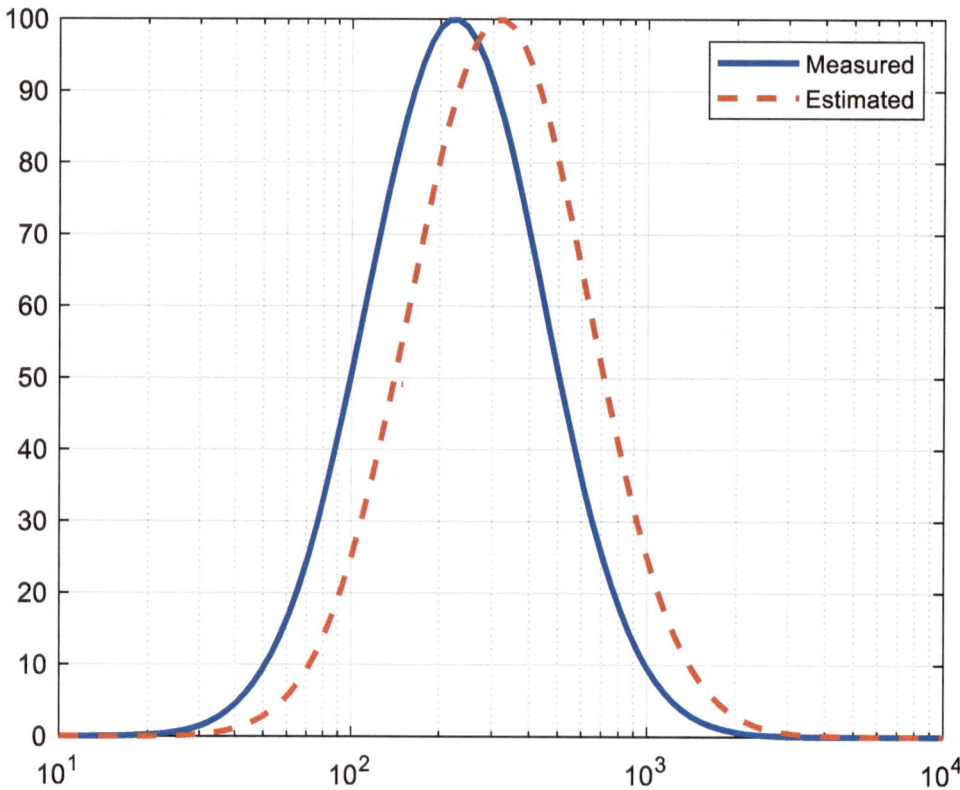

Fig. 18.9 Plotting using a specific line width and adding a legend to the specific place (northwest) of the graph

References

1. MATLAB 2023a.
2. Rahmani-Andebili, M. (2024). *Mathematics of engineering and science – Practice problems, methods, and solutions*. Springer Nature.
3. Rahmani-Andebili, M. (2022). *Differential equations – Practice problems, methods, and solutions*. Springer Nature.
4. Rahmani-Andebili, M. (2023). *Calculus III – Practice problems, methods, and solutions*. Springer Nature.
5. Rahmani-Andebili, M. (2023). *Calculus II – Practice problems, methods, and solutions*. Springer Nature.
6. Rahmani-Andebili, M. (2023). *Calculus I – Practice problems, methods, and solutions* (2nd ed.). Springer Nature.
7. Rahmani-Andebili, M. (2021). *Calculus – Practice problems, methods, and solutions*. Springer Nature.
8. Rahmani-Andebili, M. (2024). *Precalculus – Practice problems, methods, and solutions* (2nd ed.). Springer Nature.
9. Rahmani-Andebili, M. (2021). *Precalculus – Practice problems, methods, and solutions*. Springer Nature.

Abstract

In this chapter, the semilog scale plot (for y-axis) in MATLAB is presented and described. In this regard, several examples and exercises for each section of the chapter are presented. The exercises that include writing the codes, executing them, and achieving the results need to be done by students to master programming skills. In this book, the codes, outputs, and descriptions are in blue, black, and green colors, respectively. To program in MATLAB, a script file can be created and saved with an appropriate name (e.g., untitled01) in the preferred directory of a computer. The program can be run by clicking on the "Run" available on the top toolbar of the script in MATLAB or by calling the script by typing its name in Command Window or in the other scripts.

19.1 Semilog Scale Plot (for Y-Axis)

In the following, the description of semilog scale plot is presented and exemplified [1–9].

- This plot in the format semilogy(X,Y) plots x- and y-coordinates using a linear scale on the x-axis and a base-10 logarithmic scale on the y-axis.
- To plot a set of coordinates connected by line segments, specify X and Y as vectors of the same length.
- To plot multiple sets of coordinates on the same set of axes, specify at least one of X or Y as a matrix.

Example

```
x = 1:100;
y = x.^2;
semilogy(x,y)
grid on
```

Description

Plotting the function in the semilog scale (y-axis). See Fig. 19.1.

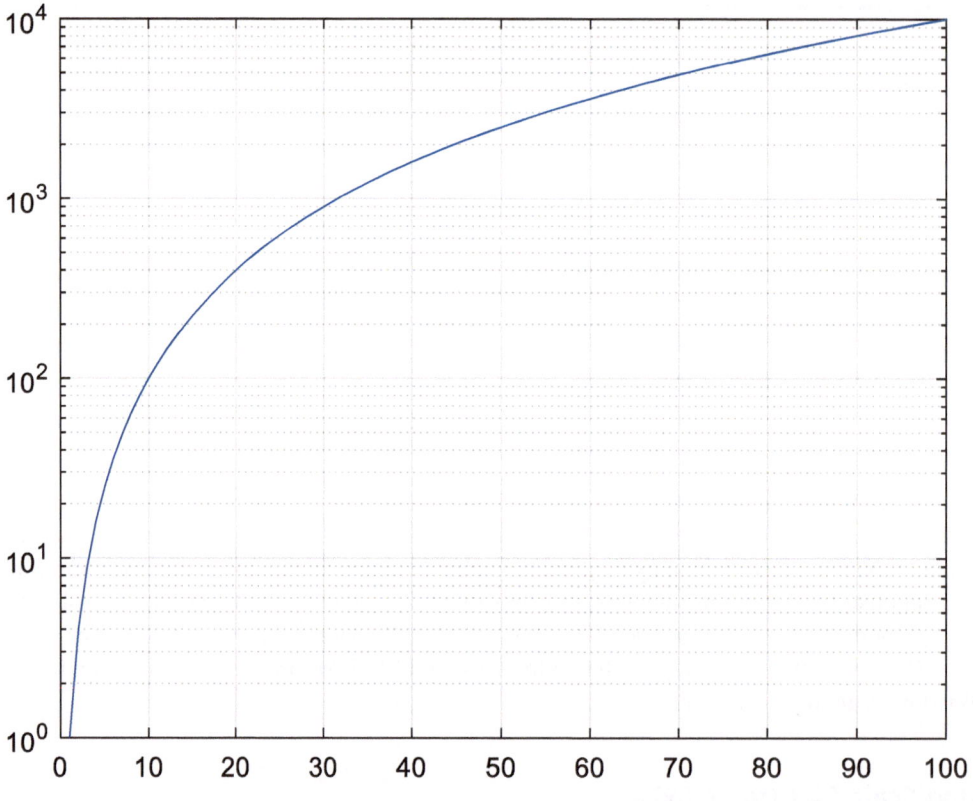

Fig. 19.1 Plotting the function in the semilog scale (y-axis)

Exercise

(a) *Write the codes to plot the graph of the function $y = x^{0.5}$ for the interval (1100).*

(b) *Execute the codes in your computer, then draw the result(s) on the axes.*

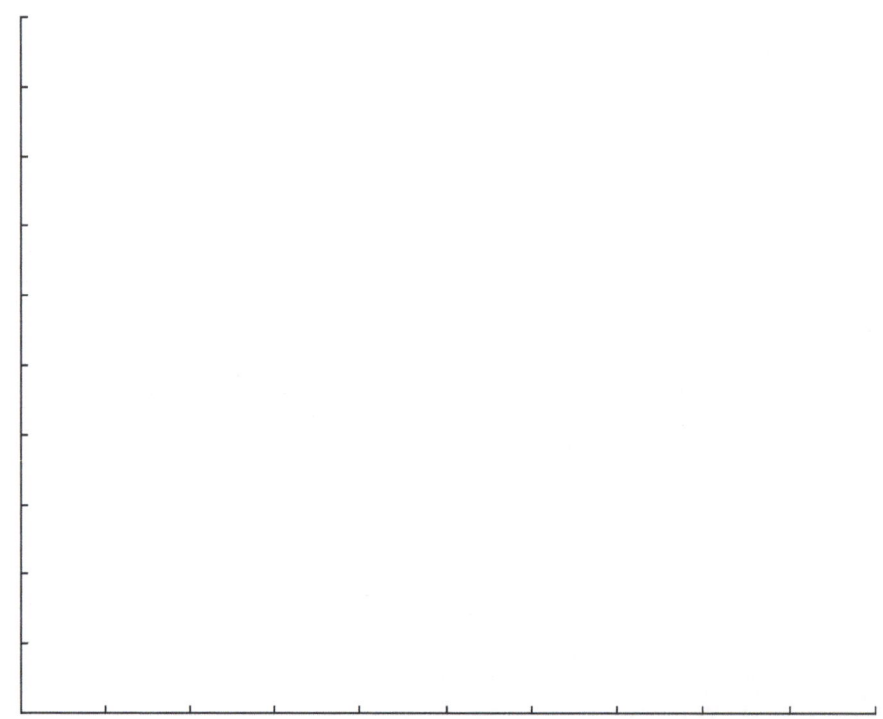

```
x = 1:100;
y1 = x.^2;
y2 = x.^3;
semilogy(x,y1,x,y2)
grid on
```

Alternative codes using a matrix:

```
x = 1:100;
y1 = x.^2;
y2 = x.^3;
y = [y1;y2];
semilogy(x,y)
grid on
```

Plotting the multiple functions in the semilog scale (y-axis). See Fig. 19.2.

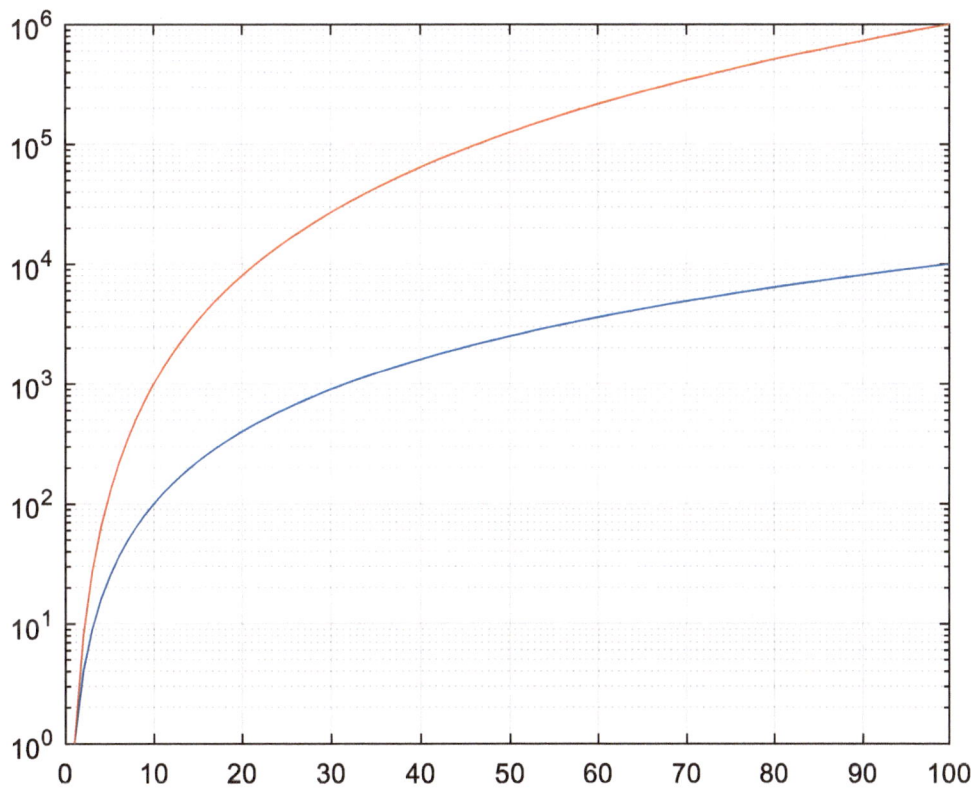

Fig. 19.2 Plotting the multiple functions in the semilog scale (y-axis)

```
P = 1000;
npayments = 240;
rate = 0.08/12;
mpayment = P*(rate*(1+rate)^npayments)/(((1+rate)^npayments) - 1);
x = 1:240;
y = x * mpayment;
semilogy(x,y,'LineWidth',2);
grid on
xlabel ('Installment')
ylabel('Cumulate Cost')
title('2-D Semilog Plot')
```

Description

Plotting the graph with a specific width and adding x and y labels and title to it. See Fig. 19.3.

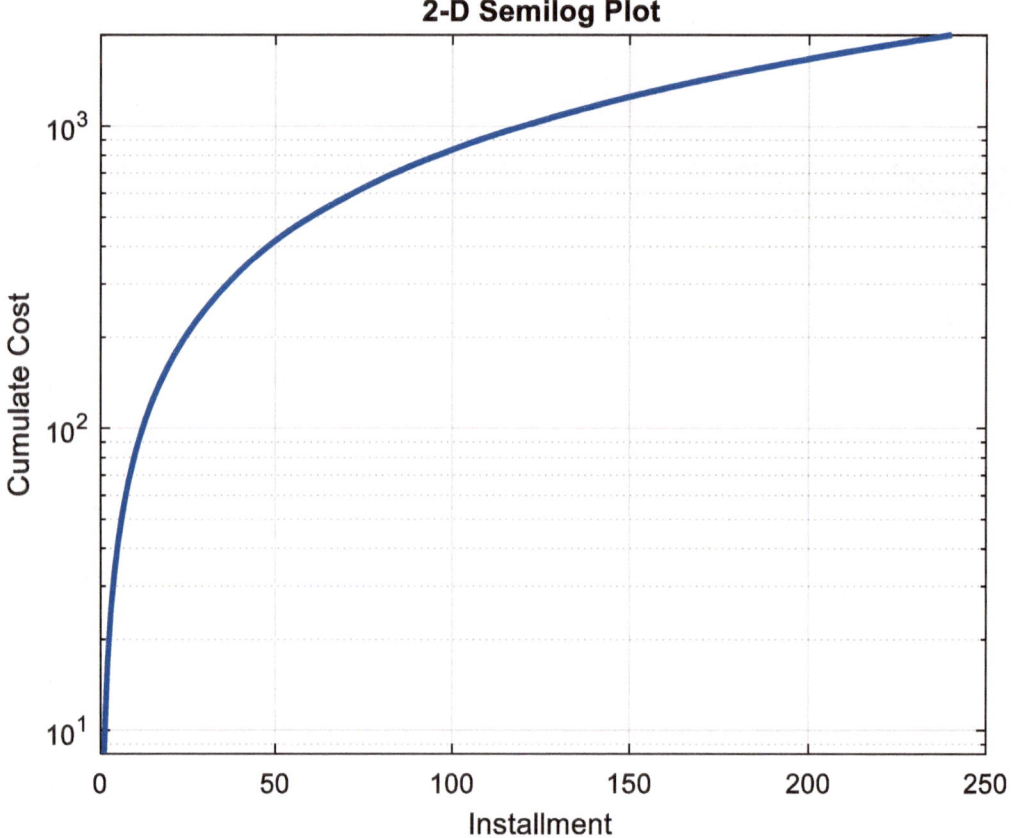

Fig. 19.3 Plotting the graph with a specific width and adding x and y labels and title to it

```
P = 1000;
npayments = 240;
rate = 0.08/12;
```

```
mpayment = P*(rate*(1+rate)^npayments)/(((1+rate)^npayments) - 1);
x = 1:240;
y = x * mpayment;
semilogy(x,y);
grid on
yticks([10 50 100 500 1000])
yticklabels({'$10','$50','$100','$500','$1000'})
```

Description

Adding tick values and tick labels to the graph. See Fig. 19.4.

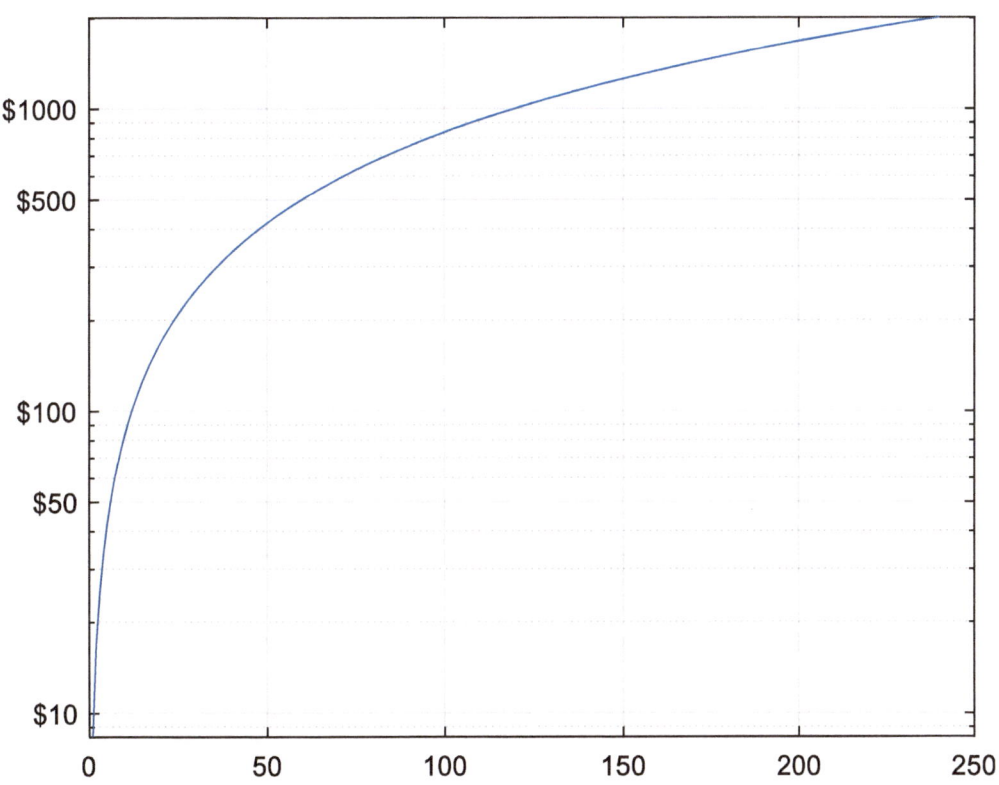

Fig. 19.4   Adding tick values and tick labels to the graph

Example

```
P = 1000;
npayments = 240;
rate = 0.08/12;
mpayment = P*(rate*(1+rate)^npayments)/(((1+rate)^npayments) - 1);
x = 1:240;
y = x * mpayment;
semilogy(x,y,'--dg');
grid on
```

Plotting the graph using a specific line style, color, and marker (See Fig. 19.5). The lists of line styles, markers, and colors available in MATLAB are illustrated in Figs. 19.6, 19.7, and 19.8, respectively.

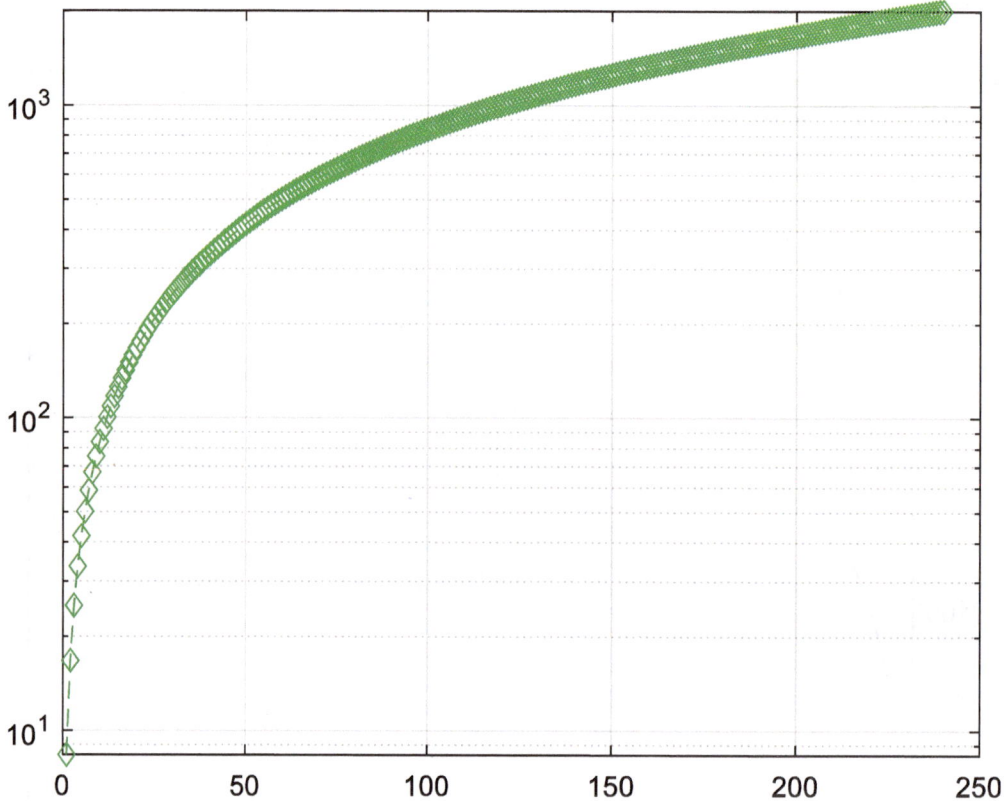

Fig. 19.5 Plotting the graph using a specific line style, color, and marker

Line Style	Description	Resulting Line
"-"	Solid line	——————
"--"	Dashed line	- — — — —
":"	Dotted line	················
"-."	Dash-dotted line	—·—·—·—··

Fig. 19.6 The list of line styles available in MATLAB

Marker	Description	Resulting Marker
"o"	Circle	○
"+"	Plus sign	+
"*"	Asterisk	✳
"."	Point	•
"x"	Cross	×
"_"	Horizontal line	—
"|"	Vertical line	|
"square"	Square	□
"diamond"	Diamond	◇
"^"	Upward-pointing triangle	△
"v"	Downward-pointing triangle	▽
">"	Right-pointing triangle	▷
"<"	Left-pointing triangle	◁
"pentagram"	Pentagram	☆
"hexagram"	Hexagram	✡

Fig. 19.7 The list of line markers available in MATLAB

Color Name	Short Name	Appearance
"red"	"r"	
"green"	"g"	
"blue"	"b"	
"cyan"	"c"	
"magenta"	"m"	
"yellow"	"y"	
"black"	"k"	
"white"	"w"	

Fig. 19.8 The list of line colors available in MATLAB

Example

```
x = 1:100;
y1 = x.^2;
y2 = x.^3;
semilogy(x,y1,'--',x,y2)
legend('x^2','x^3','Location','northwest')
grid on
```

Description

Adding a legend to the specific place (northwest) of the graph. See Fig. 19.9.

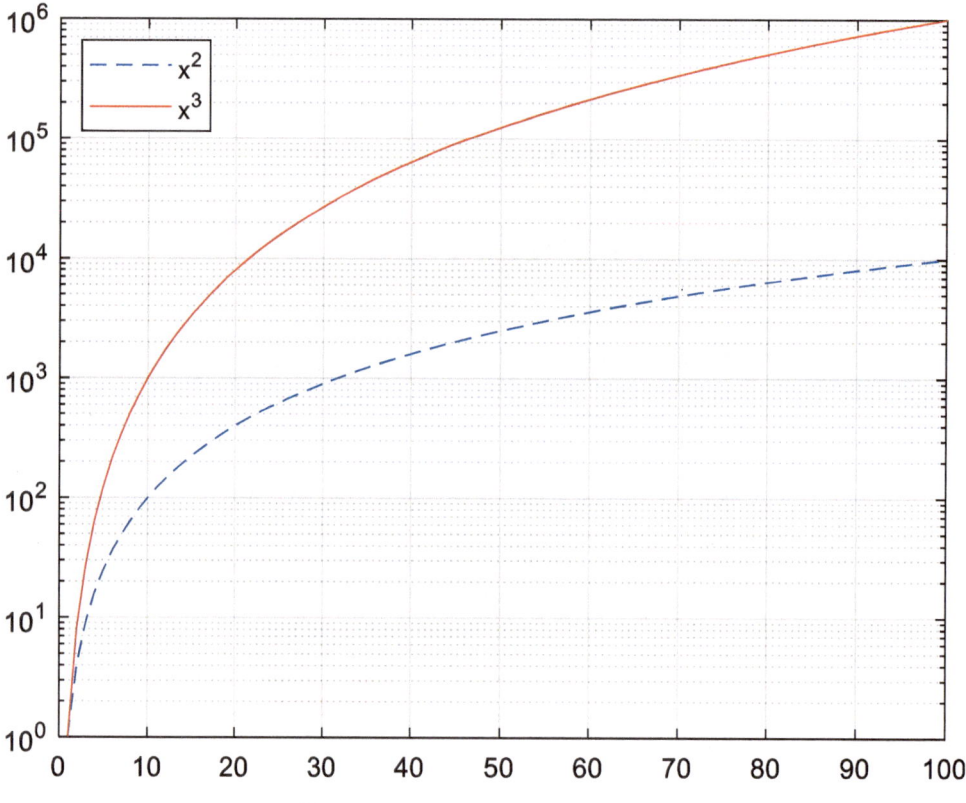

Fig. 19.9 Adding a legend to the specific place (northwest) of the graph

Exercise

(a) *Write the codes to plot the graph of the functions $y = x^4$ and $y = x^5$ for the interval (1100). Add appropriate x and y labels and legend. Also, select the dashed line style for one of them.*

(b) *Execute the codes in your computer, then draw the result(s) on the axes.*

References

1. MATLAB 2023a.
2. Rahmani-Andebili, M. (2024). *Mathematics of engineering and science – Practice problems, methods, and solutions*. Springer Nature.
3. Rahmani-Andebili, M. (2022). *Differential equations – Practice problems, methods, and solutions*. Springer Nature.
4. Rahmani-Andebili, M. (2023). *Calculus III – Practice problems, methods, and solutions*. Springer Nature.
5. Rahmani-Andebili, M. (2023). *Calculus II – Practice problems, methods, and solutions*. Springer Nature.
6. Rahmani-Andebili, M. (2023). *Calculus I – Practice problems, methods, and solutions* (2nd ed.). Springer Nature.
7. Rahmani-Andebili, M. (2021). *Calculus – Practice problems, methods, and solutions*. Springer Nature.
8. Rahmani-Andebili, M. (2024). *Precalculus – Practice problems, methods, and solutions* (2nd ed.). Springer Nature.
9. Rahmani-Andebili, M. (2021). *Precalculus – Practice problems, methods, and solutions*. Springer Nature.

Abstract

In this chapter, the area plot in MATLAB is presented and described. In this regard, several examples and exercises for each section of the chapter are presented. The exercises that include writing the codes, executing them, and achieving the results need to be done by students to master programming skills. In this book, the codes, outputs, and descriptions are in blue, black, and green colors, respectively. To program in MATLAB, a script file can be created and saved with an appropriate name (e.g., untitled01) in the preferred directory of a computer. The program can be run by clicking on the "Run" available on the top toolbar of the script in MATLAB or by calling the script by typing its name in Command Window or in the other scripts.

20.1   Area Plot

In the following, the description of area plot is presented and exemplified [1].

- This plot in the format area(Y) plots Y against an implicit set of x-coordinates and fills the areas between the curves. If Y is a vector, the x-coordinates range from 1 to length(Y). Moreover, if Y is a matrix, the x-coordinates range from 1 to the number of rows in Y.
- This plot in the format area(X,Y) plots the values in Y against the x-coordinates X. The function then fills the areas between the curves based on the shape of Y. If Y is a vector, the plot contains one curve. The area plot fills the area between the curve and the horizontal axis. In addition, if Y is a matrix, the plot contains one curve for each column in Y. The area plot fills the areas between the curves and stacks them, showing the relative contribution of each row element to the total height at each x-coordinate.
- This plot in the format area(,basevalue) specifies the baseline value for the area plot. The basevalue corresponds to a horizontal baseline. The area plot fills the area confined between the curves and this line.

© The Author(s), under exclusive license to Springer Nature Switzerland AG 2024

M. Rahmani-Andebili, *MATLAB Lessons, Examples, and Exercises*, https://doi.org/10.1007/978-3-031-76177-5_20

Example

y = [1 0.5 5 8 7 4];
area(y)

Description

Displaying the elements of the vector using the area plot. See Fig. 20.1.

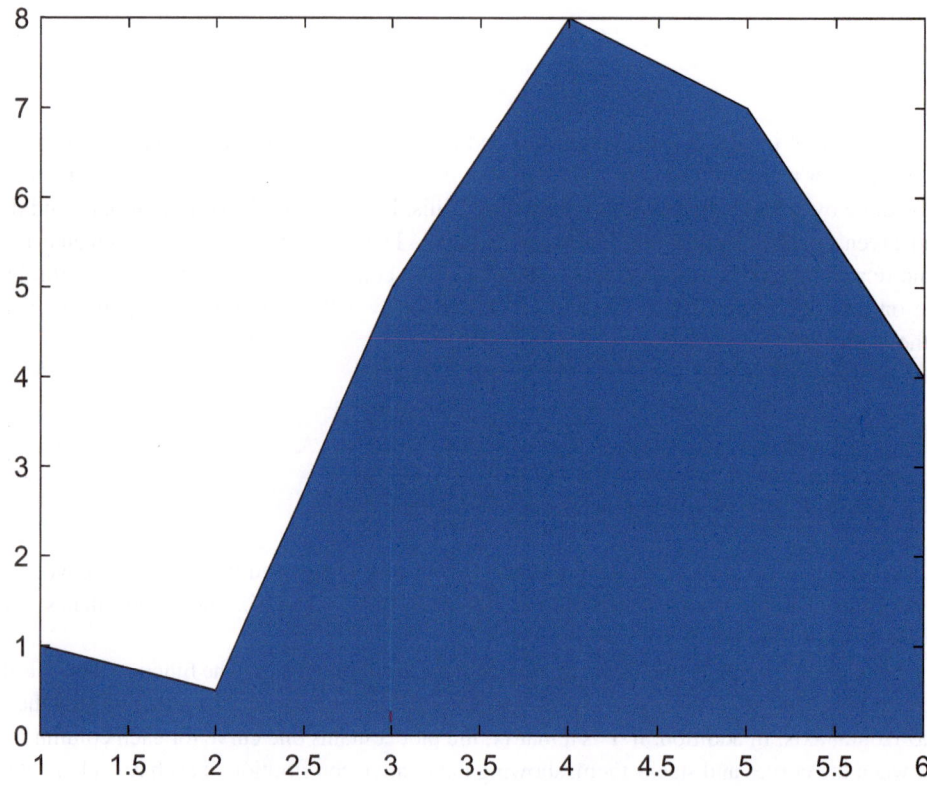

Fig. 20.1   Displaying the elements of the vector using the area plot

Example

Y = [1 5 3; 3 2 7; 1 5 3; 2 6 1];
area(Y)
legend({'Model A','Model B','Model C'})

Description

Displaying the elements of the matrix using the area plot and adding a legend to that. See Fig. 20.2.

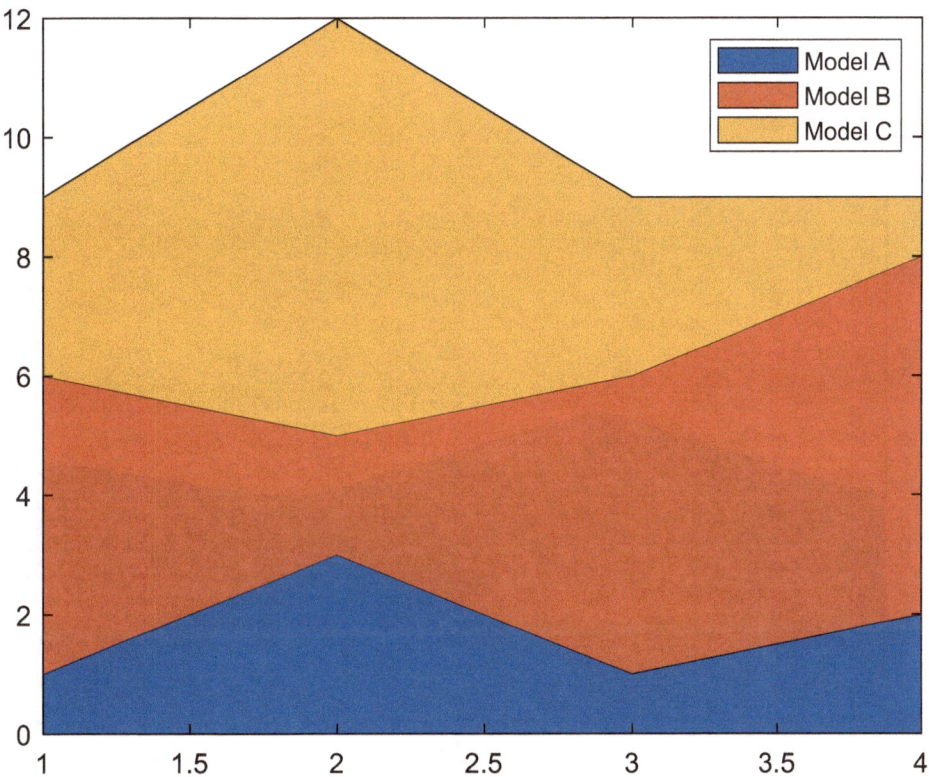

Fig. 20.2 Displaying the elements of the matrix using the area plot and adding a legend to that

Y = [1 5 3; 3 2 7; 1 5 3; 2 6 1];
basevalue = -4;
area(Y,basevalue)

Description

Displaying the elements of the matrix using the area plot while applying a baseline value. See Fig. 20.3.

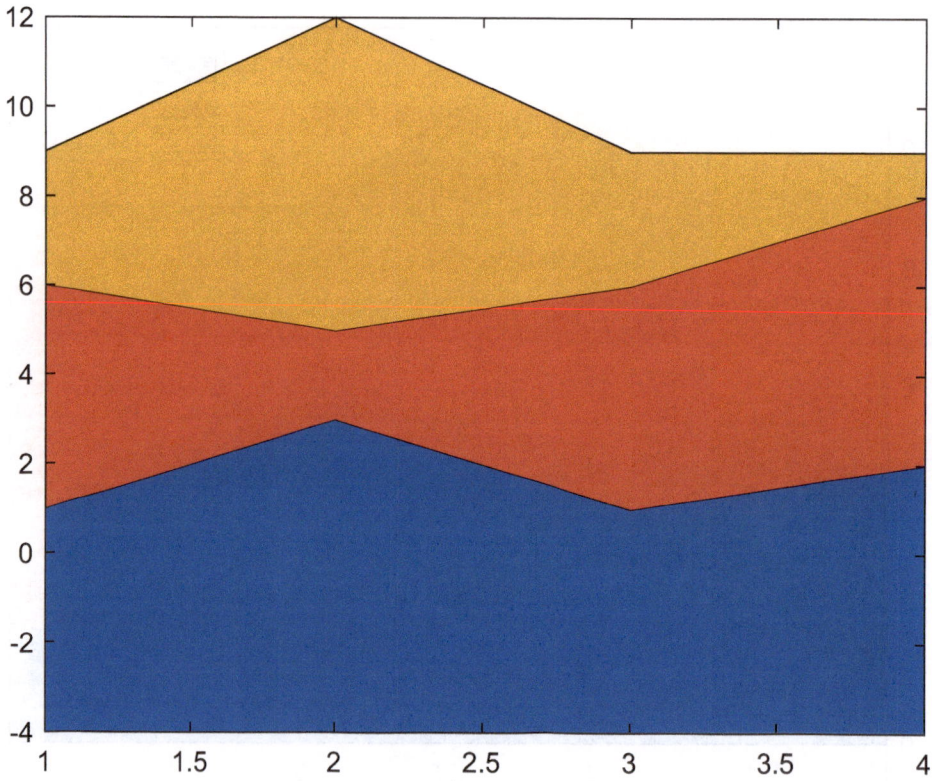

Fig. 20.3 Displaying the elements of the matrix using the area plot while applying a baseline value

Exercise

(a) *Write the codes to plot the graph of the matrix below. Add a legend and apply the baseline value of −2.*

(b) *Execute the codes in your computer, then draw the result(s) on the axes.*

Y =

1	5	3	1
3	2	7	2
1	5	3	3
2	6	1	4

Reference

1. MATLAB 2023a.

Abstract

In this chapter, the polar coordinates plot in MATLAB is presented and described. In this regard, several examples and exercises for each section of the chapter are presented. The exercises that include writing the codes, executing them, and achieving the results need to be done by students to master programming skills. In this book, the codes, outputs, and descriptions are in blue, black, and green colors, respectively. To program in MATLAB, a script file can be created and saved with an appropriate name (e.g., untitled01) in the preferred directory of a computer. The program can be run by clicking on the "Run" available on the top toolbar of the script in MATLAB or calling the script by typing its name in Command Window or in the other scripts.

21.1 Polar Coordinates Plot

In the following, the description of polar coordinates plot is presented and exemplified [1–9].

- This plot in the format polarplot(theta,rho) plots a line in polar coordinates, with theta indicating the angle in radians and rho indicating the radius value for each point. The inputs must be vectors of equal length or matrices of equal size.
- If the inputs are matrices, then it plots columns of rho versus columns of theta. Alternatively, one of the inputs can be a vector and the other a matrix as long as the vector is the same length as one dimension of the matrix.

Example

```
theta = 0:0.01:2*pi;
rho = sin(theta).*cos(theta);
polarplot(theta,rho)
```

Description

Plotting the line in polar coordinates. See Fig. 21.1.

M. Rahmani-Andebili, *MATLAB Lessons, Examples, and Exercises*, https://doi.org/10.1007/978-3-031-76177-5_21

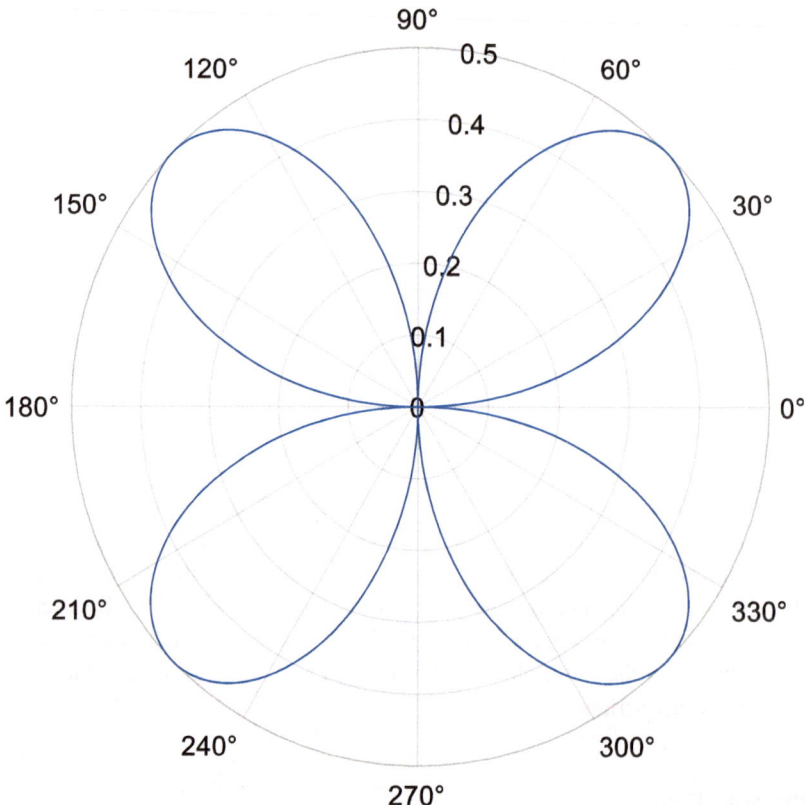

Fig. 21.1 Plotting the line in polar coordinates

Execute the codes in your computer. Then, draw the result(s) on the polar coordinates below.

```
theta = 0:0.01:2*pi;
rho = sin(3*theta).*cos(3*theta);
polarplot(theta,rho)
```

Example

```
theta = linspace(0,360,50);
theta_radians = deg2rad(theta);
rho = 0.005*theta/10;
polarplot(theta_radians,rho)
```

Description

Plotting the line in polar coordinates after converting the values in theta from degrees to radians. See Fig. 21.2.

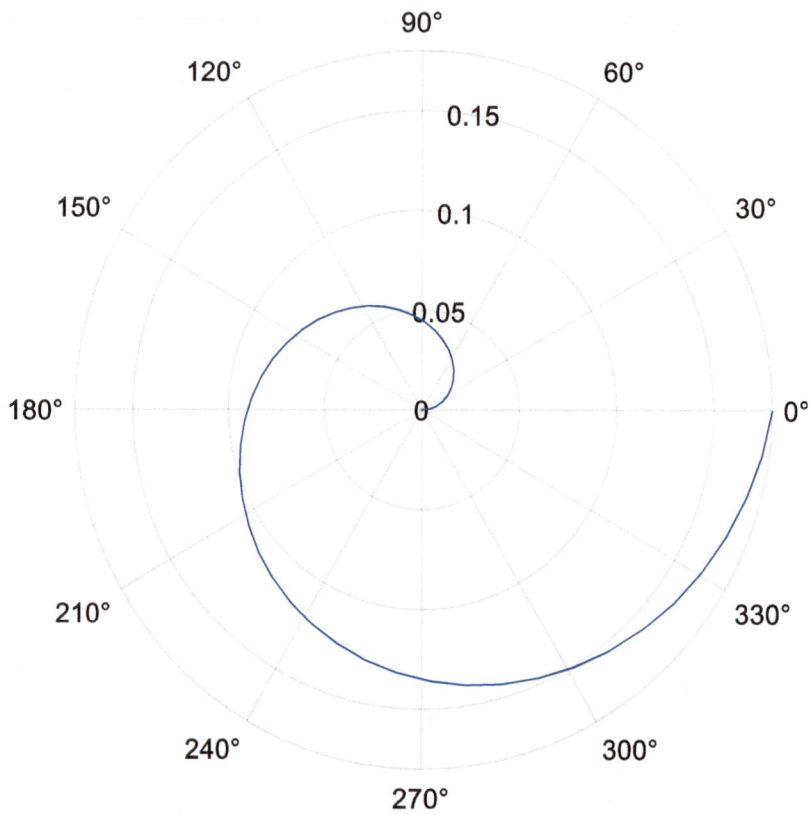

Fig. 21.2 Plotting the line in polar coordinates

Exercise

Execute the codes in your computer. Then, draw the result(s) on the polar coordinates below.

```
theta = linspace(0,720,50);
theta_radians = deg2rad(theta);
rho = 0.005*theta/10;
polarplot(theta_radians,rho)
```

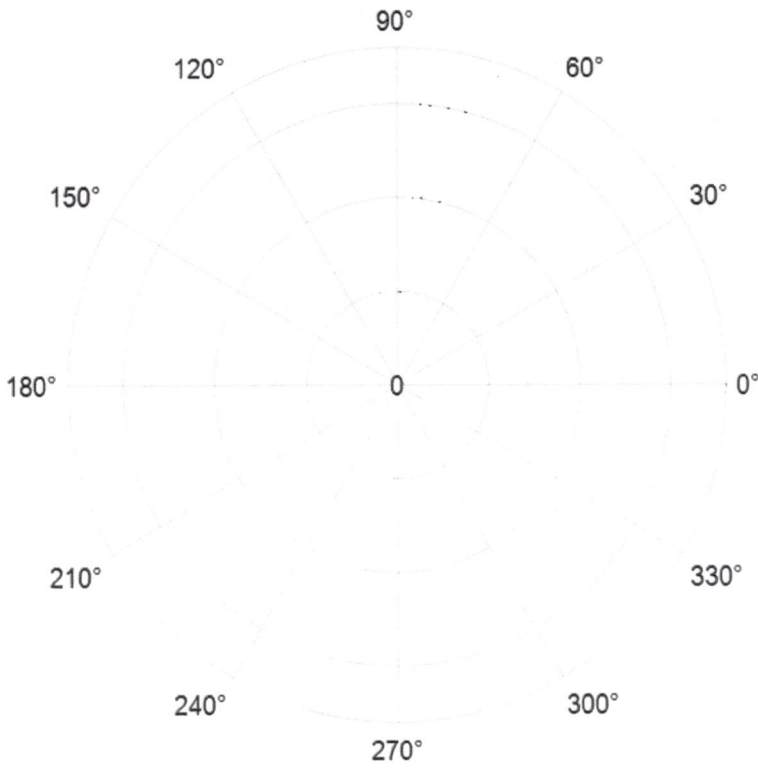

Example

theta = linspace(0,6*pi);
rho1 = theta/10;
polarplot(theta,rho1,'LineWidth',2)
hold on
rho2 = theta/12;
polarplot(theta,rho2,'--','LineWidth',2)
hold off

Description

Plotting the multiple lines in polar coordinates with the specific line width. See Fig. 21.3.

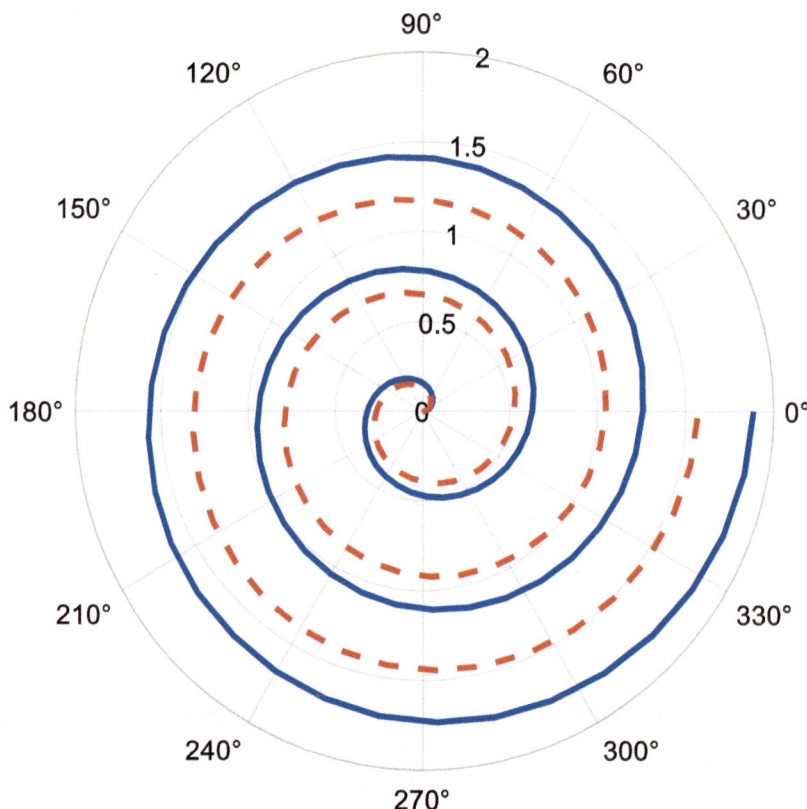

Fig. 21.3 Plotting the multiple lines in polar coordinates with the specific line width

Example

```
theta = linspace(0,360,50);
theta_radians = deg2rad(theta);
rho = 0.005*theta/10;
polarplot(theta_radians,rho,'-om')
```

Description

Plotting using the specific line marker, color, and style (See Fig. 21.4). The lists of line styles, markers, and colors available in MATLAB are illustrated in Figs. 21.5, 21.6, and 21.7, respectively.

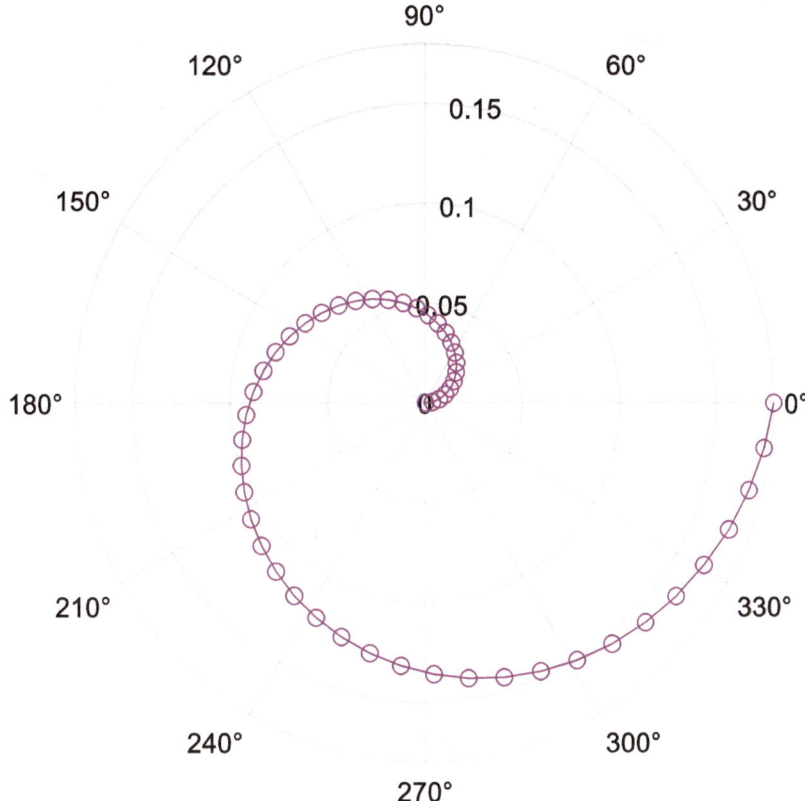

Fig. 21.4 Plotting using the specific line marker, color, and style

Line Style	Description	Resulting Line
" - "	Solid line	————————
" - - "	Dashed line	– — — —
" : "	Dotted line	················
" - . "	Dash-dotted line	—·—·—·—··

Fig. 21.5 The list of line styles available in MATLAB

Marker	Description	Resulting Marker
"o"	Circle	○
"+"	Plus sign	+
"*"	Asterisk	✳
"."	Point	•
"x"	Cross	×
"_"	Horizontal line	—
"|"	Vertical line	|
"square"	Square	□
"diamond"	Diamond	◇
"^"	Upward-pointing triangle	△
"v"	Downward-pointing triangle	▽
">"	Right-pointing triangle	▷
"<"	Left-pointing triangle	◁
"pentagram"	Pentagram	☆
"hexagram"	Hexagram	✡

Fig. 21.6 The list of line markers available in MATLAB

Color Name	Short Name	Appearance
"red"	"r"	
"green"	"g"	
"blue"	"b"	
"cyan"	"c"	
"magenta"	"m"	
"yellow"	"y"	
"black"	"k"	
"white"	"w"	

Fig. 21.7 The list of line colors available in MATLAB

References

1. MATLAB 2023a.
2. Rahmani-Andebili, M. (2024). *Mathematics of engineering and science – Practice problems, methods, and solutions*. Springer Nature.
3. Rahmani-Andebili, M. (2022). *Differential equations – Practice problems, methods, and solutions*. Springer Nature.
4. Rahmani-Andebili, M. (2023). *Calculus III – Practice problems, methods, and solutions*. Springer Nature.
5. Rahmani-Andebili, M. (2023). *Calculus II – Practice problems, methods, and solutions*. Springer Nature.
6. Rahmani-Andebili, M. (2023). *Calculus I – Practice problems, methods, and solutions* (2nd ed.). Springer Nature.
7. Rahmani-Andebili, M. (2021). *Calculus – Practice problems, methods, and solutions*. Springer Nature.
8. Rahmani-Andebili, M. (2024). *Precalculus – Practice problems, methods, and solutions* (2nd ed.). Springer Nature.
9. Rahmani-Andebili, M. (2021). *Precalculus – Practice problems, methods, and solutions*. Springer Nature.

Abstract

In this chapter, the contour plot in MATLAB is presented and described. In this regard, several examples and exercises for each section of the chapter are presented. The exercises that include writing the codes, executing them, and achieving the results need to be done by students to master programming skills. In this book, the codes, outputs, and descriptions are in blue, black, and green colors, respectively. To program in MATLAB, a script file can be created and saved with an appropriate name (e.g., untitled01) in the preferred directory of a computer. The program can be run by clicking on the "Run" available on the top toolbar of the script in MATLAB or calling the script by typing its name in Command Window or in the other scripts.

22.1 Contour Plot

In the following, the description of contour plot is presented and exemplified [1].

- This plot in the format contour(X,Y,Z) creates a contour plot containing the isolines of matrix Z, where Z contains height values on the x–y plane. The column and row indices of Z are the x and y coordinates in the plane, respectively. Herein, X and Y are matrices in the x–y plane.

Example

```
x = linspace(-2*pi,2*pi);
y = linspace(0,4*pi);
[X,Y] = meshgrid(x,y);
Z = sin(X)+cos(Y);
contour(X,Y,Z)
```

Description

Plotting the contours of Z as the heights above the grid defined by the matrices X and Y in the x–y plane. See Fig. 22.1.

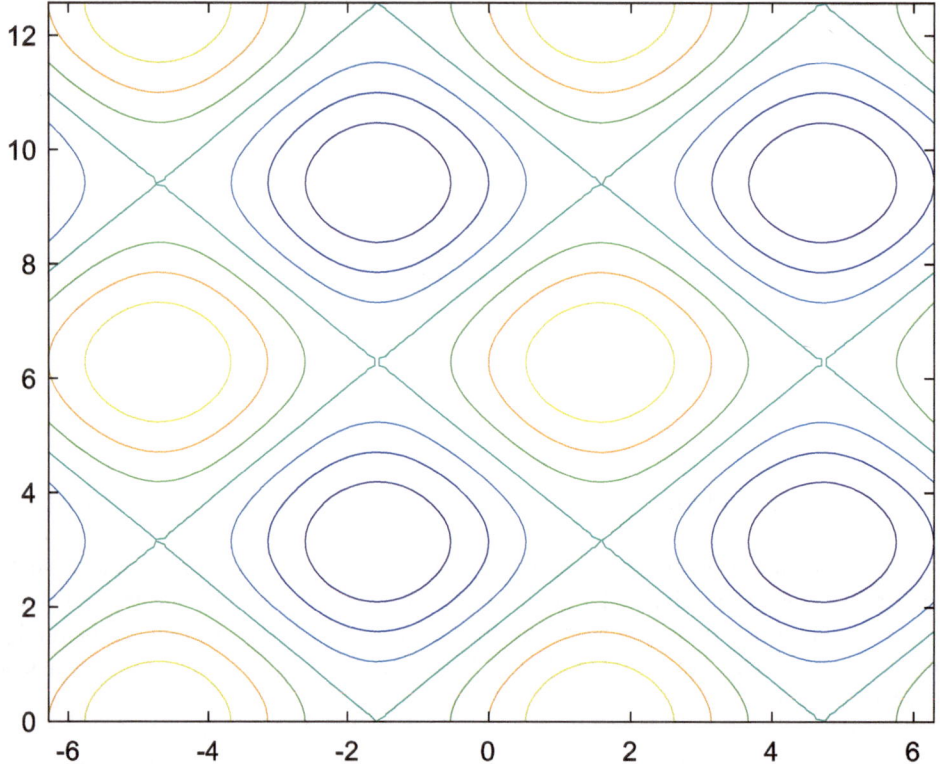

Fig. 22.1 Plotting the contours as the heights above the grid in the x–y plane

Example

```
x = linspace(-2*pi,2*pi);
y = linspace(0,4*pi);
[X,Y] = meshgrid(x,y);
Z = sin(X)+cos(Y);
contour(X,Y,Z,'--')
```

Description

Plotting the contours by using a dashed line style. See Fig. 22.2. The list of line styles available in MATLAB is illustrated in Fig. 22.3.

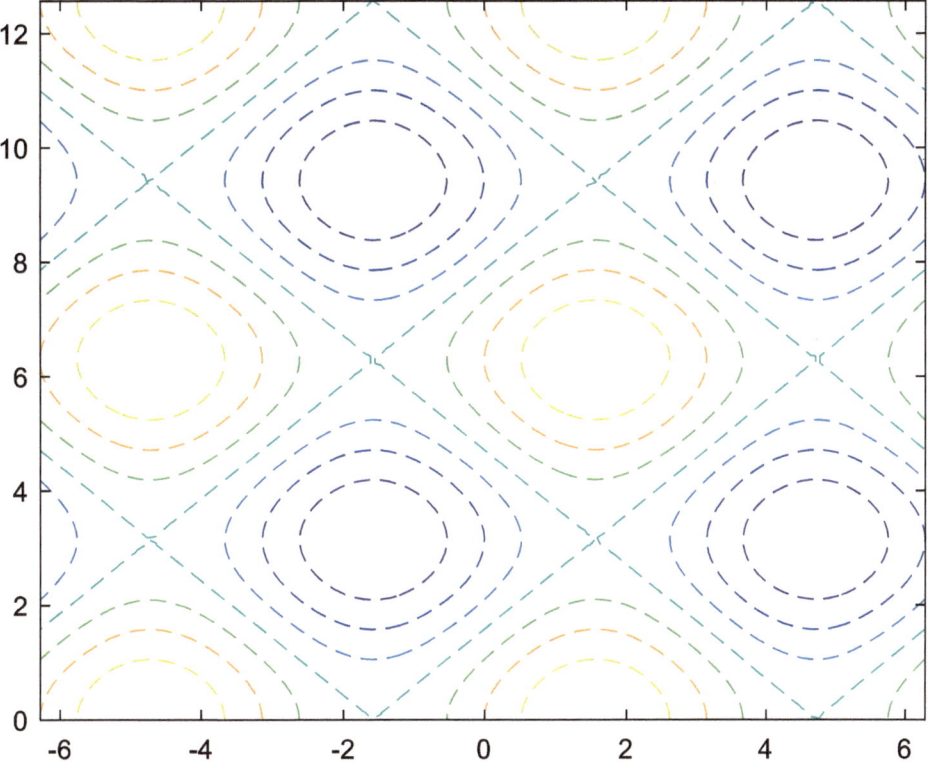

Fig. 22.2 Plotting the contours by using a dashed line style

Line Style	Description	Resulting Line
" - "	Solid line	————
" -- "	Dashed line	– — — — —
" : "	Dotted line	················
" -. "	Dash-dotted line	–·–·–·–···

Fig. 22.3 The list of line styles available in MATLAB

Example

```
x = linspace(-2*pi,2*pi);
y = linspace(0,4*pi);
[X,Y] = meshgrid(x,y);
Z = sin(X)+cos(Y);
contour(X,Y,Z,'ShowText','on')
```

Description

Adding labels (values) to the contours. See Fig. 22.4.

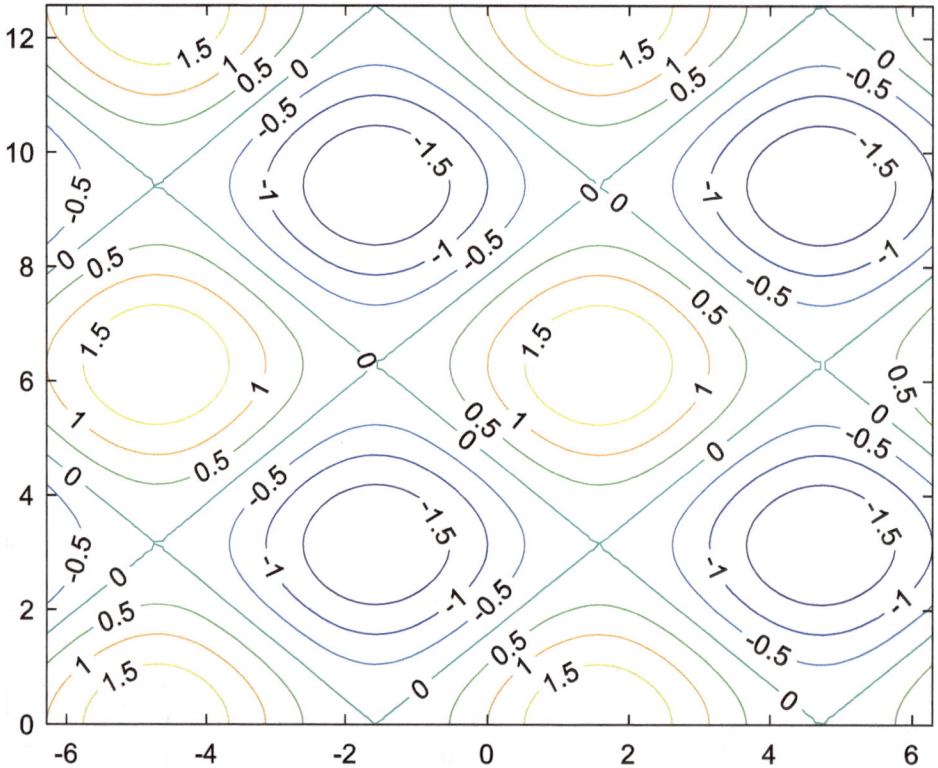

Fig. 22.4 Adding labels (values) to the contours

Example

```
x = linspace(-2*pi,2*pi);
y = linspace(0,4*pi);
[X,Y] = meshgrid(x,y);
Z = sin(X)+cos(Y);
[M,c] = contour(X,Y,Z);
c.LineWidth = 2;
```

Description

Changing the contours width. See Fig. 22.5.

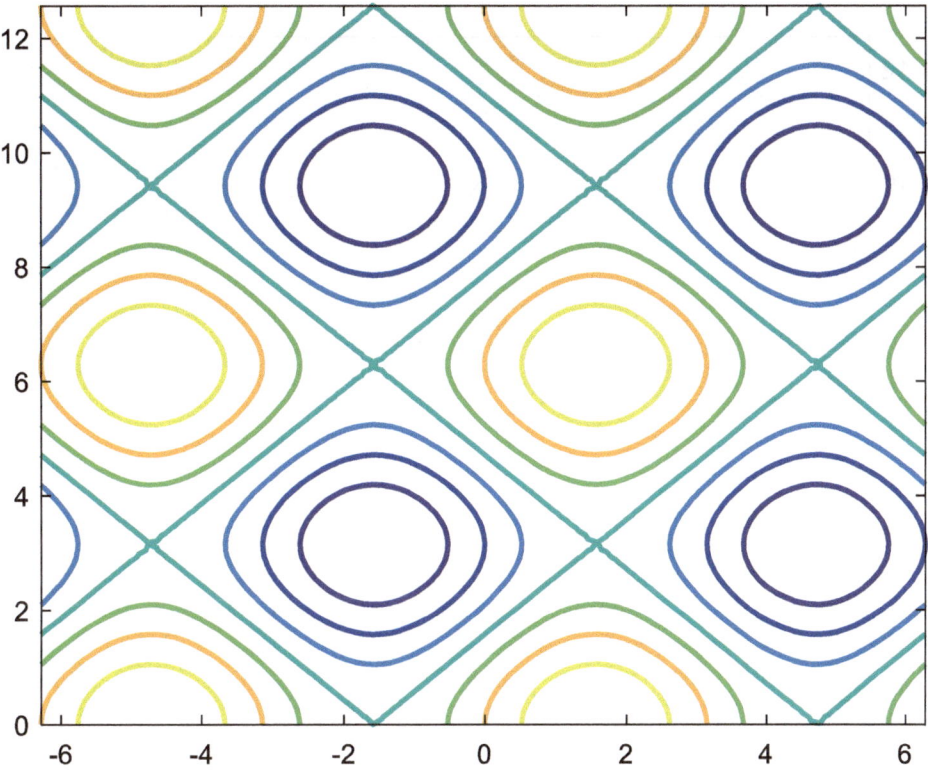

Fig. 22.5 Changing the contours width

Example

```
x = linspace(-2*pi,2*pi);
y = linspace(0,4*pi);
[X,Y] = meshgrid(x,y);
Z = sin(X)+cos(Y);
contour(X,Y,Z)
title('Contours')
xlabel('Your xlabel')
ylabel('Your ylabel')
grid on
```

Description

Adding title, axis labels, and grid to the graph. See Fig. 22.6.

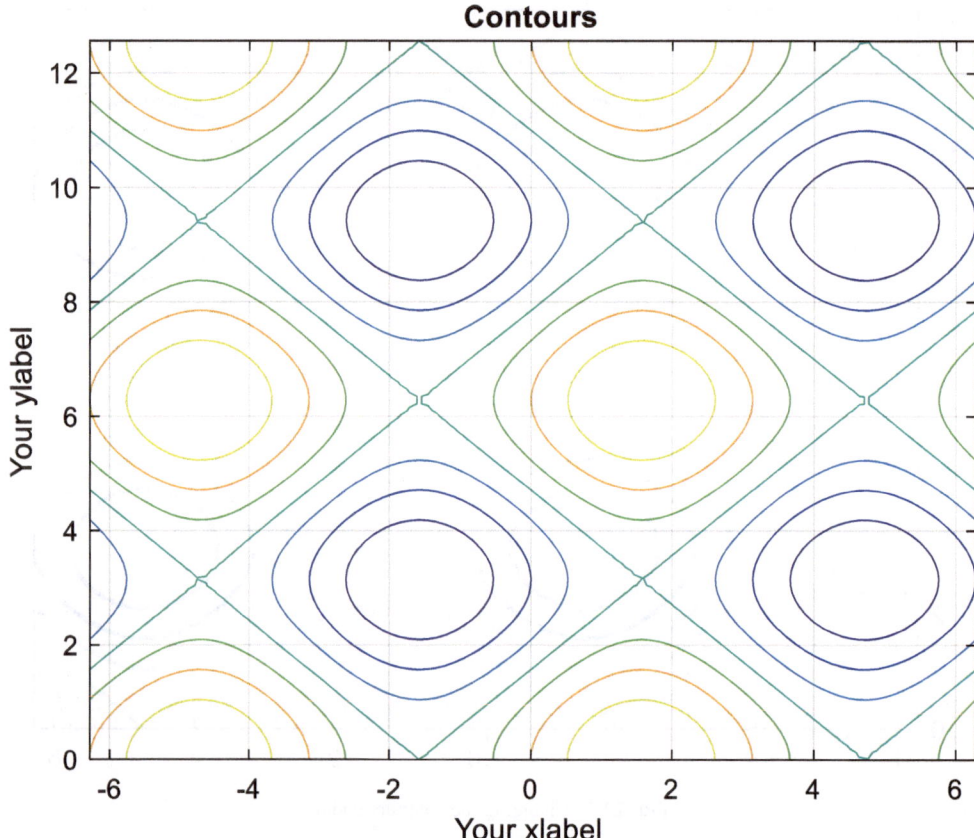

Fig. 22.6 Adding title, axis labels, and grid to the graph

Reference

1. MATLAB 2023a.

Abstract

In this chapter, the vector plot (quiver) in MATLAB is presented and described. In this regard, several examples and exercises for each section of the chapter are presented. The exercises that include writing the codes, executing them, and achieving the results need to be done by students to master programming skills. In this book, the codes, outputs, and descriptions are in blue, black, and green colors, respectively. To program in MATLAB, a script file can be created and saved with an appropriate name (e.g., untitled01) in the preferred directory of a computer. The program can be run by clicking on the "Run" available on the top toolbar of the script in MATLAB or calling the script by typing its name in Command Window or in the other scripts.

23.1   Vector Plot (Quiver)

In the following, the description of quiver plot is presented and exemplified [1].

- This plot in the format quiver(X,Y,U,V) plots arrows with directional components U and V at the Cartesian coordinates specified by X and Y. For example, the first arrow originates from the point X(1) and Y(1), extends horizontally according to U(1), and extends vertically according to V(1). By default, the quiver function scales the arrow lengths so that they do not overlap.

Example

```
load('wind','x','y','u','v')
X = x(11:22,11:22,1);
Y = y(11:22,11:22,1);
U = u(11:22,11:22,1);
V = v(11:22,11:22,1);
quiver(X,Y,U,V)
axis equal
```

Description

Plotting the air currents by using the vector plot. Herein, the vectors X and Y represent the location of the base of each arrow, and U and V represent the directional components of each arrow. Moreover, the "axis equal" is applied to use equal data unit lengths along each axis. This makes the arrows point in the correct direction. See Fig. 23.1.

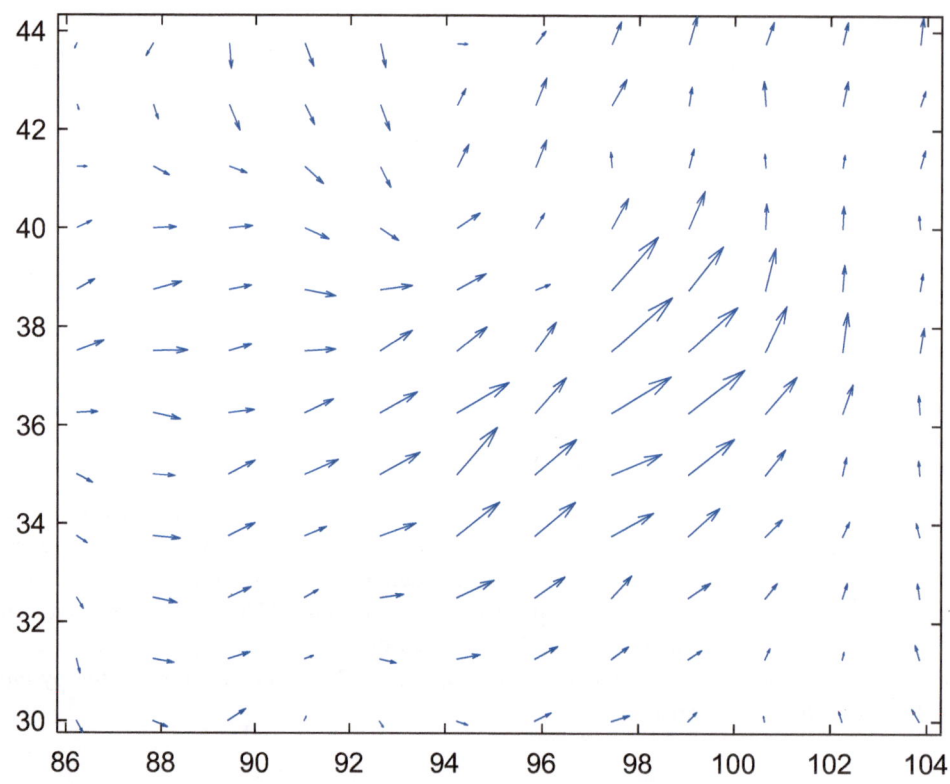

Fig. 23.1 Plotting the air currents by using the vector plot

Example

```
[X,Y] = meshgrid(0:6,0:6);
U = 0.25*X;
V = 0.5*Y;
quiver(X,Y,U,V,0)
```

Description

Disabling automatic scaling so that arrow lengths are determined entirely by U and V by setting the scale argument to 0. See Fig. 23.2.

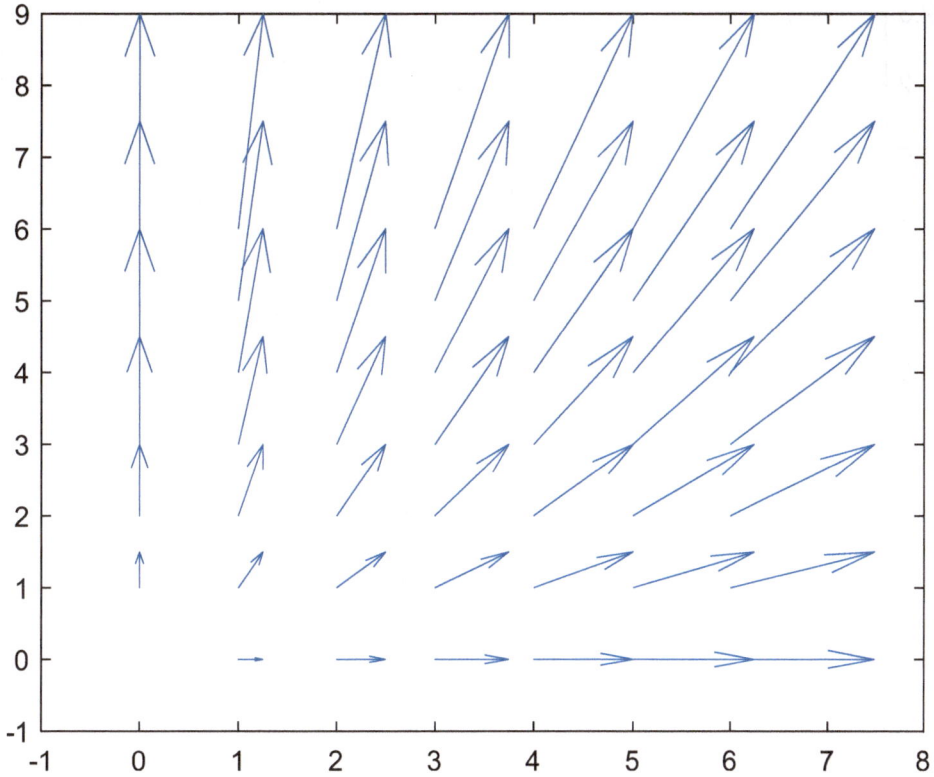

Fig. 23.2 Plotting the air currents by using the vector plot while disabling automatic scaling

spacing = 0.2;
[X,Y] = meshgrid(-2:spacing:2);
Z = X.*exp(-X.^2 - Y.^2);
[DX,DY] = gradient(Z,spacing);
quiver(X,Y,DX,DY)

Plotting the gradient of the function. See Fig. 23.3.

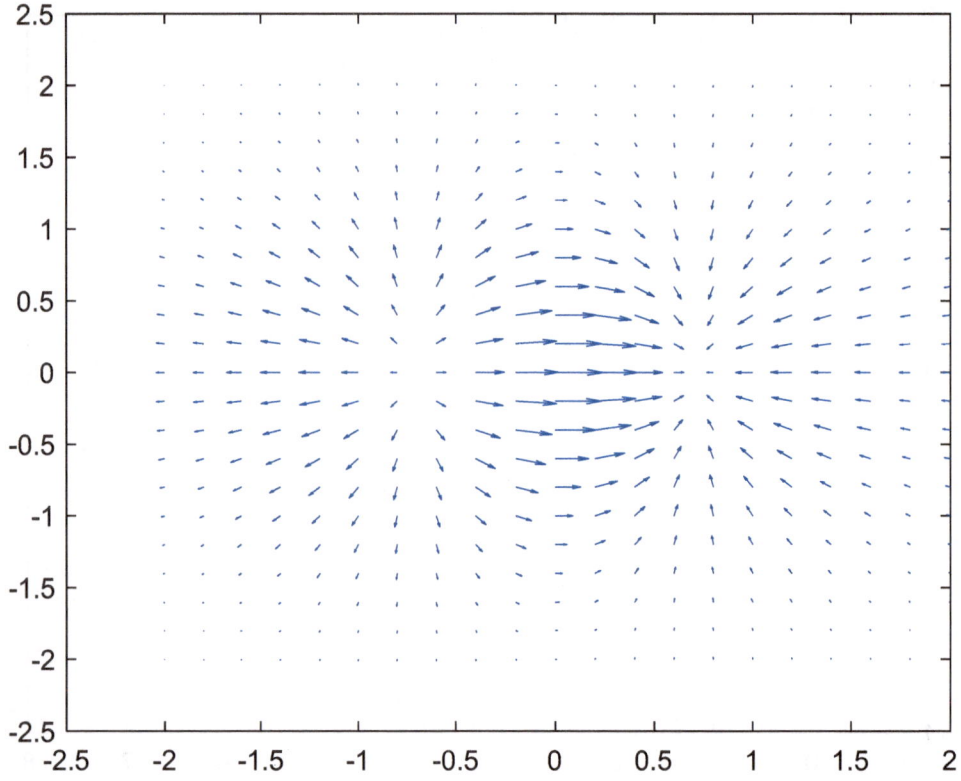

Fig. 23.3 Plotting the gradient of the function

[X,Y] = meshgrid(0:6,0:6);
U = 0.25*X;
V = 0.5*Y;
quiver(X,Y,U,V,'m','LineWidth',2)

Plotting the gradient of the function using the specific line width and color (See Fig. 23.4). The list of line colors available in MATLAB is shown in Fig. 23.5.

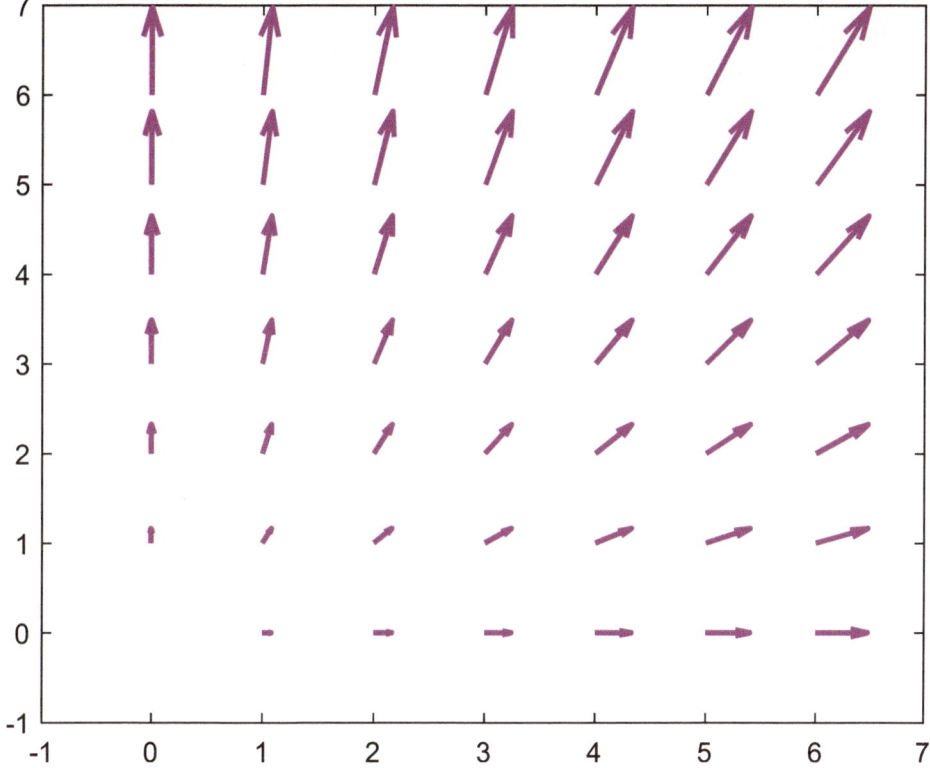

Fig. 23.4 Plotting the gradient of the function using the specific line width and color

Color Name	Short Name	Appearance
"red"	"r"	
"green"	"g"	
"blue"	"b"	
"cyan"	"c"	
"magenta"	"m"	
"yellow"	"y"	
"black"	"k"	
"white"	"w"	

Fig. 23.5 The list of line colors available in MATLAB

Exercise

Execute the codes in your computer. Then, draw the result(s) on the axes.

```
[X,Y] = meshgrid(-pi:pi/8:pi,-pi:pi/8:pi);
U = sin(Y);
V = cos(X);
quiver(X,Y,U,V)
```

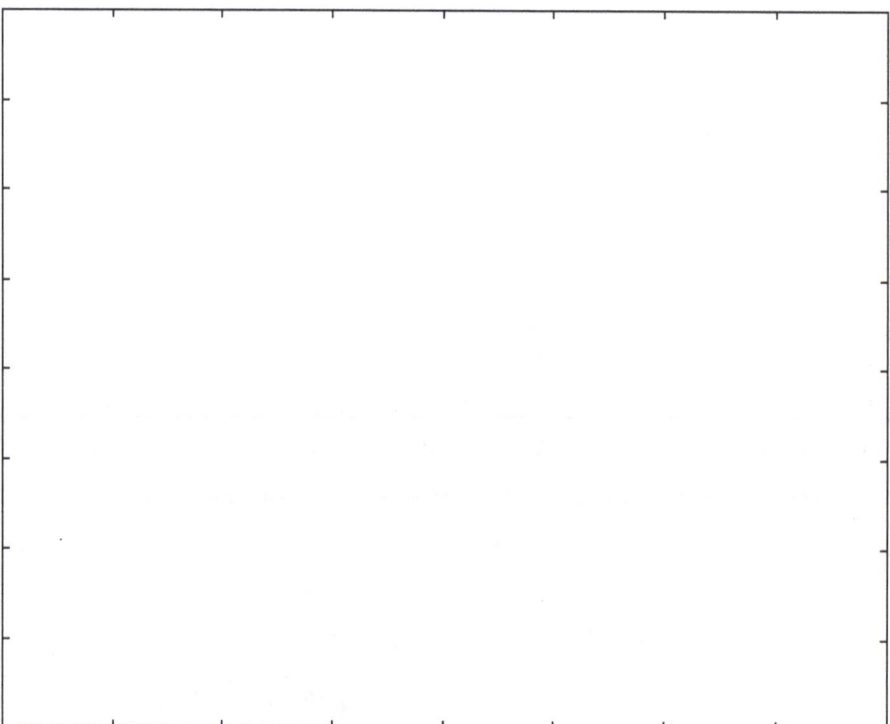

Reference

1. MATLAB 2023a.

Abstract

In this chapter, the 2D and 3D streamline plots in MATLAB are presented and described. In this regard, several examples and exercises for each section of the chapter are presented. The exercises that include writing the codes, executing them, and achieving the results need to be done by students to master programming skills. In this book, the codes, outputs, and descriptions are in blue, black, and green colors, respectively. To program in MATLAB, a script file can be created and saved with an appropriate name (e.g., untitled01) in the preferred directory of a computer. The program can be run by clicking on the "Run" available on the top toolbar of the script in MATLAB or calling the script by typing its name in Command Window or in the other scripts.

24.1 2D and 3D Streamline Plots

In the following, the description of streamline plots is presented and exemplified [1].

- This plot in the format streamline(X,Y,Z,U,V,W,startX,startY,startZ) returns plotted streamlines for 3D vector data. The inputs X, Y, and Z are vector data coordinates; U, V, and W are vector data; and startX, startY, and startZ are the starting positions of the streamlines.
- This plot in the format streamline(X,Y,U,V,startX,startY) returns plotted streamlines for 2D vector data. The inputs X and Y are vector data coordinates, U and V are vector data, and startX and startY are the starting positions of the streamlines.

Example

```
load wind
[startX,startY,startZ] = meshgrid(80,20:10:50,0:5:15);
verts = stream3(x,y,z,u,v,w,startX,startY,startZ);
lineobj = streamline(verts);
view(3)
grid on
```

Description

Plotting 3D streamlines. Herein, the 3D arrays x, y, and z represent the locations of air current measurements. Moreover, the 3D arrays u, v, and w represent the velocity of the air current in 3D vector fields. See Fig. 24.1.

M. Rahmani-Andebili, *MATLAB Lessons, Examples, and Exercises*, https://doi.org/10.1007/978-3-031-76177-5_24

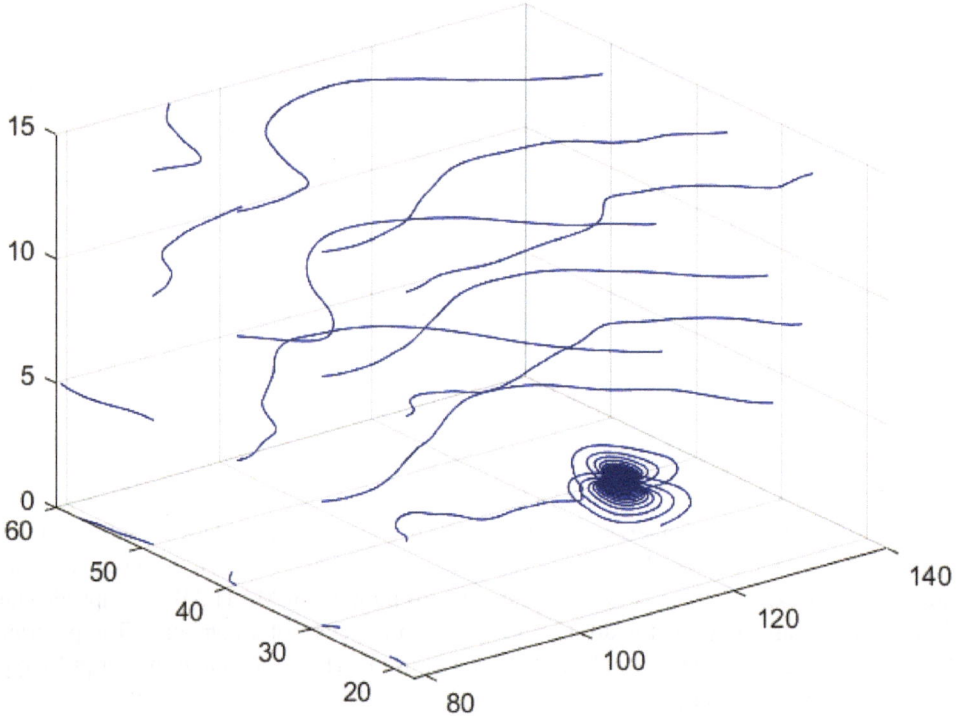

Fig. 24.1   Plotting the locations of air current measurements by using 3D streamlines

Example

```
load wind
[startX,startY,startZ] = meshgrid(80,20:10:50,0:5:15);
verts = stream3(x,y,z,u,v,w,startX,startY,startZ);
lineobj = streamline(verts);
view(3)
grid on
lineobj(10).Color = "m";
lineobj(10).LineStyle = "--";
lineobj(10).LineWidth = 2;
```

Description

Changing the aspects of a particular line of the 3D streamlines by setting the properties (style, color, width) on one of the returned line objects (see Fig. 24.2). The lists of line colors and styles available in MATLAB are illustrated in Figs. 24.3 and 24.4, respectively.

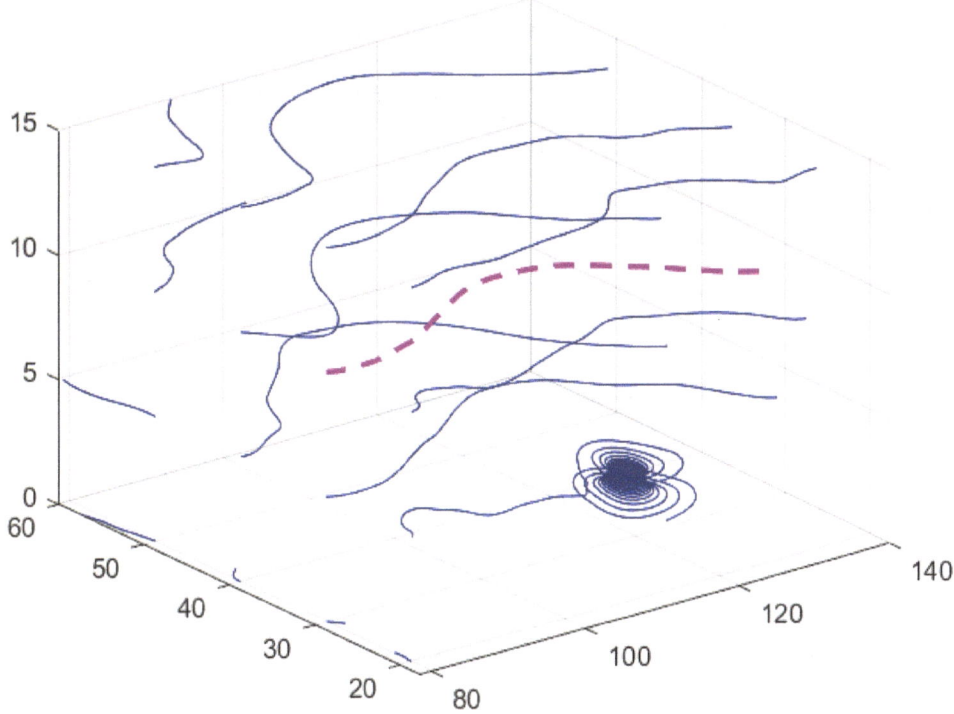

Fig. 24.2 Changing the aspects of a particular line of the 3D streamlines

Color Name	Short Name	Appearance
"red"	"r"	
"green"	"g"	
"blue"	"b"	
"cyan"	"c"	
"magenta"	"m"	
"yellow"	"y"	
"black"	"k"	
"white"	"w"	

Fig. 24.3 The list of line colors available in MATLAB

Line Style	Description	Resulting Line
"-"	Solid line	———————
"--"	Dashed line	- — — — —
":"	Dotted line	·················
"-."	Dash-dotted line	—·—·—·—··

Fig. 24.4 The list of line styles available in MATLAB

Example

load wind
x5 = x(:,:,5);
y5 = y(:,:,5);
u5 = u(:,:,5);
v5 = v(:,:,5);
[startX,startY] = meshgrid(80,20:10:50);
verts = stream2(x5,y5,u5,v5,startX,startY);
lineobj = streamline(verts);
grid on

Description

Plotting 2D streamlines. Herein, the 2D arrays x and y represent the locations of air current measurements. Moreover, the 2D arrays u and v represent the velocity of the air current in 2D vector fields. See Fig. 24.5.

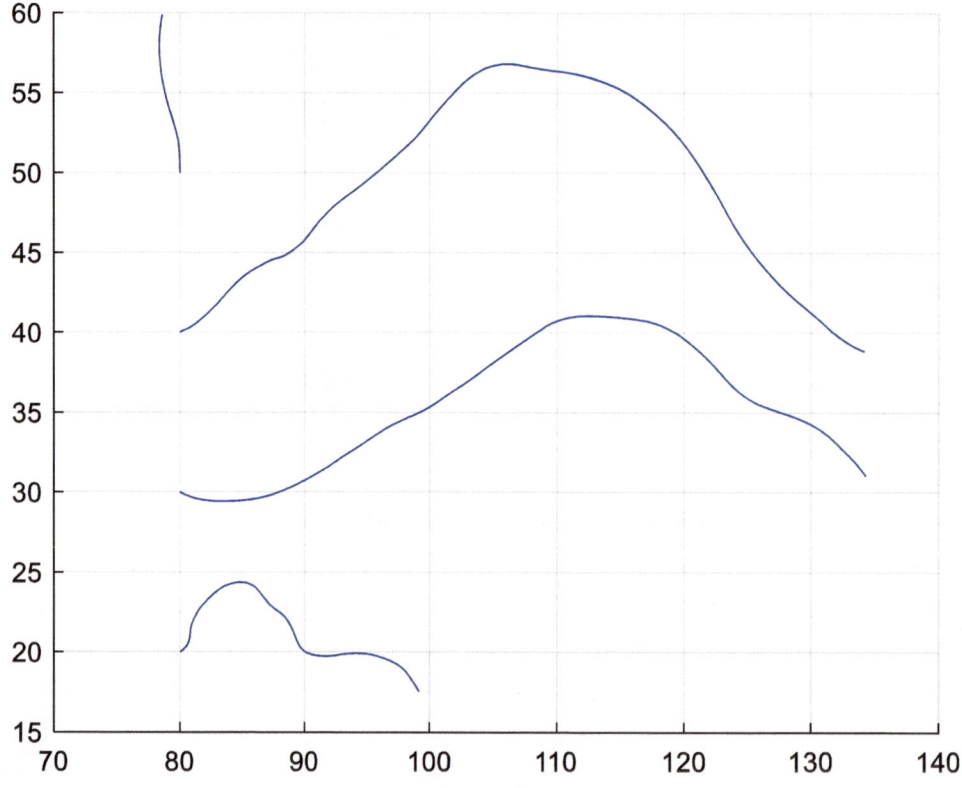

Fig. 24.5 Plotting the locations of air current measurements by using 2D streamlines

Reference

1. MATLAB 2023a.

Abstract

In this chapter, the surface plot in MATLAB is presented and described. In this regard, several examples and exercises for each section of the chapter are presented. The exercises that include writing the codes, executing them, and achieving the results need to be done by students to master programming skills. In this book, the codes, outputs, and descriptions are in blue, black, and green colors, respectively. To program in MATLAB, a script file can be created and saved with an appropriate name (e.g., untitled01) in the preferred directory of a computer. The program can be run by clicking on the "Run" available on the top toolbar of the script in MATLAB or calling the script by typing its name in Command Window or in the other scripts.

25.1 Surface Plot

In the following, the description of surface plot is presented and exemplified [1–9].
- This plot in the format surf(X,Y,Z) creates a three-dimensional surface plot. The function plots the values in matrix Z as heights above a grid in the x–y plane defined by X and Y. The color of the surface varies according to the heights specified by Z. Alternatively, surf(Z) can be used.
- This plot in the format surf(X,Y,Z,C) specifies the surface color. Alternatively, surf(Z,C) can be used.

Example

```
[X,Y] = meshgrid(1:0.2:20,1:0.2:20);
Z = sin(X) + cos(Y);
surf(X,Y,Z)
```

Description

Plotting the surface defined by the matrix Z. See Fig. 25.1.

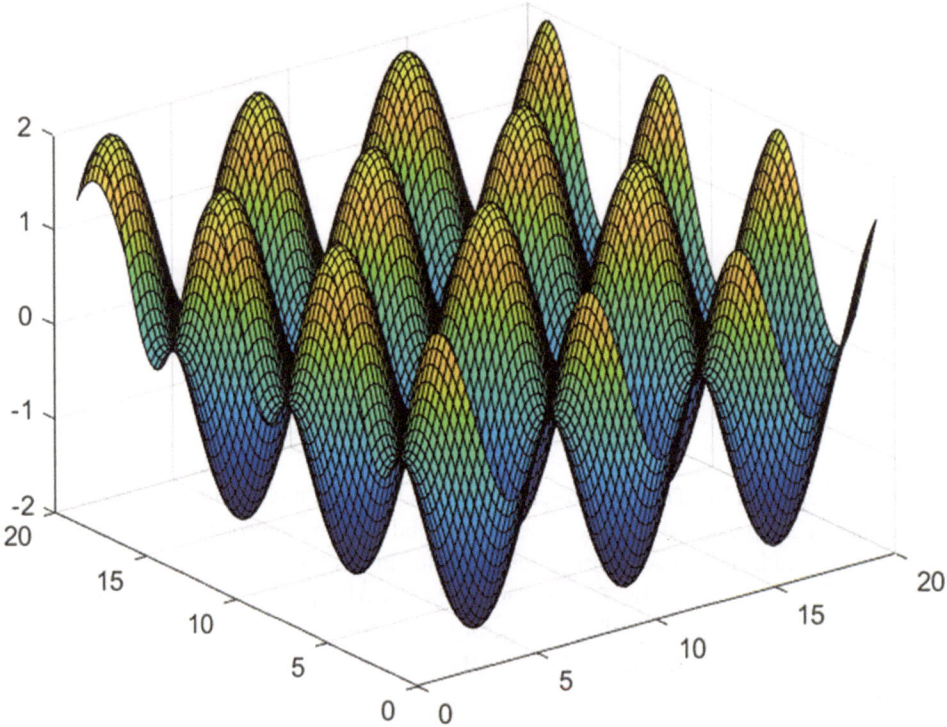

Fig. 25.1 Plotting the surface defined by the matrix

Demonstration

Execute the codes in your computer and observe the result(s).

[X,Y] = meshgrid(1:20,1:20);
Z = sin(X) + cos(Y);
surf(X,Y,Z)

Demonstration

Execute the codes in your computer and observe the result(s).

[X,Y] = meshgrid(1:0.2:20,1:0.2:20);
Z = (sin(X)).^2 + cos(Y);
surf(X,Y,Z)

Example

[X,Y] = meshgrid(1:0.2:20,1:0.2:20);
Z = sin(X) + cos(Y);
C = X.*Y;
surf(X,Y,Z,C)
colorbar

Description

Plotting the surface with specific colors using a colormap. See Fig. 25.2.

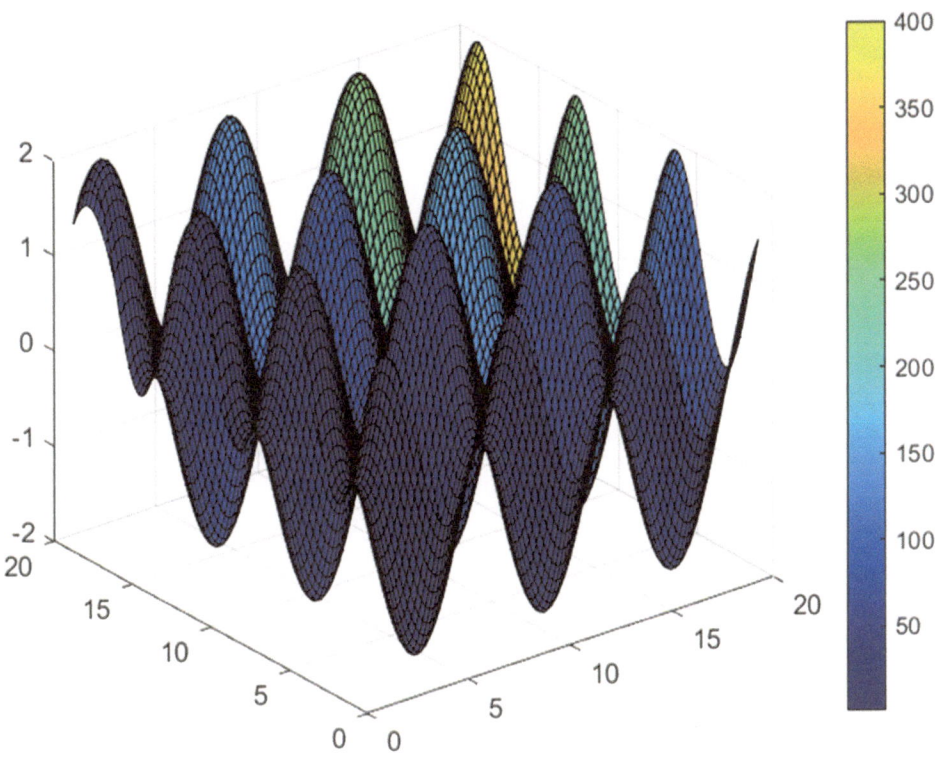

Fig. 25.2 Plotting the surface with specific colors using a colormap

References

1. MATLAB 2023a.
2. Rahmani-Andebili, M. (2024). *Mathematics of engineering and science – Practice problems, methods, and solutions*. Springer Nature.
3. Rahmani-Andebili, M. (2022). *Differential equations – Practice problems, methods, and solutions*. Springer Nature.
4. Rahmani-Andebili, M. (2023). *Calculus III – Practice problems, methods, and solutions*. Springer Nature.
5. Rahmani-Andebili, M. (2023). *Calculus II – Practice problems, methods, and solutions*. Springer Nature.
6. Rahmani-Andebili, M. (2023). *Calculus I – Practice problems, methods, and solutions* (2nd ed.). Springer Nature.
7. Rahmani-Andebili, M. (2021). *Calculus – Practice problems, methods, and solutions*. Springer Nature.
8. Rahmani-Andebili, M. (2024). *Precalculus – Practice problems, methods, and solutions* (2nd ed.). Springer Nature.
9. Rahmani-Andebili, M. (2021). *Precalculus – Practice problems, methods, and solutions*. Springer Nature.

Abstract

In this chapter, the mesh surface plot in MATLAB is presented and described. In this regard, several examples and exercises for each section of the chapter are presented. The exercises that include writing the codes, executing them, and achieving the results need to be done by students to master programming skills. In this book, the codes, outputs, and descriptions are in blue, black, and green colors, respectively. To program in MATLAB, a script file can be created and saved with an appropriate name (e.g., untitled01) in the preferred directory of a computer. The program can be run by clicking on the "Run" available on the top toolbar of the script in MATLAB or calling the script by typing its name in Command Window or in the other scripts.

26.1 Mesh Surface Plot

In the following, the description of mesh surface plot is presented and exemplified [1–9].

- This plot in the format mesh(X,Y,Z) creates a mesh plot which is a three-dimensional surface including solid edge colors and no face colors. The function plots the values in matrix Z as heights above a grid in the x–y plane defined by X and Y. The edge colors vary according to the heights specified by Z. Alternatively, mesh(Z) can be used.
- This plot in the format mesh(X,Y,Z,C) specifies the color of the edges. Alternatively, mesh(Z,C) can be used.

Example

```
[X,Y] = meshgrid(-8:.5:8);
R = sqrt(X.^2 + Y.^2) + eps;
Z = sin(R)./R;
mesh(X,Y,Z)
```

Description

Plotting the mesh surface plot defined by the matrix Z. See Fig. 26.1.

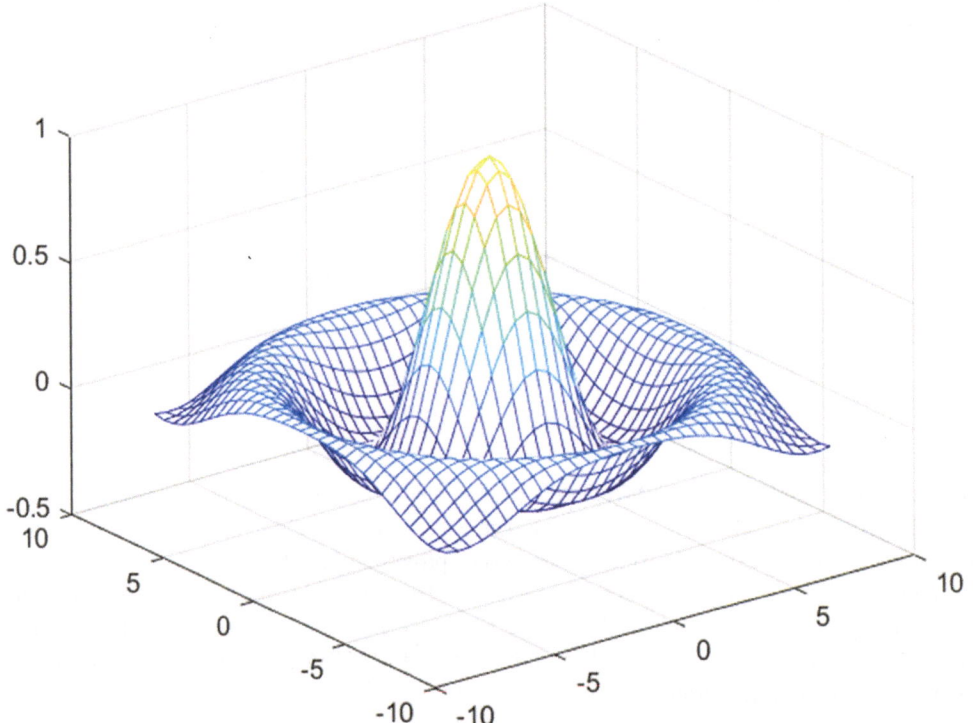

Fig. 26.1 Plotting the mesh surface plot defined by the matrix

[X,Y] = meshgrid(-8:0.1:8);
R = sqrt(X.^2 + Y.^2) + eps;
Z = sin(R)./R;
mesh(X,Y,Z)

Description

Plotting the mesh surface plot with more resolution. See Fig. 26.2.

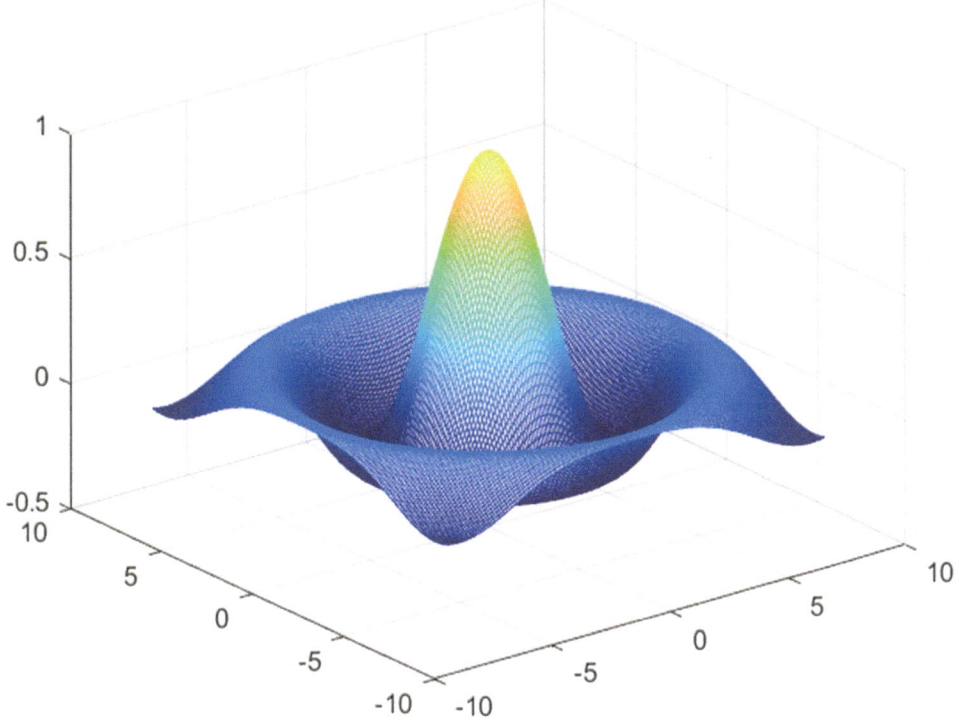

Fig. 26.2 Plotting the mesh surface plot with more resolution

Example

```
[X,Y] = meshgrid(-8:.5:8);
R = sqrt(X.^2 + Y.^2) + eps;
Z = sin(R)./R;
C = X.*Y;
mesh(X,Y,Z,C)
colorbar
```

Description

Plotting the mesh surface plot with specific colors using a colormap. See Fig. 26.3.

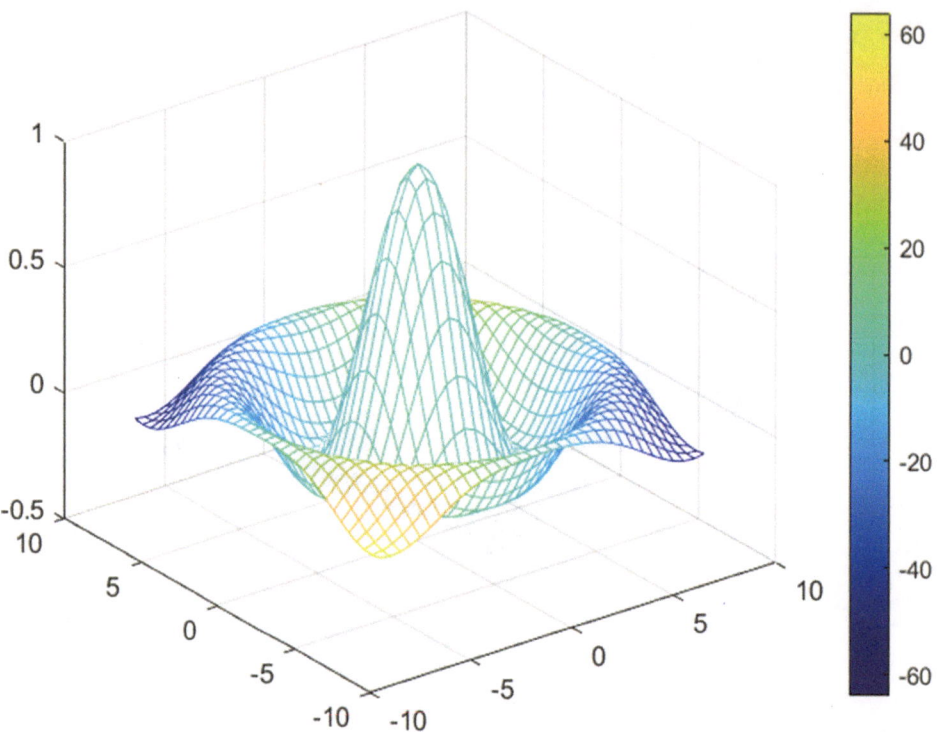

Fig. 26.3 Plotting the mesh surface plot with specific colors using a colormap

Demonstration

Execute the codes in your computer and observe the result(s).

```
[X,Y] = meshgrid(-8:0.1:8);
R = sqrt(X.^2 + Y.^2) + eps;
Z = sin(R)./R;
C = X.*Y;
mesh(X,Y,Z,C)
colorbar
```

References

1. MATLAB 2023a.
2. Rahmani-Andebili, M. (2024). *Mathematics of engineering and science – Practice problems, methods, and solutions*. Springer Nature.
3. Rahmani-Andebili, M. (2022). *Differential equations – Practice problems, methods, and solutions*. Springer Nature.
4. Rahmani-Andebili, M. (2023). *Calculus III – Practice problems, methods, and solutions*. Springer Nature.
5. Rahmani-Andebili, M. (2023). *Calculus II – Practice problems, methods, and solutions*. Springer Nature.
6. Rahmani-Andebili, M. (2023). *Calculus I – Practice problems, methods, and solutions* (2nd ed.). Springer Nature.
7. Rahmani-Andebili, M. (2021). *Calculus – Practice problems, methods, and solutions*. Springer Nature.
8. Rahmani-Andebili, M. (2024). *Precalculus – Practice problems, methods, and solutions* (2nd ed.). Springer Nature.
9. Rahmani-Andebili, M. (2021). *Precalculus – Practice problems, methods, and solutions*. Springer Nature.

Abstract

In this chapter, the display image in MATLAB is presented and described. In this regard, several examples and exercises for each section of the chapter are presented. The exercises that include writing the codes, executing them, and achieving the results need to be done by students to master programming skills. In this book, the codes, outputs, and descriptions are in blue, black, and green colors, respectively. To program in MATLAB, a script file can be created and saved with an appropriate name (e.g., untitled01) in the preferred directory of a computer. The program can be run by clicking on the "Run" available on the top toolbar of the script in MATLAB or calling the script by typing its name in Command Window or in the other scripts.

27.1 Display Image

In the following, the description of display image is presented and exemplified [1].

- This function in the format imshow(I) displays the grayscale image I in a figure. It uses the default display range for the image data type, and optimizes figure, axes, and image object properties for image display.

Example

```
rgbImage = imread("peppers.png");
imshow(rgbImage)
```

Description

Reading a sample RGB image (peppers.png) into the MATLAB workspace and then displaying the image using imshow. See Fig. 27.1.

M. Rahmani-Andebili, *MATLAB Lessons, Examples, and Exercises*, https://doi.org/10.1007/978-3-031-76177-5_27

Fig. 27.1 Displaying an RGB image

grayImage = im2gray(rgbImage);
imshow(grayImage)

Converting the RGB image to a grayscale image by using the im2gray function and then displaying the image using imshow. See Fig. 27.2.

Fig. 27.2 Displaying a grayscale image

Example

meanVal = mean(grayImage,"all");
binaryImage = grayImage >= meanVal;
imshow(binaryImage)

Description

Converting the grayscale image to a binary image by using thresholding and then displaying the image using imshow. See Fig. 27.3.

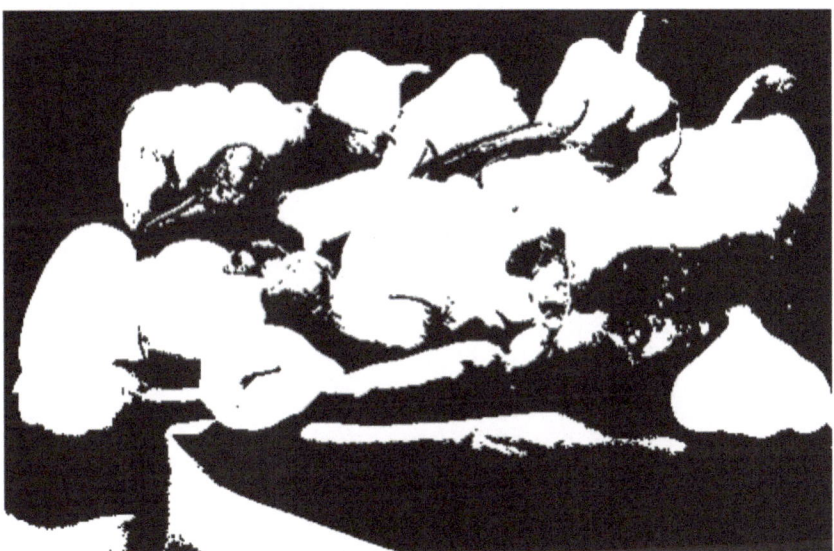

Fig. 27.3 Displaying a binary image

Exercise

Store an image in your computer with your favorite name (e.g., MyImage) with png format. Then, execute the codes below in your computer to display it.

filename = "MyImage.png";
imshow(filename)

Reference

1. MATLAB 2023a.

Abstract

In this chapter, the bar graph which is one of the plot types in MATLAB is presented and described. In this regard, several examples and exercises for each section of the chapter are presented. The exercises that include writing the codes, executing them, and achieving the results need to be done by students to master programming skills. In this book, the codes, outputs, and descriptions are in blue, black, and green colors, respectively. To program in MATLAB, a script file can be created and saved with an appropriate name (e.g., untitled01) in the preferred directory of a computer. The program can be run by clicking on the "Run" available on the top toolbar of the script in MATLAB or calling the script by typing its name in Command Window or in the other scripts.

28.1 Bar Graph

In the following, the description of bar graph is presented and exemplified [1].
- This plot in the format bar(y) creates a bar graph with one bar for each element in y.
- To plot a single series of bars, specify y as a vector of length m. The bars are positioned from 1 to m along the x-axis.
- To plot multiple series of bars, specify y as a matrix with one column for each series.

Example

```
y = [91 87 69 55 44 39 32 29 22 11 5];
bar(y)
```

Description

Displaying a bar graph. See Fig. 28.1.

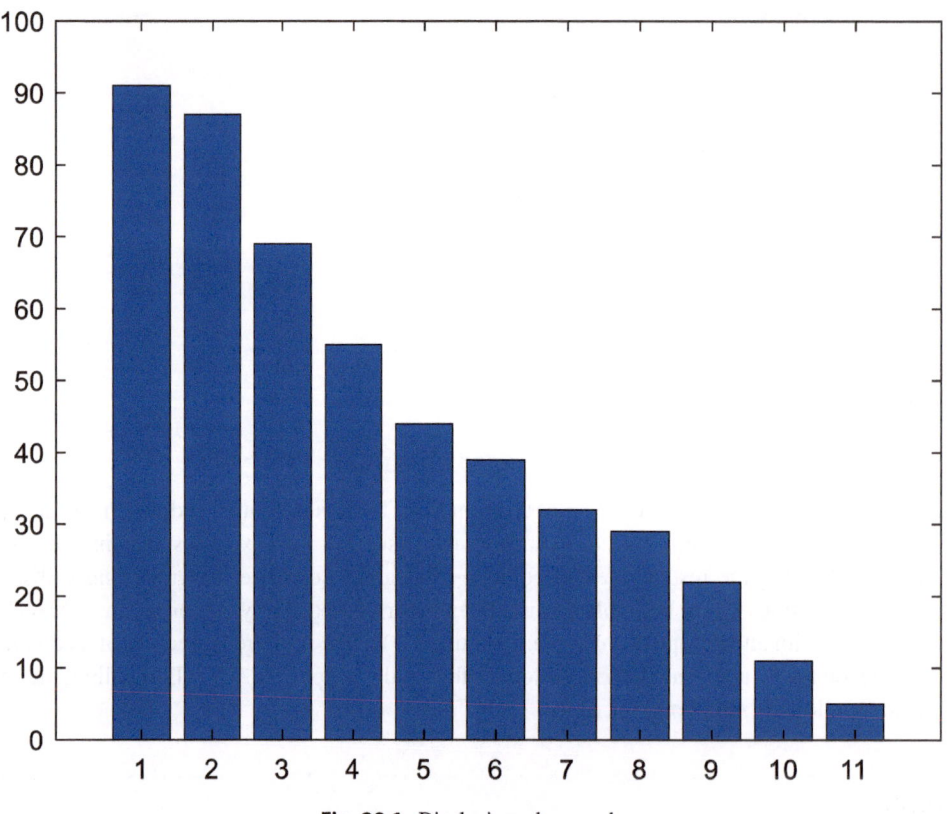

Fig. 28.1 Displaying a bar graph

Example

x = 2000:2010;
y = [91 87 69 55 44 39 32 29 22 11 5];
bar(x,y)

Description

Displaying a bar graph while specifying the bar locations along the x-axis. See Fig. 28.2.

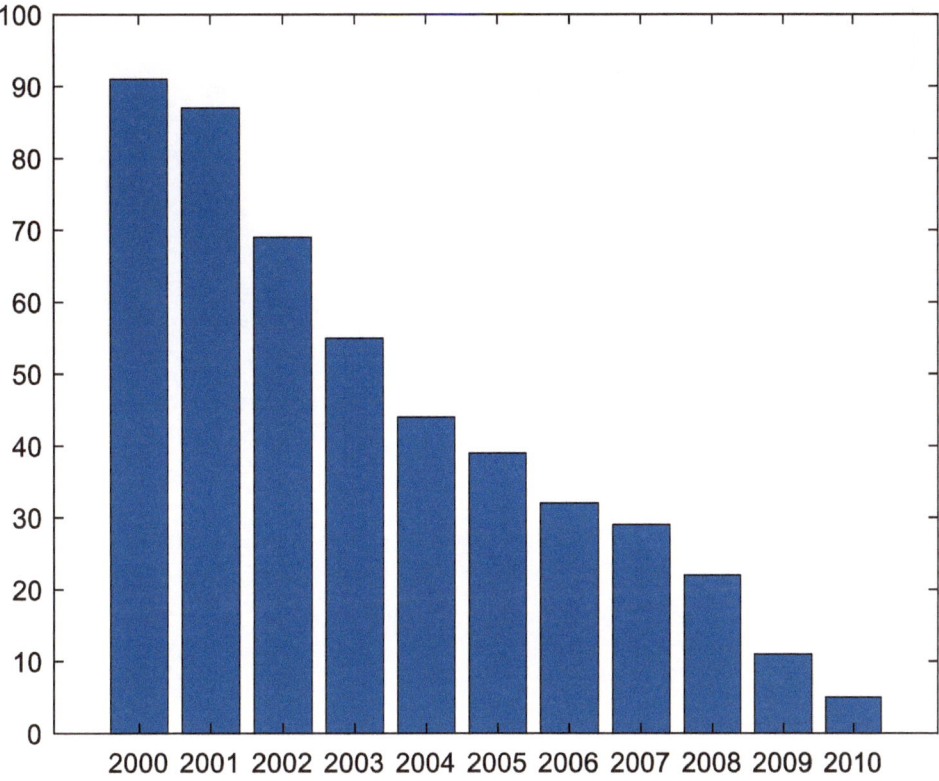

Fig. 28.2 Displaying a bar graph while specifying the bar locations along the x-axis

Exercise

Execute the codes in your computer, then draw the result(s) on the axes.

```
x = 2020:2024;
y = [0.98 0.87 0.83 0.88 0.91];
bar(x,y)
title('The World Population Growth Rate During 2020-2024')
xlabel('Year')
ylabel(' Growth Rate (%)')
```

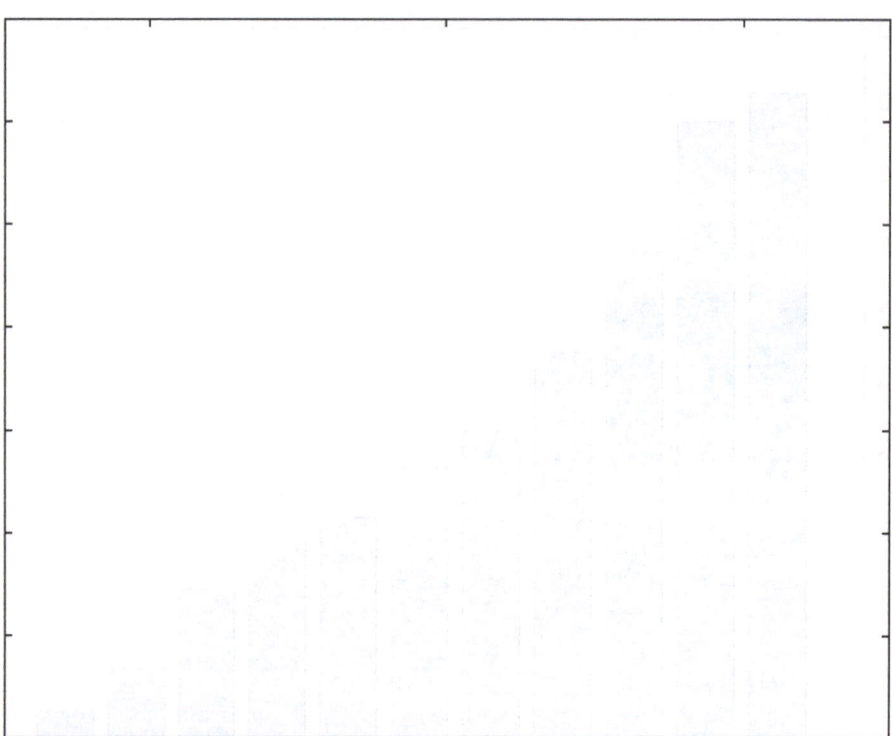

y = [91 87 69 55 44 39 32 29 22 11 5];
bar(y,0.3)

Displaying a bar graph with a specific bar width. See Fig. 28.3.

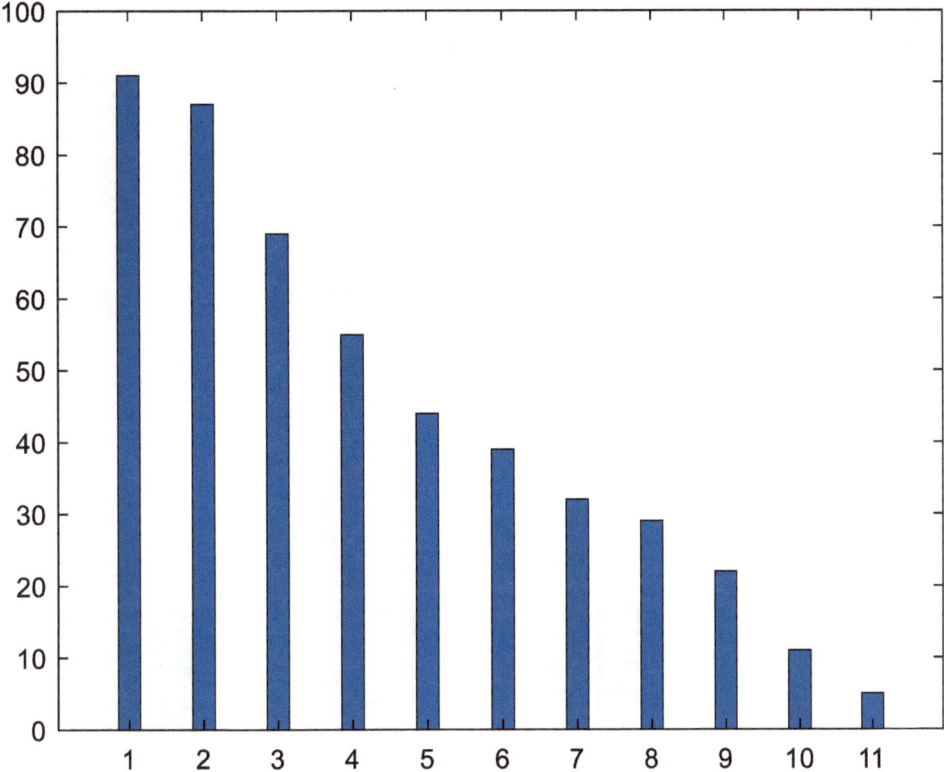

Fig. 28.3 Displaying a bar graph with a specific bar width

Example

y = [2 2 3; 2 5 6; 2 8 9; 2 11 12];
bar(y)

Description

Displaying four groups of three bars. See Fig. 28.4.

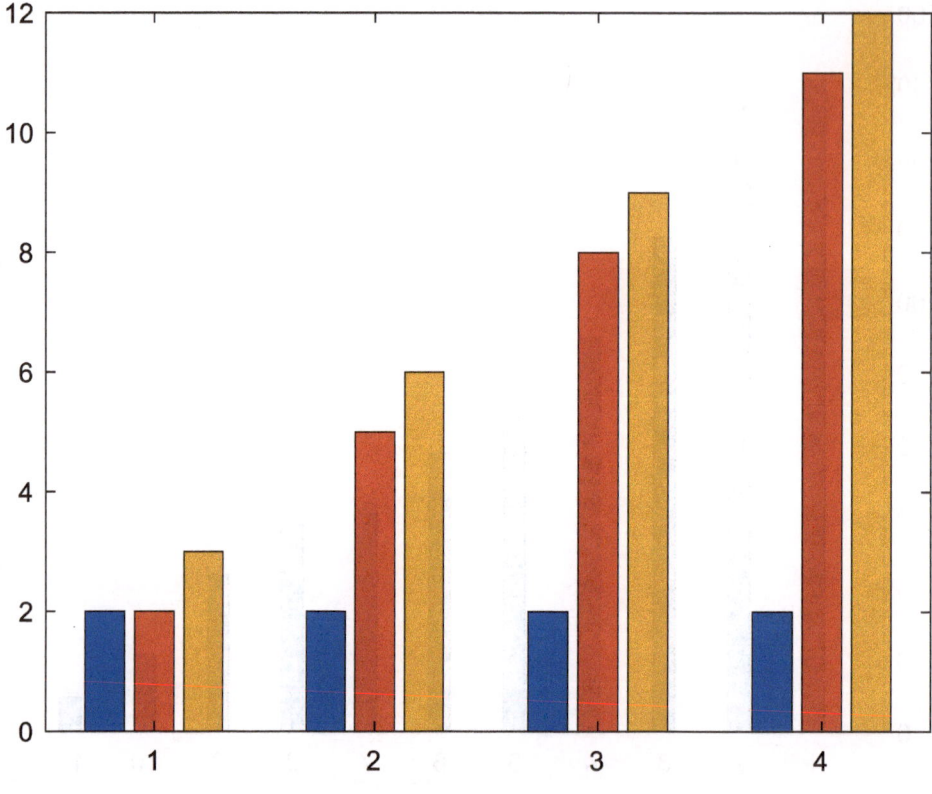

Fig. 28.4 Displaying grouped bars

y = [2 2 3; 2 5 6; 2 8 9; 2 11 12];
bar(y,'stacked')

Displaying stacked bars. See Fig. 28.5.

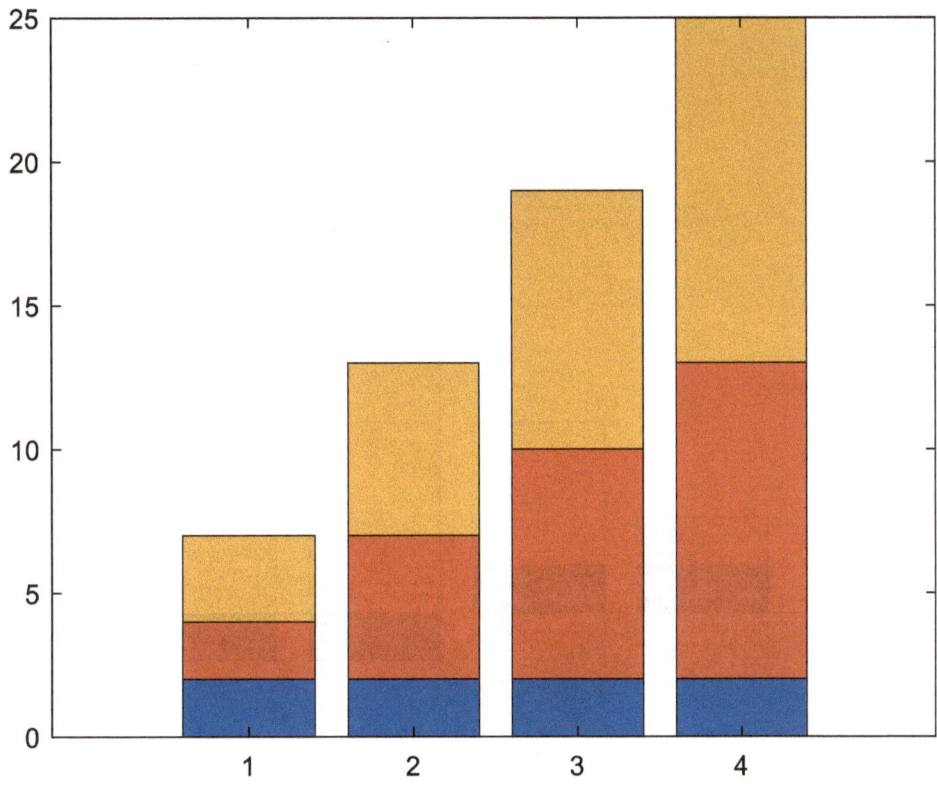

Fig. 28.5 Displaying stacked bars

Description

Displaying stacked bars with negative quantities. See Fig. 28.6.

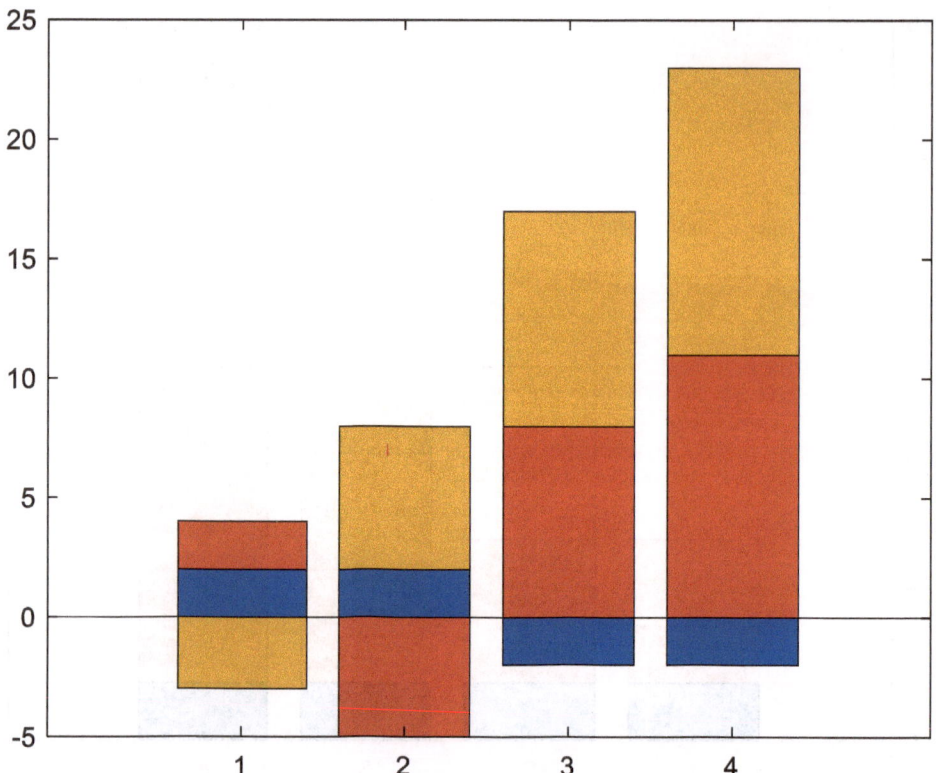

Fig. 28.6 Displaying stacked bars including negative values

```
X = categorical({'January','February','March','April'});
Y = [3 1 2 6];
bar(X,Y)
```

Indicating the categories for the bars. Herein, the bar function alphabetically sorts the categories. See Fig. 28.7.

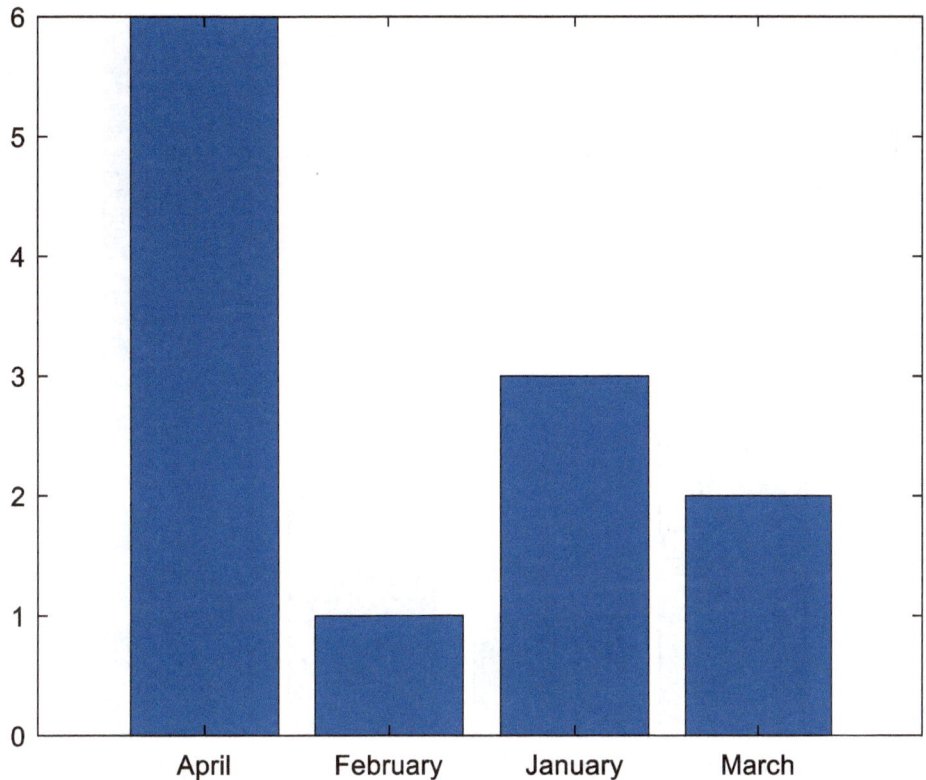

Fig. 28.7 Indicating the categories for the bars

X = categorical({'January','February','March','April'});
X = reordercats(X,{'January','February','March','April'});
Y = [3 1 2 6];
bar(X,Y)

Applying the "reordercats" function to specify the order for the bars. See Fig. 28.8.

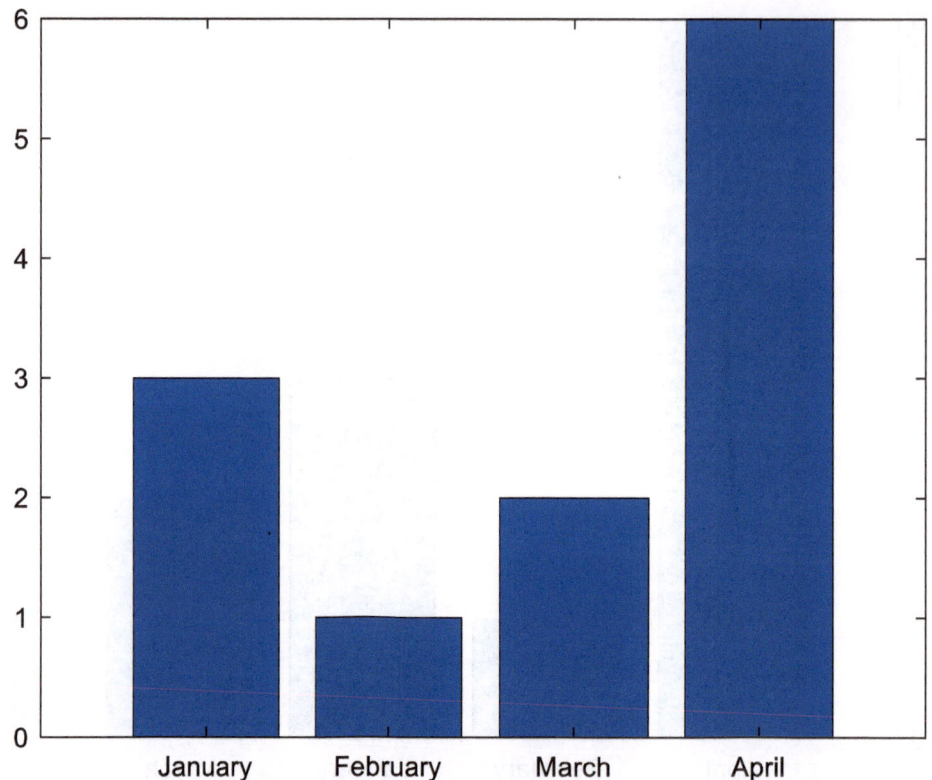

Fig. 28.8 Displaying the bars with the specific order

```
y = [1 2 3; 4 5 6];
tiledlayout(2,1)
nexttile;
bar(y)
nexttile;
bar(y,'stacked')
```

Displaying a bar graph in tiled chart layout. See Fig. 28.9.

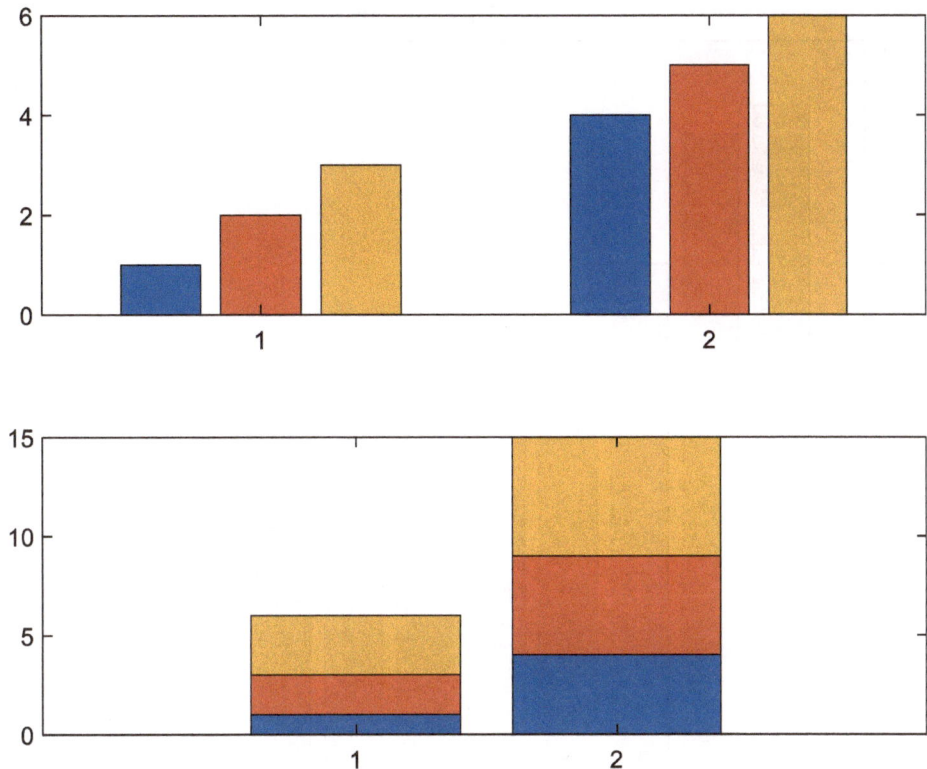

Fig. 28.9 Displaying bars in the tiled chart layout

y = [91 87 69 55 44 39 32 29 22 11 5];
bar(y,'m')
title('Bar Graph')
xlabel('Your X-Label')
ylabel('Your Y-Label')

Description

Displaying a bar graph using a specific color as well as adding title and x and y labels (See Fig. 28.10). The list of line colors available in MATLAB is illustrated in Fig. 28.11.

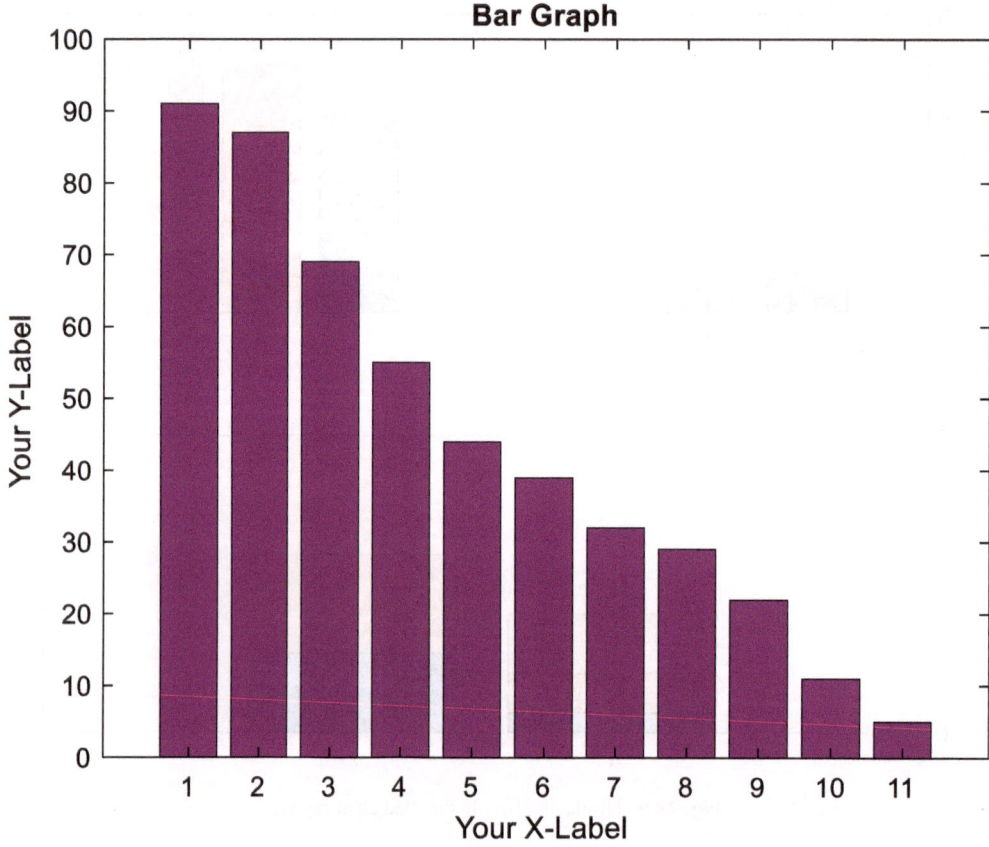

Fig. 28.10 Displaying a bar graph using a specific color as well as adding title and x and y labels to it

Color Name	Short Name	Appearance
"red"	"r"	
"green"	"g"	
"blue"	"b"	
"cyan"	"c"	
"magenta"	"m"	
"yellow"	"y"	
"black"	"k"	
"white"	"w"	

Fig. 28.11 The list of line colors available in MATLAB

Exercise

(a) *Write the codes to plot the bar graph of the normalized monthly rain, temperature, and humidity of your city or your favourite city. Display the results in 12 groups of three bars. The data can be accessed from the internet.*

(b) *Execute the codes in your computer. Then, draw the result(s) on the axes.*

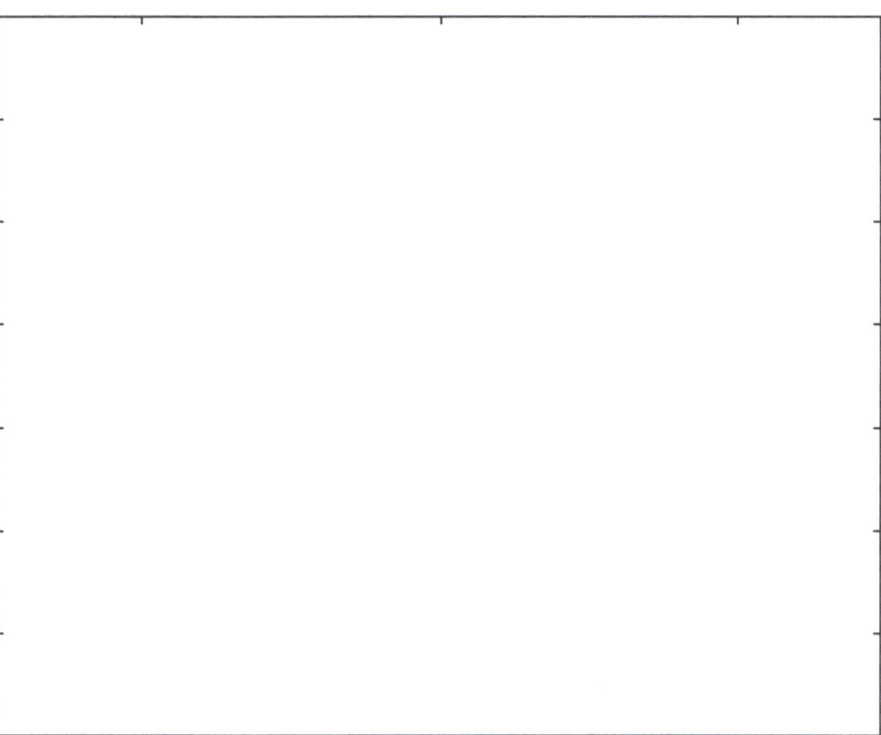

Reference

1. MATLAB 2023a.

Abstract

In this chapter, the discrete sequence plot in MATLAB is presented and described. In this regard, several examples and exercises for each section of the chapter are presented. The exercises that include writing the codes, executing them, and achieving the results need to be done by students to master programming skills. In this book, the codes, outputs, and descriptions are in blue, black, and green colors, respectively. To program in MATLAB, a script file can be created and saved with an appropriate name (e.g., untitled01) in the preferred directory of a computer. The program can be run by clicking on the "Run" available on the top toolbar of the script in MATLAB or calling the script by typing its name in Command Window or in the other scripts.

29.1 Discrete Sequence Plot

In the following, the description of discrete sequence plot is presented and exemplified [1–9].

- This plot in the format stem(Y) plots the data sequence, Y, as stems that extend from a baseline along the x-axis. If Y is a vector, then the x-axis scale ranges from 1 to length(Y). If Y is a matrix, then stem plots all elements in a row against the same x value, and the x-axis scale ranges from 1 to the number of rows in Y.
- This plot in the format stem(X,Y) plots the data sequence, Y, at values specified by X. The X and Y inputs must be vectors or matrices of the same size. Additionally, X can be a row or column vector and Y must be a matrix with length(X) rows. If X and Y are both vectors, then stem plots entries in Y against corresponding entries in X. If X is a vector and Y is a matrix, then stem plots each column of Y against the set of values specified by X, such that all elements in a row of Y are plotted against the same value. If X and Y are both matrices, then stem plots columns of Y against corresponding columns of X.

Example

X = linspace(0,2*pi,50)';
Y = cos(X);
stem(Y)

Description

Plotting a single discrete sequence graph. See Fig. 29.1.

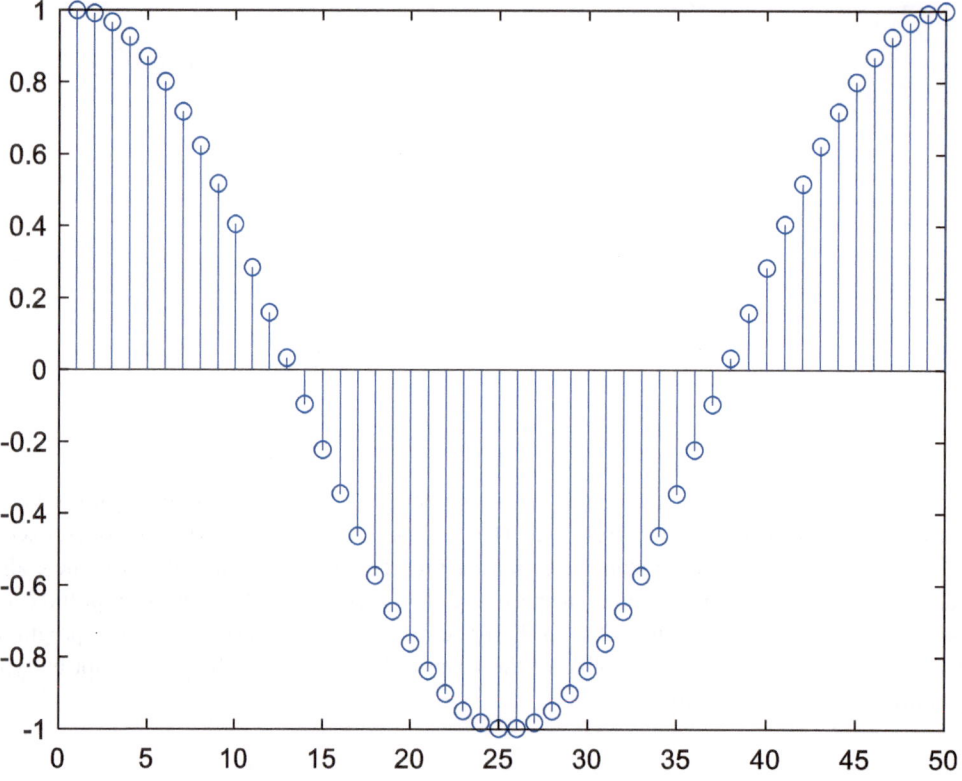

Fig. 29.1 Plotting a single discrete sequence graph

Example

```
X = linspace(0,2*pi,50)';
Y = [cos(X), 2*sin(X)];
stem(Y)
xlabel('x')
ylabel('cos(x) and 2cos(x)')
legend("cos(x)", "2sin(x)")
title('Trigonometric Functions')
```

Description

Plotting multiple discrete sequence graphs as well as adding x and y labels, legend, and a title. See Fig. 29.2.

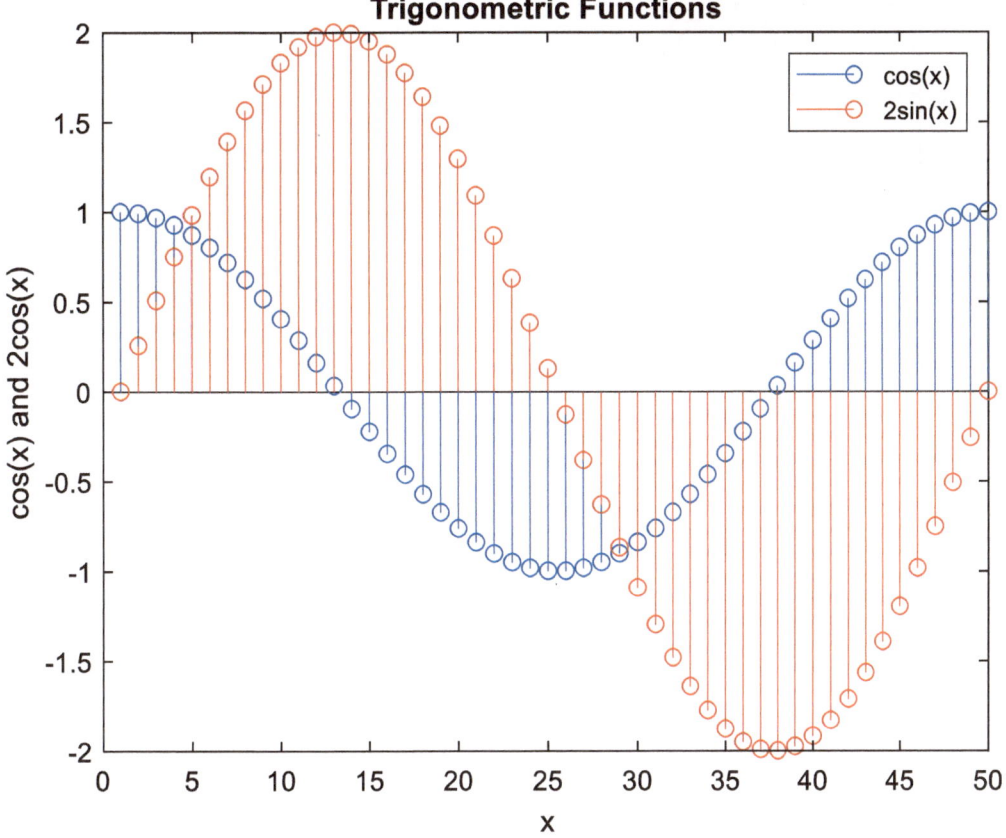

Fig. 29.2 Plotting multiple discrete sequence graphs and adding x and y labels, legend, and a title to it

Example

```
X = linspace(0,2*pi,50)';
Y = [cos(X), 2*sin(X)];
stem(X,Y)
```

Description

Plotting multiple discrete sequence graphs at the specified set of x values. See Fig. 29.3.

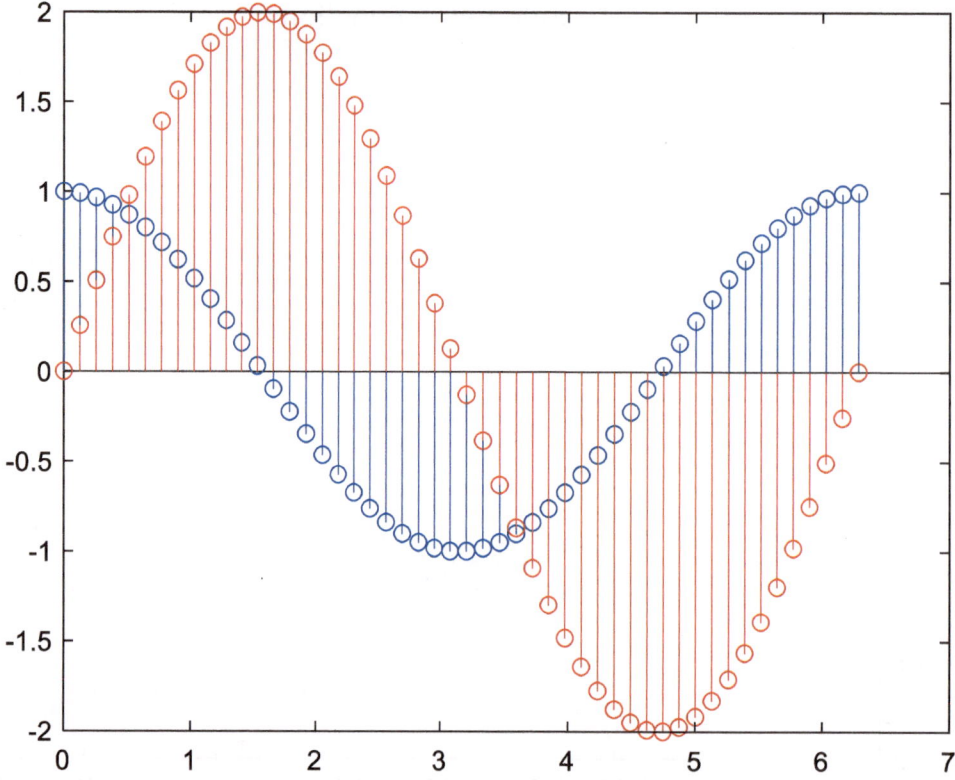

Fig. 29.3 Plotting multiple discrete sequence graphs at the specified set of x values

Exercise

Execute the codes in your computer. Then, draw the result(s) on the axes.

```
X = linspace(0,4*pi,15)';
Y = sin(X);
stem(X,Y)
```

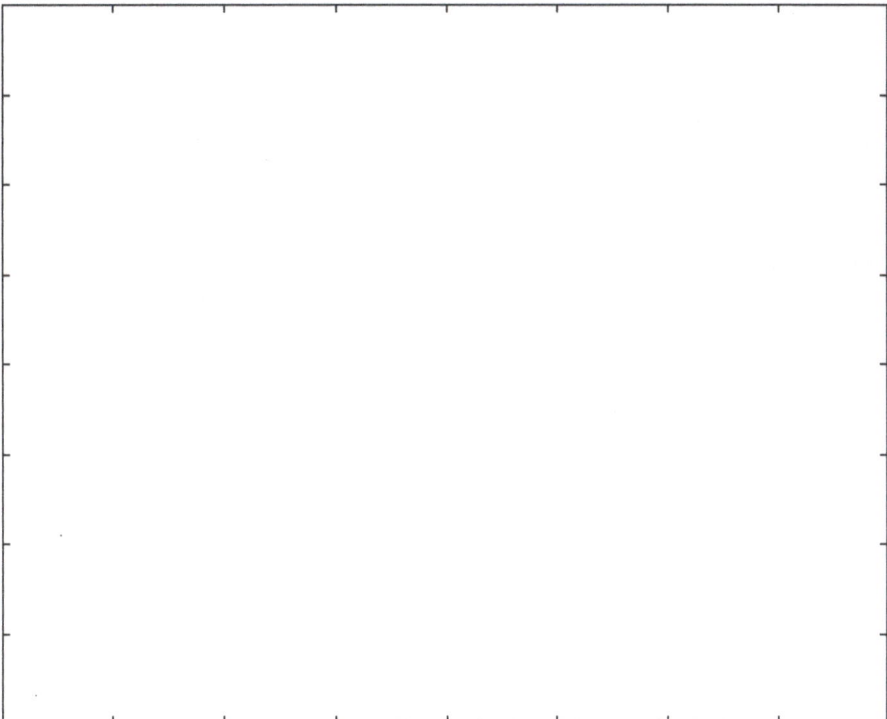

Example

```
x1 = linspace(0,2*pi,50)';
x2 = linspace(pi,3*pi,50)';
X = [x1, x2];
Y = [cos(x1), 2*sin(x2)];
stem(X,Y)
```

Description

Plotting multiple discrete sequence graphs at the corresponding sets of x values for each series. See Fig. 29.4.

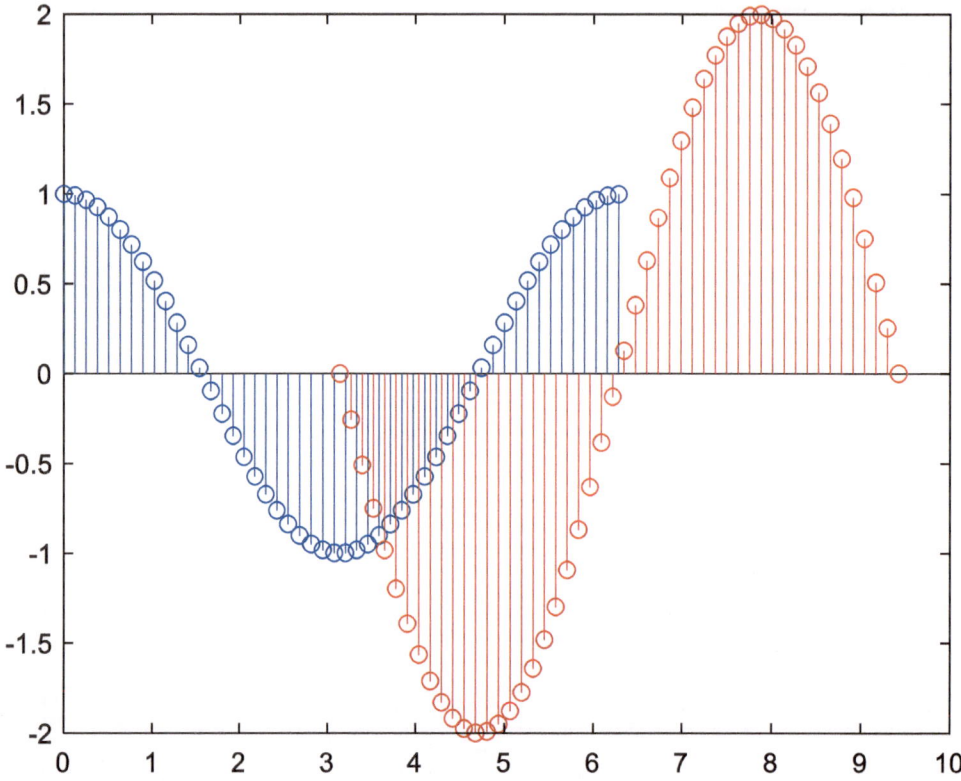

Fig. 29.4 Plotting multiple discrete sequence graphs at the corresponding sets of x values for each series

Example

```
x1 = linspace(0,2*pi,50)';
x2 = linspace(pi,3*pi,50)';
X = [x1, x2];
Y = [cos(x1), 2*sin(x2)];
stem(X,Y ,':sm')
```

Description

Plotting the graphs by using a specific style, color, and marker (See Fig. 29.5). The lists of line styles, markers, and colors available in MATLAB are illustrated in Figs. 29.6, 29.7, and 29.8, respectively.

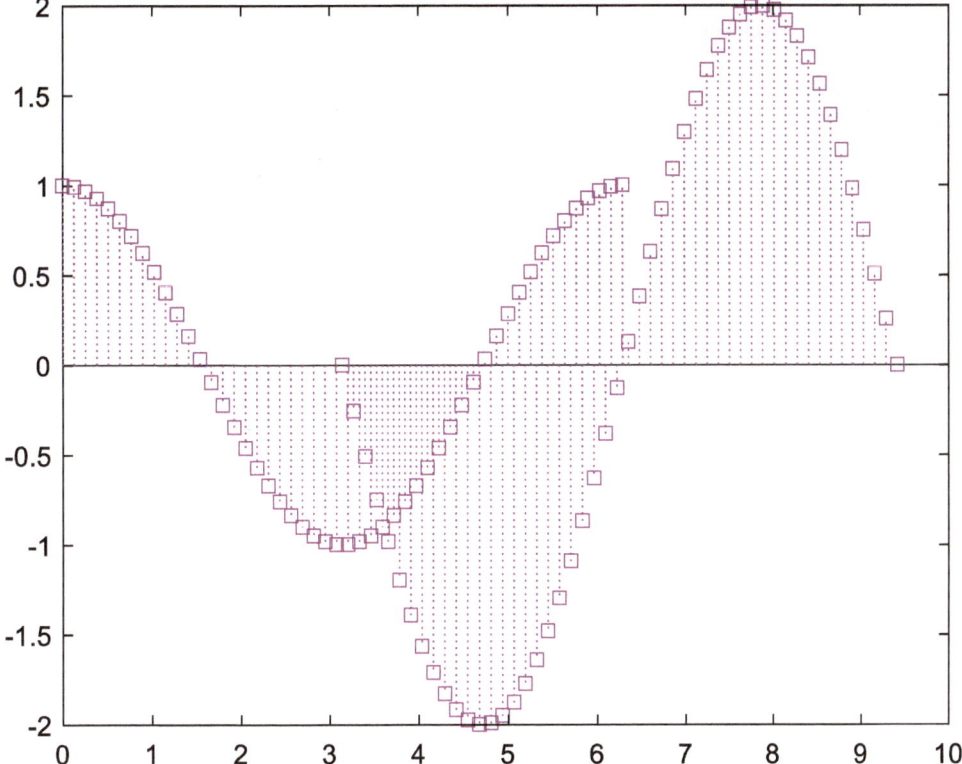

Fig. 29.5 Plotting the graphs by using a specific style, color, and marker

Line Style	Description	Resulting Line
" - "	Solid line	———————
" - - "	Dashed line	– — — — —
" : "	Dotted line	....
" -. "	Dash-dotted line	–·–·–·–···

Fig. 29.6 The list of line styles available in MATLAB

Marker	Description	Resulting Marker
"o"	Circle	○
"+"	Plus sign	+
"*"	Asterisk	＊
"."	Point	•
"x"	Cross	×
"_"	Horizontal line	—
"|"	Vertical line	|
"square"	Square	□
"diamond"	Diamond	◇
"^"	Upward-pointing triangle	△
"v"	Downward-pointing triangle	▽
">"	Right-pointing triangle	▷
"<"	Left-pointing triangle	◁
"pentagram"	Pentagram	☆
"hexagram"	Hexagram	✡

Fig. 29.7 The list of line markers available in MATLAB

Color Name	Short Name	Appearance
"red"	"r"	
"green"	"g"	
"blue"	"b"	
"cyan"	"c"	
"magenta"	"m"	
"yellow"	"y"	
"black"	"k"	
"white"	"w"	

Fig. 29.8 The list of line colors available in MATLAB

Exercise

Execute the codes in your computer. Then, draw the result(s) on the axes.

```
X = linspace(0,2*pi,50)';
Y = (exp(X).*sin(X));
stem(X,Y,'-.*g')
```

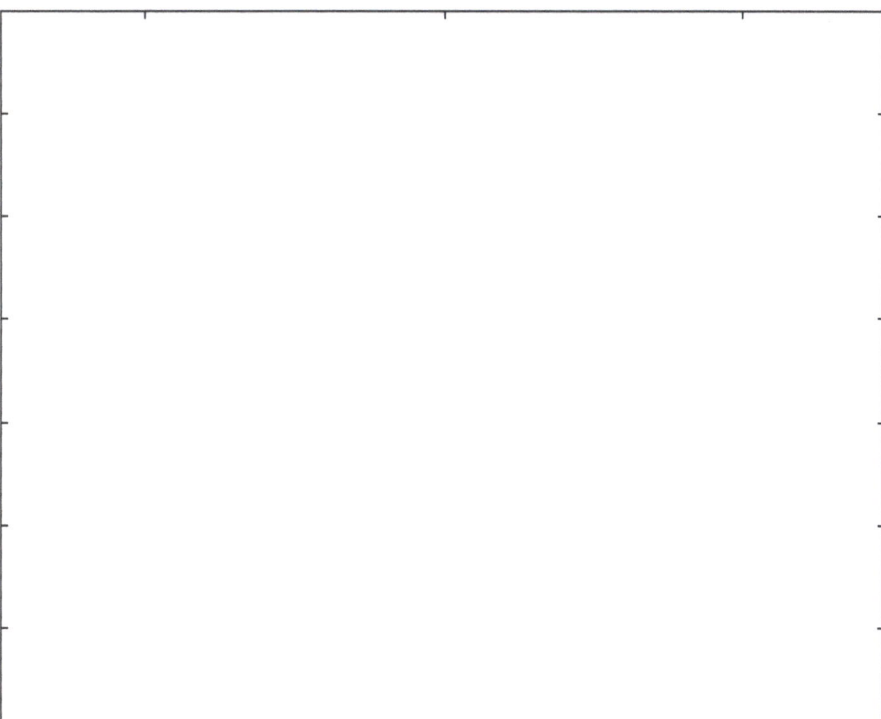

Exercise

(a) *Write the codes to plot the discrete sequence graphs of the daily temperature your city or your favorite city in the last 10 days. Plot it using a specific style, color, and marker as well as add x and y labels and a title. The data can be accessed from the internet.*

(b) *Execute the codes in your computer. Then, draw the result(s) on the axes.*

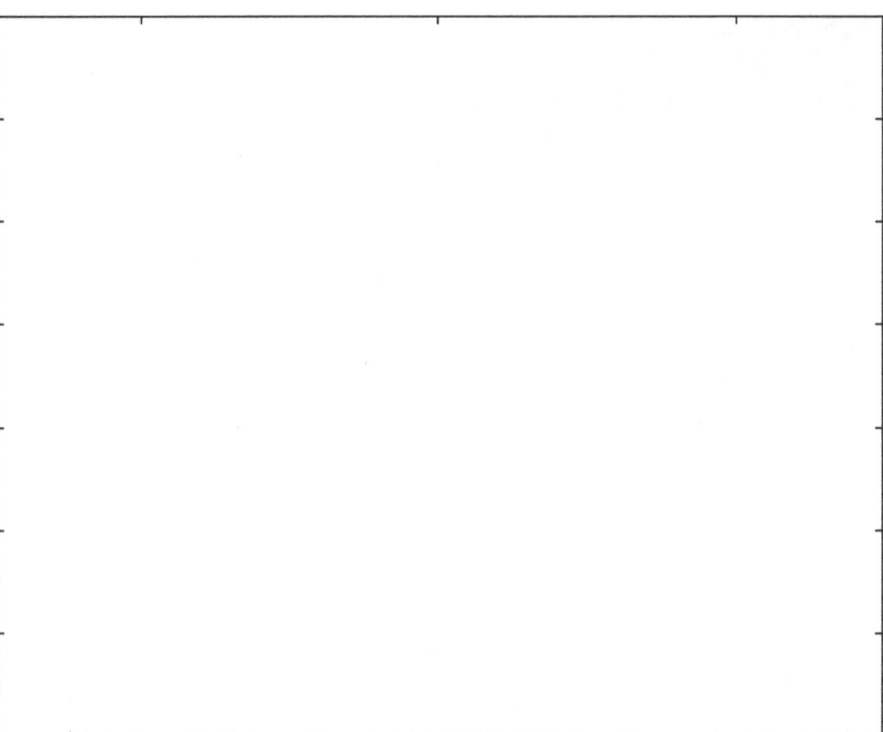

References

1. MATLAB 2023a.
2. Rahmani-Andebili, M. (2024). *Mathematics of engineering and science – Practice problems, methods, and solutions*. Springer Nature.
3. Rahmani-Andebili, M. (2022). *Differential equations – Practice problems, methods, and solutions*. Springer Nature.
4. Rahmani-Andebili, M. (2023). *Calculus III – Practice problems, methods, and solutions*. Springer Nature.
5. Rahmani-Andebili, M. (2023). *Calculus II – Practice problems, methods, and solutions*. Springer Nature.
6. Rahmani-Andebili, M. (2023). *Calculus I – Practice problems, methods, and solutions* (2nd ed.). Springer Nature.
7. Rahmani-Andebili, M. (2021). *Calculus – Practice problems, methods, and solutions*. Springer Nature.
8. Rahmani-Andebili, M. (2024). *Precalculus – Practice problems, methods, and solutions* (2nd ed.). Springer Nature.
9. Rahmani-Andebili, M. (2021). *Precalculus – Practice problems, methods, and solutions*. Springer Nature.

Abstract

In this chapter, the histogram plot in MATLAB is presented and described. In this regard, several examples and exercises for each section of the chapter are presented. The exercises that include writing the codes, executing them, and achieving the results need to be done by students to master programming skills. In this book, the codes, outputs, and descriptions are in blue, black, and green colors, respectively. To program in MATLAB, a script file can be created and saved with an appropriate name (e.g., untitled01) in the preferred directory of a computer. The program can be run by clicking on the "Run" available on the top toolbar of the script in MATLAB or calling the script by typing its name in Command Window or in the other scripts.

30.1 Histogram Plot

In the following, the description of histogram plot is presented and exemplified [1].

- This plot in the format histogram(X,nbins) uses a number of bins specified by the scalar, nbins.

Example

```
x = randn(1000,1);
h = histogram(x)
```

Output

```
h =
  Histogram with properties:
    Data: [1000×1 double]
    Values: [1 3 9 12 19 38 44 70 83 109 122 112 113 86 68 46 35 12 14 2 1 1]
    NumBins: 22
    BinEdges: [-3.3000 -3.0000 -2.7000 -2.4000 -2.1000 -1.8000 -1.5000 -1.2000 -0.9000 ...]
    BinWidth: 0.3000
    BinLimits: [-3.3000 3.3000]
    Normalization: 'count'
    FaceColor: 'auto'
    EdgeColor: [0 0 0]
```

M. Rahmani-Andebili, *MATLAB Lessons, Examples, and Exercises*, https://doi.org/10.1007/978-3-031-76177-5_30

Description

Plotting the histogram function of 1000 random numbers. Herein, the histogram function automatically chooses an appropriate number of bins to cover the range of values in x and show the shape of the underlying distribution. See Fig. 30.1.

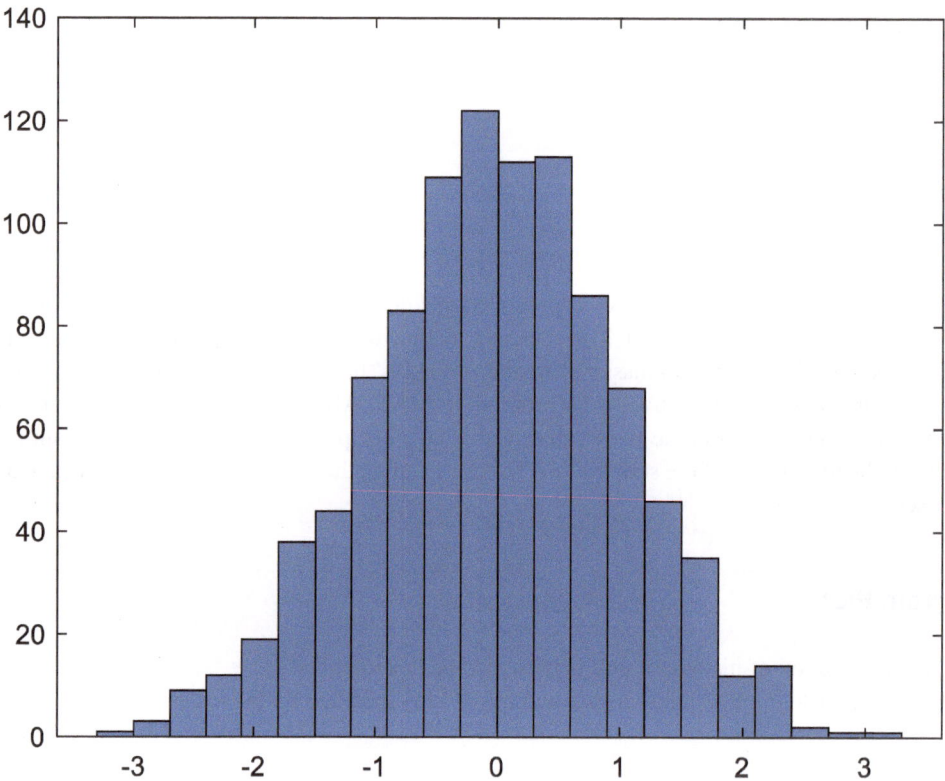

Fig. 30.1   Plotting the histogram function of 1000 random numbers

Example

x = randn(1000,1);
nbins = 10;
h = histogram(x,nbins)

Output

h =
   Histogram with properties
      Data: [1000×1 double]
      Values: [3 11 66 165 250 246 160 65 28 6]
      NumBins: 10
      BinEdges: [-3.6000 -2.9000 -2.2000 -1.5000 -0.8000 -0.1000 0.6000 1.3000 2 2.7000]
      BinWidth: 0.7000
      BinLimits: [-3.6000 3.4000]
      Normalization: 'count'
      FaceColor: 'auto'
      EdgeColor: [0 0 0]

Description

Plotting the histogram function of 1000 random numbers sorted into 10 equally spaced bins. See Fig. 30.2.

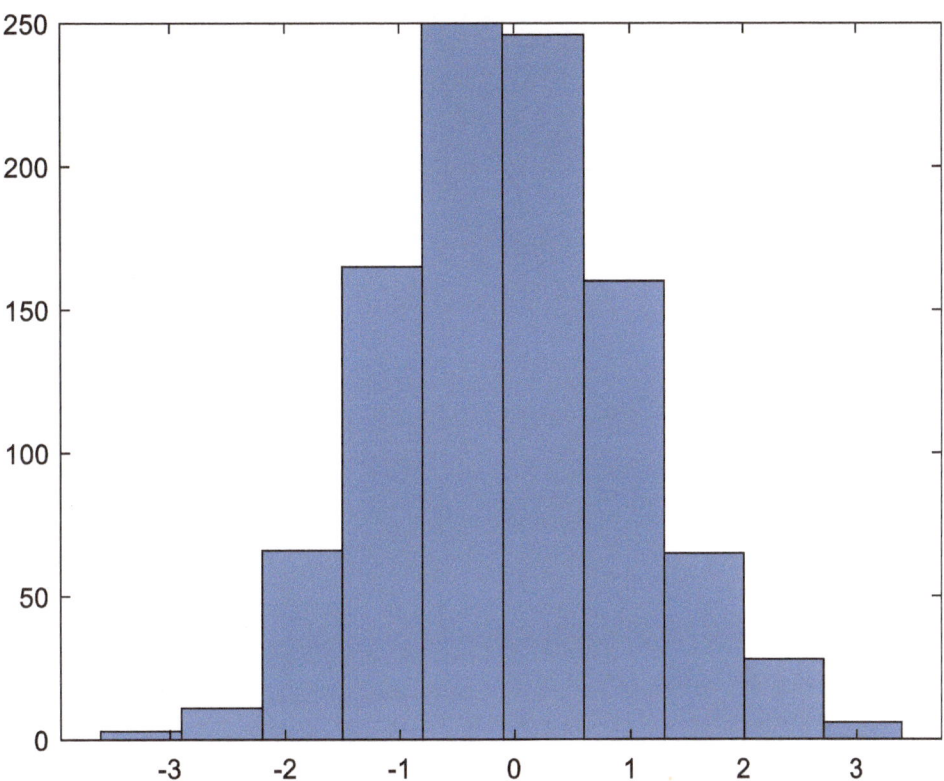

Fig. 30.2 Plotting the histogram function of 1000 random numbers sorted into 10 equally spaced bins

Exercise

Execute the codes in your computer. Then, draw the result(s) on the axes.

x = [
 -1.7
  0.8
  0.2
 -0.5
  1.6
  0.6
 -0.2
  0.6
 -0.4
 -1.1
  0.3
  0.4
 -0.5
  1.0
 -0.9

```
      2.2
      0.0
      0.5
     -0.7
      0.5];
nbins = 10;
h = histogram(x,nbins)
h.FaceColor = 'k';
```

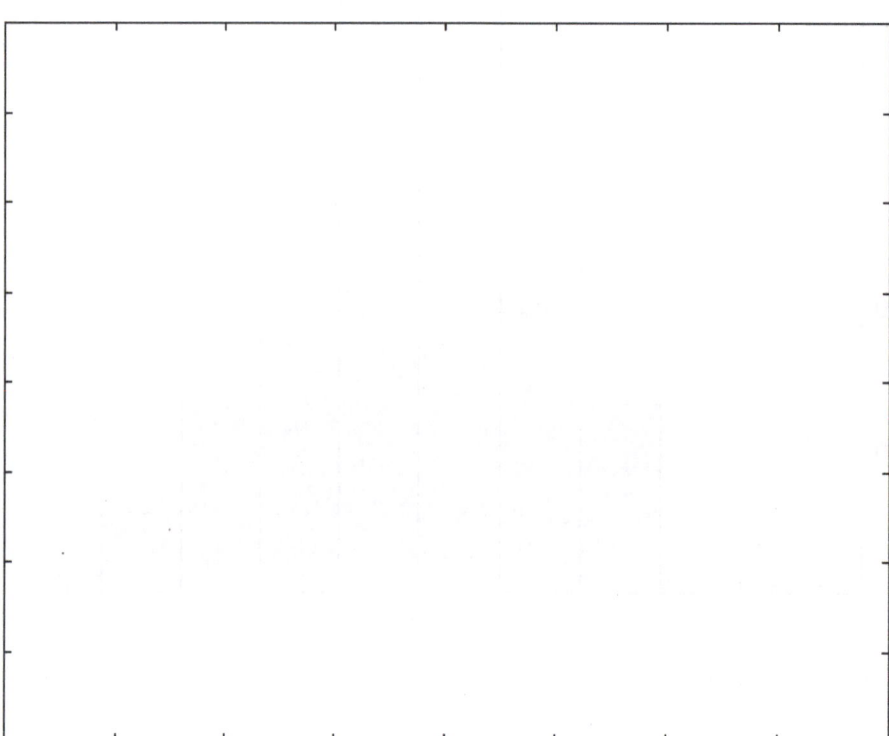

Example

```
x = randn(2000,1);
y = 1 + randn(5000,1);
h1 = histogram(x);
hold on
h2 = histogram(y);
```

Description

Plotting the histogram function of two vectors of random numbers in the same figure. See Fig. 30.3.

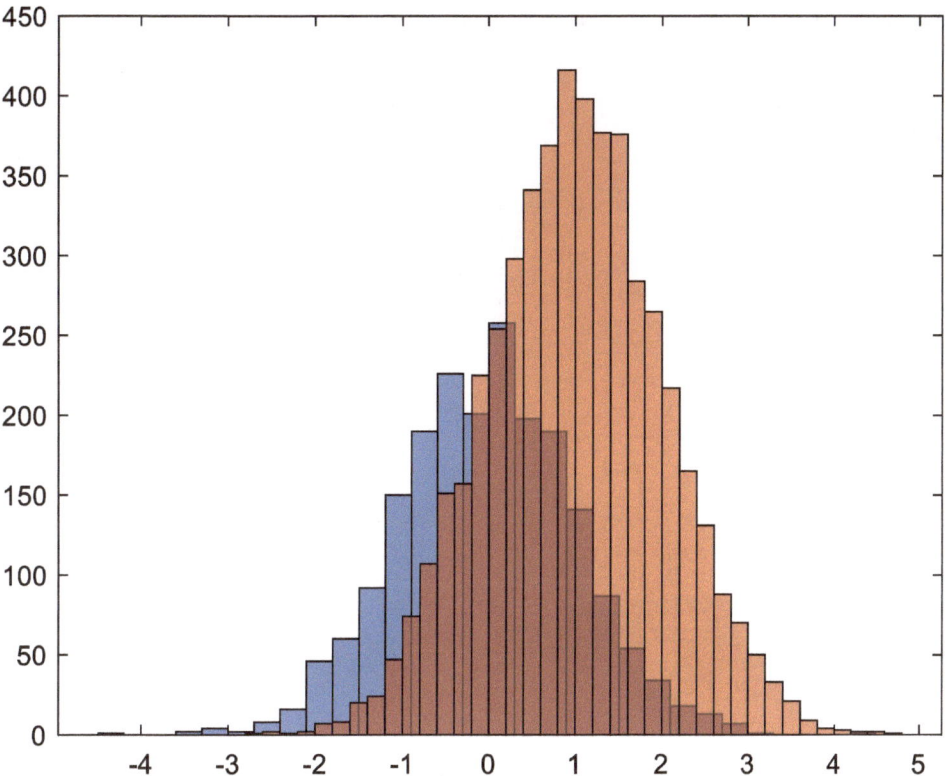

Fig. 30.3 Plotting the histogram function of two vectors of random numbers in the same figure

```
x = randn(1000,1);
h = histogram(x)
h.FaceColor = 'r';
```

Plotting a histogram function with a specific color (See Fig. 30.4). The list of line colors available in MATLAB is illustrated in Fig. 30.5.

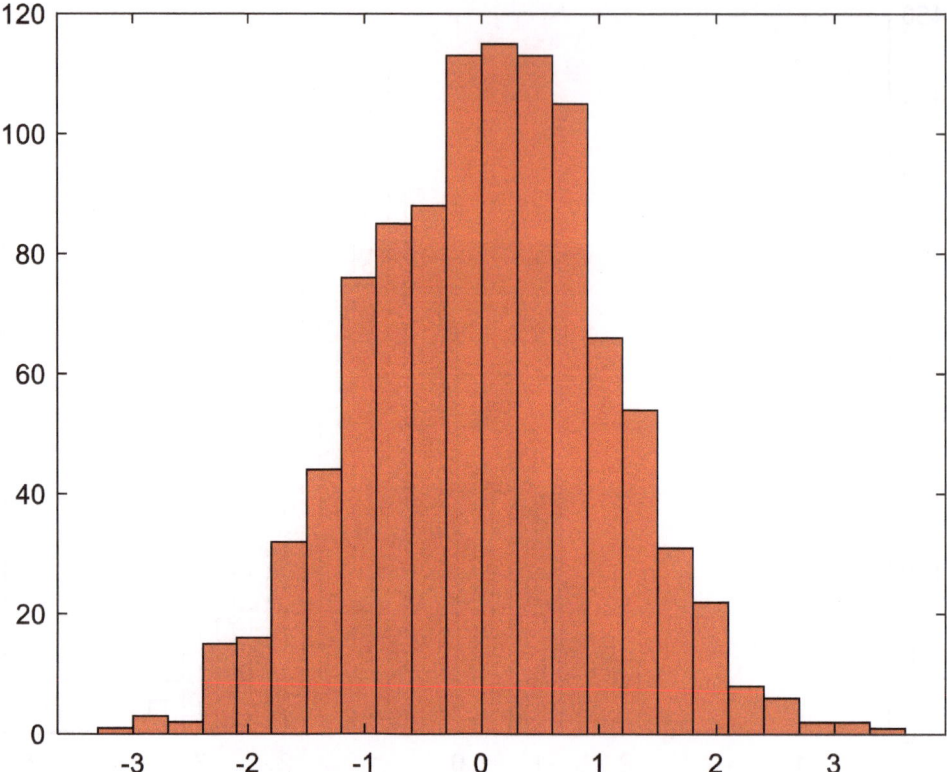

Fig. 30.4 Plotting a histogram function with a specific color

Color Name	Short Name	Appearance
"red"	"r"	
"green"	"g"	
"blue"	"b"	
"cyan"	"c"	
"magenta"	"m"	
"yellow"	"y"	
"black"	"k"	
"white"	"w"	

Fig. 30.5 The list of line colors available in MATLAB

Reference

1. MATLAB 2023a.

Abstract

In this chapter, the 2D and 3D pie charts in MATLAB are presented and described. In this regard, several examples and exercises for each section of the chapter are presented. The exercises that include writing the codes, executing them, and achieving the results need to be done by students to master programming skills. In this book, the codes, outputs, and descriptions are in blue, black, and green colors, respectively. To program in MATLAB, a script file can be created and saved with an appropriate name (e.g., untitled01) in the preferred directory of a computer. The program can be run by clicking on the "Run" available on the top toolbar of the script in MATLAB or calling the script by typing its name in Command Window or in the other scripts.

31.1   2D and 3D Pie Charts

In the following, the description of 2D and 3D pie charts is presented and exemplified [1].

- The pie chart in the format pie(X) draws a pie chart using the data in X. Each slice of the pie chart represents an element in X. If sum(X) = 1, then the values in X directly specify the areas of the pie slices. If sum(X) < 1, pie draws only a partial pie. If sum(X) > 1, then pie normalizes the values by X/sum(X) to determine the area of each slice of the pie.
- The pie chart in the format pie(X,explode) offsets slices from the pie. Herein, the explode is a vector or matrix of zeros and nonzeros that correspond to X. The pie function offsets slices for the nonzero elements only.
- The pie chart in the format pie3(X) draws a three-dimensional pie chart using the data in X. Each slice of the pie chart represents an element in X.
- The pie chart in the format pie3(X,explode) specifies which slices to offset from the center of the pie chart.

Example

```
X = [2 0.5 1.5 3 .9];
pie(X)
```

Description

Plotting the pie chart of vector X. See Fig. 31.1.

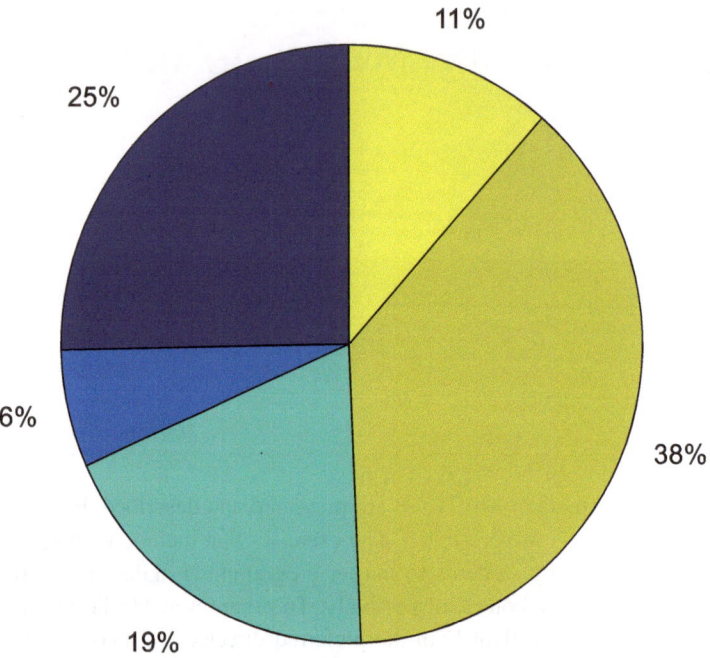

Fig. 31.1 Plotting the pie chart of the vector

X = [2 0.5 1.5 3 .9];
explode = [1 1 1 1 1];
pie(X,explode)

Plotting the pie chart of vector X in a sliced mode. See Fig. 31.2.

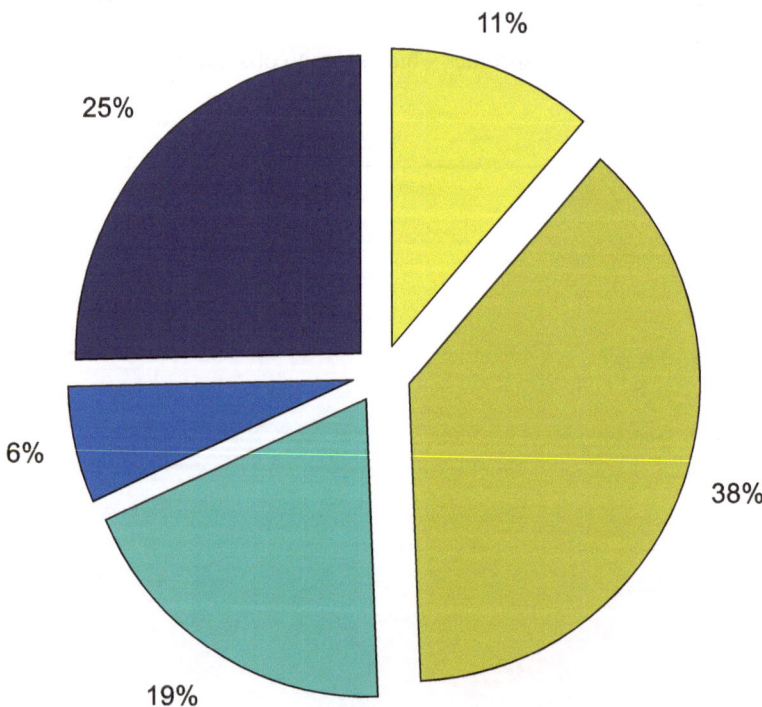

Fig. 31.2 Plotting the pie chart of the vector in a sliced mode

Execute the codes in your computer. Then, draw the result(s) in the assigned box.

X = [3 7 4 2 1];
explode = [0 0 0 0 1];
pie(X,explode)

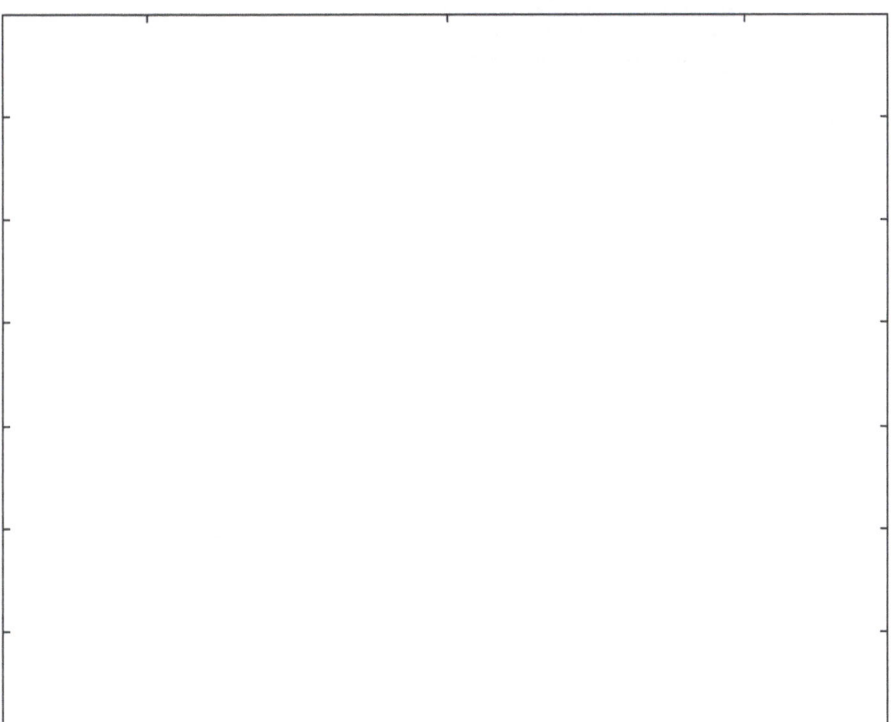

Example

X = [2 0.5 1.5 3 .9];
labels = {'Type I','Type II','Type III','Type IV','Type V'};
pie(X,labels)

Description

Adding text labels to a pie chart. See Fig. 31.3.

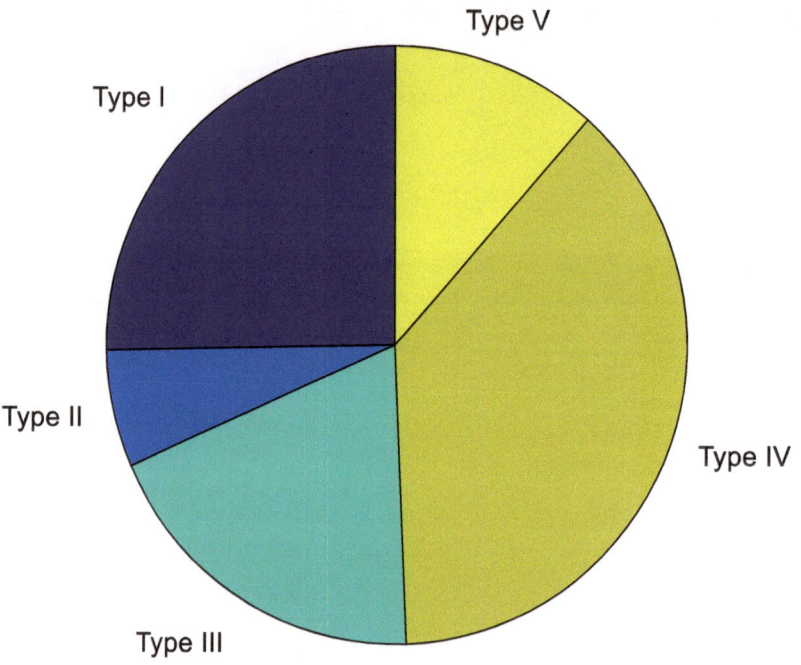

Fig. 31.3 Adding text labels to the pie chart

X = [0.19 0.22 0.41];
pie(X)

Description

Plotting a partial pie chart. Herein, the sum of the elements is less than 1. See Fig. 31.4.

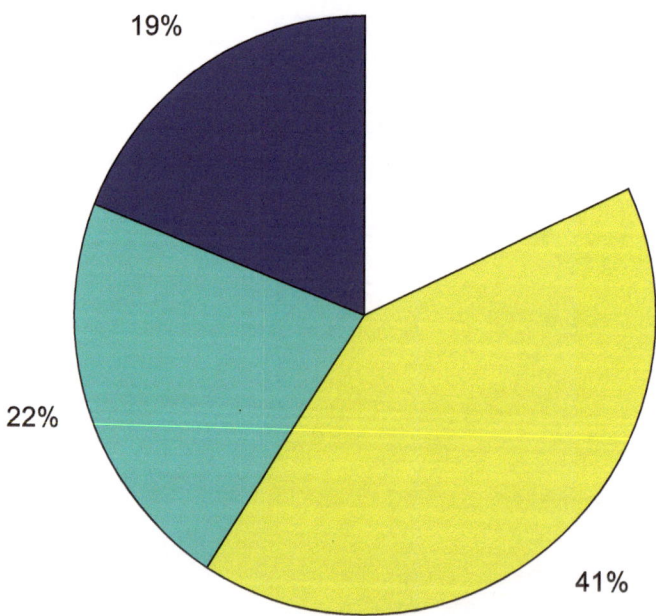

Fig. 31.4 Plotting a partial pie chart

```
y1 = [50 0 100 95];
y2 = [65 22 97 120];
t = tiledlayout(1,2);
nexttile;
pie(y1)
title('2010')
nexttile;
pie(y2)
title('2011')
labels = {'Investments','Cash','Operations','Sales'};
lgd = legend(labels);
lgd.Layout.Tile = 'east';
```

Applying tiled chart layout to compare two pie charts as well as adding title and legend to a specific location (east). See Fig. 31.5.

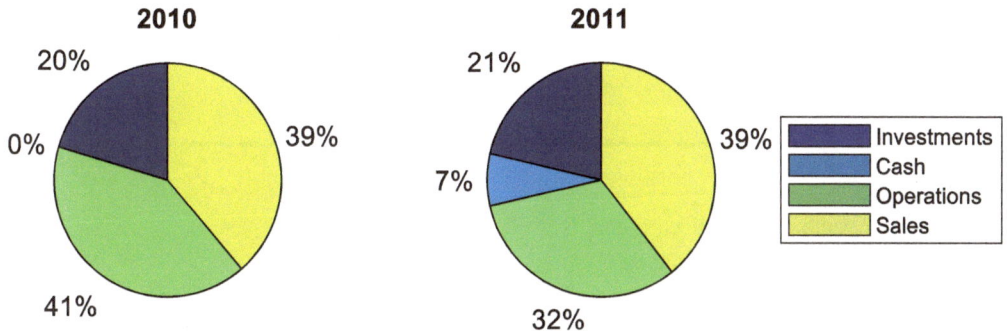

Fig. 31.5 Applying tiled chart layout to compare two pie charts and adding title and legend to a specific location (east)

```
X = [2 0.5 1.5 3 .9];
pie3(X)
```

Plotting the 3D pie chart of vector X. See Fig. 31.6.

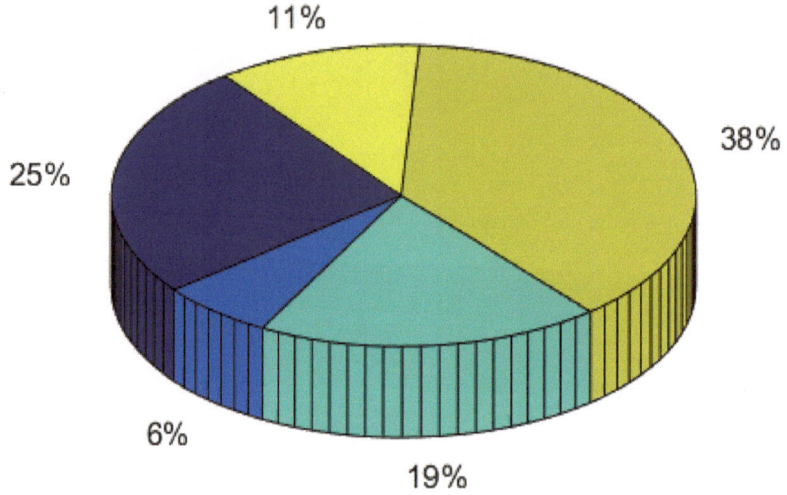

Fig. 31.6 Plotting the 3D pie chart of the vector

Example

X = [2 0.5 1.5 3 .9];
explode = [1 1 1 1 1];
pie3(X,explode)

Description

Plotting the 3D pie chart of vector X in a sliced mode. See Fig. 31.7.

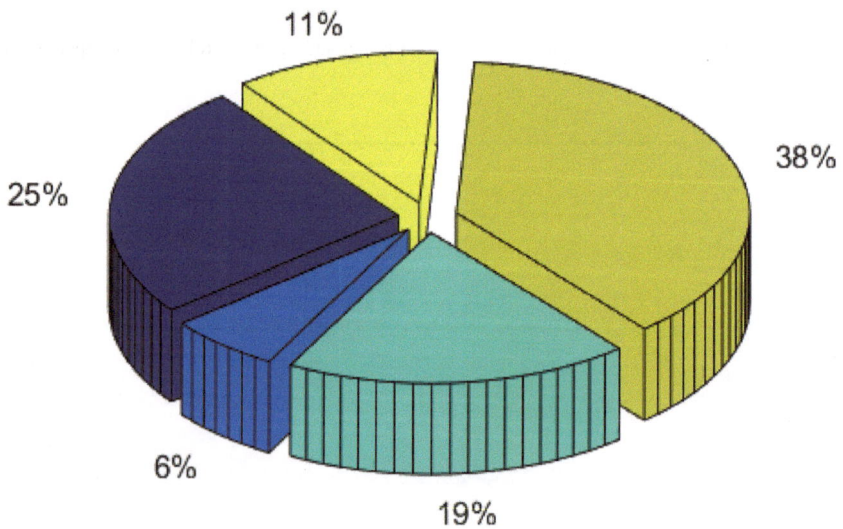

Fig. 31.7 Plotting the 3D pie chart of the vector in a sliced mode

Example

X = [2 0.5 1.5 3 .9];
labels = {'Type I','Type II','Type III','Type IV','Type V'};
pie3(X,labels)

Description

Adding text labels to a 3D pie chart. See Fig. 31.8.

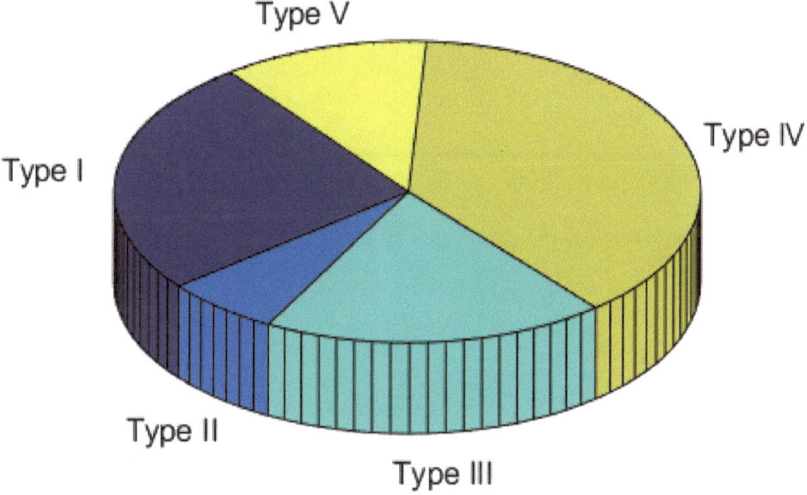

Fig. 31.8   Adding text labels to the 3D pie chart

Example

X = [0.19 0.22 0.41];
pie3(X)

Description

Plotting a partial 3D pie chart. Herein, the sum of the elements is less than 1. See Fig. 31.9.

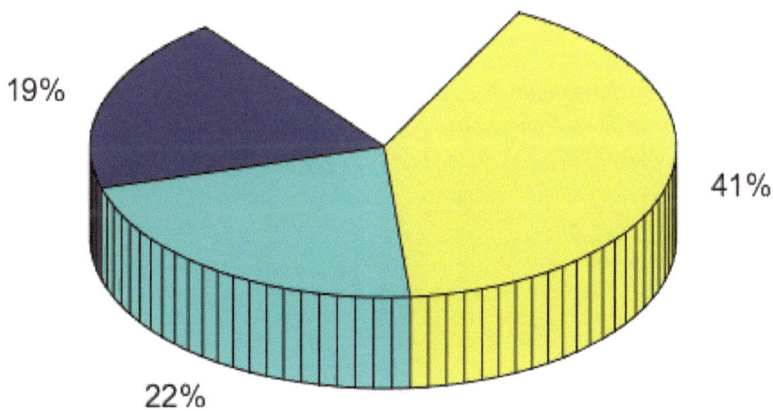

Fig. 31.9   Plotting a partial 3D pie chart

```
y1 = [50 0 100 95];
y2 = [65 22 97 120];
t = tiledlayout(1,2);
nexttile;
pie3(y1)
title('2010')
nexttile;
pie3(y2)
title('2011')
labels = {'Investments','Cash','Operations','Sales'};
lgd = legend(labels);
lgd.Layout.Tile = 'south';
```

Description

Applying tiled chart layout to compare two 3D pie charts as well as adding title and legend to a specific location (south). See Fig. 31.10.

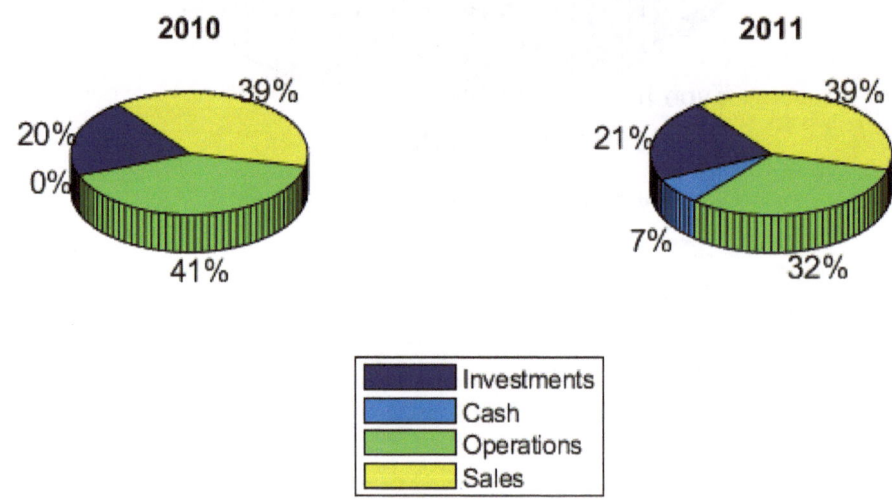

Fig. 31.10 Applying tiled chart layout to compare two 3D pie charts and adding title and legend to a specific location (south)

Exercise

(a) *Write the codes to plot the 3D pie chart regarding the contribution percentage of you and your classmates in two course projects. Add an appropriate title to each pie chart.*

(b) *Execute the codes in your computer. Then, draw the result(s) in the assigned box.*

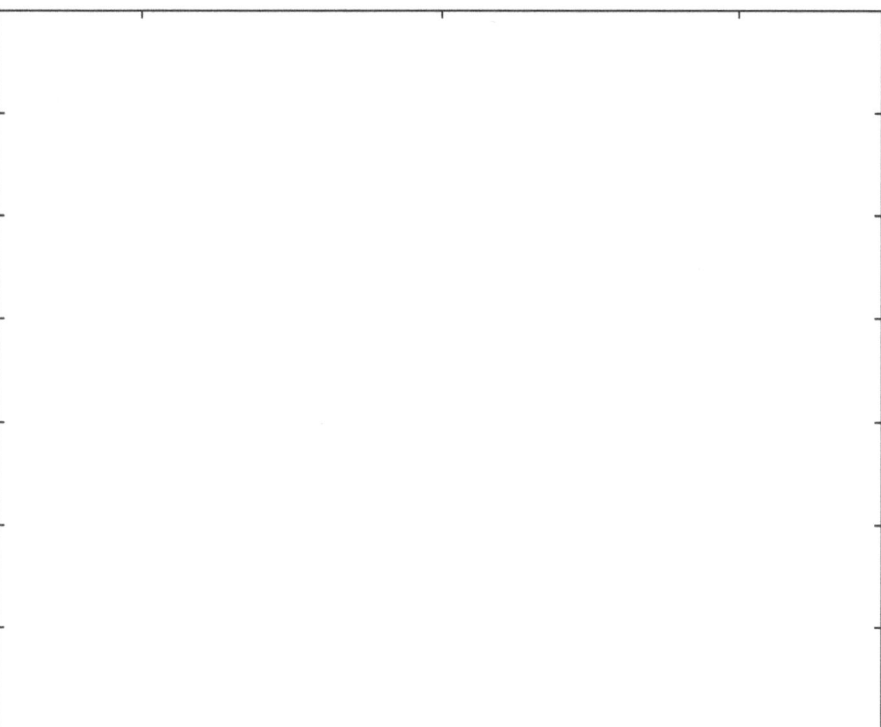

Reference

1. MATLAB 2023a.

Abstract

In this chapter, the scatter plot in MATLAB is presented and described. In this regard, several examples and exercises for each section of the chapter are presented. The exercises that include writing the codes, executing them, and achieving the results need to be done by students to master programming skills. In this book, the codes, outputs, and descriptions are in blue, black, and green colors, respectively. To program in MATLAB, a script file can be created and saved with an appropriate name (e.g., untitled01) in the preferred directory of a computer. The program can be run by clicking on the "Run" available on the top toolbar of the script in MATLAB or calling the script by typing its name in Command Window or in the other scripts.

32.1   Scatter Plot

In the following, the description of scatter plot is presented and exemplified [1].
- This plot in the format scatter(x,y) creates a scatter plot with circular markers at the locations specified by the vectors x and y. To plot one set of coordinates, specify x and y as vectors of equal length. To plot multiple sets of coordinates on the same set of axes, specify at least one of x or y as a matrix.
- This plot in the format scatter(x,y,sz) specifies the circle sizes. To use the same size for all the circles, specify sz as a scalar. To plot each circle with a different size, specify sz as a vector or a matrix.
- This plot in the format scatter(x,y,sz,c) specifies various colors to the circles.
- This plot in the format scatter(x,y,sz,'r') plots the circles in red. Herein, the other colors can be used instead of the red color.
- This plot in the format scatter(...,"filled") fills in the circles. Use the "filled" option with any of the input argument combinations in the previous syntaxes.
- This plot in the format scatter(...,mkr) specifies the marker type.

Example

```
theta = linspace(0,1,500);
x = exp(theta).*sin(100*theta);
y = exp(theta).*cos(100*theta);
s = scatter(x,y);
```

Description

Plotting the scatter plot. See Fig. 32.1.

© The Author(s), under exclusive license to Springer Nature Switzerland AG 2024
M. Rahmani-Andebili, *MATLAB Lessons, Examples, and Exercises*, https://doi.org/10.1007/978-3-031-76177-5_32

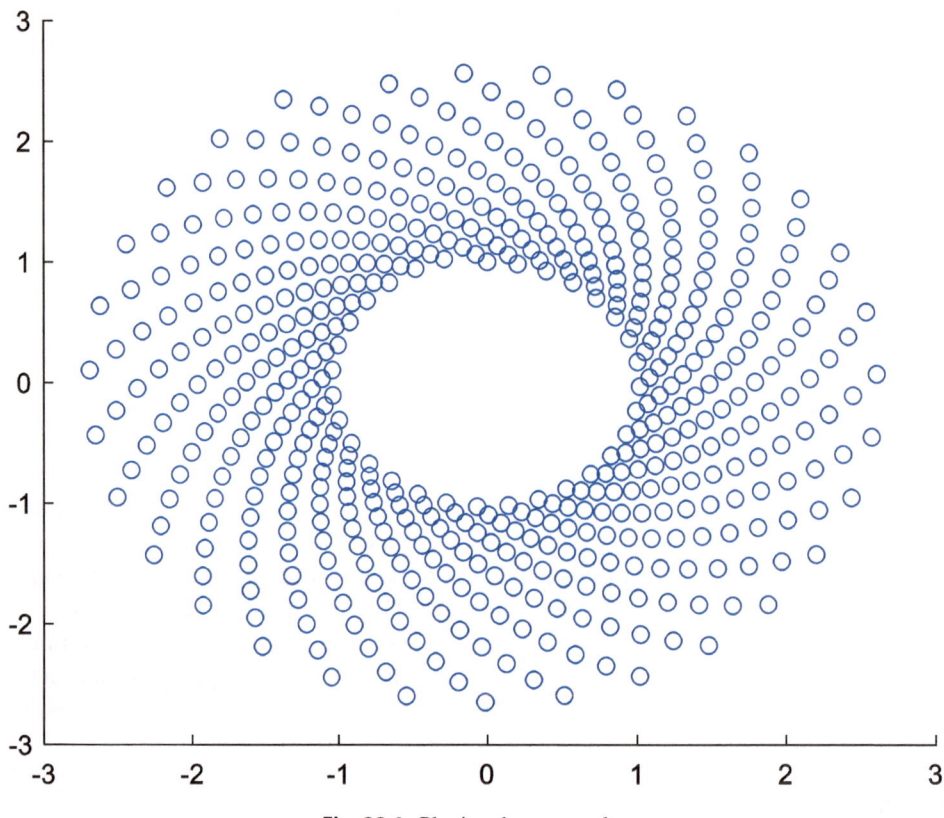

Fig. 32.1 Plotting the scatter plot

Exercise

Execute the codes in your computer. Then, draw the result(s) on the axes.

theta = linspace(0,1,500);
x = exp(theta).*sin(10*theta);
y = exp(theta).*cos(10*theta);
s = scatter(x,y);

Example

```
x = linspace(0,3*pi,200);
y = cos(x) + rand(1,200);
scatter(x,y)
```

Description

Plotting the scatter plot. See Fig. 32.2.

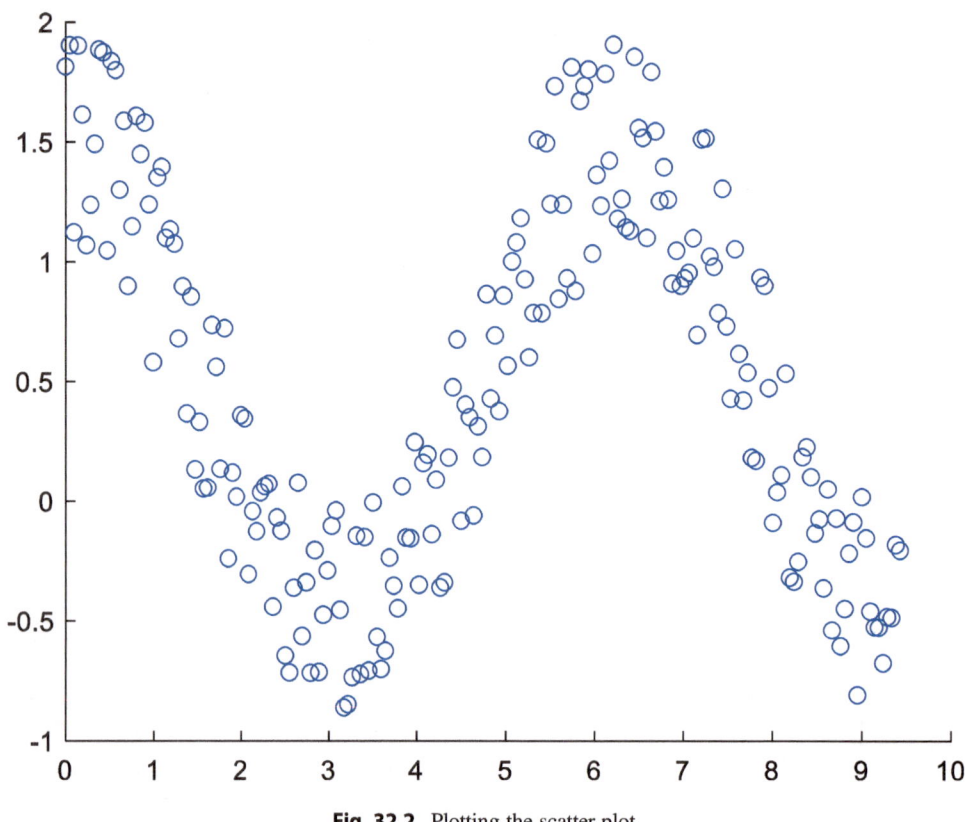

Fig. 32.2 Plotting the scatter plot

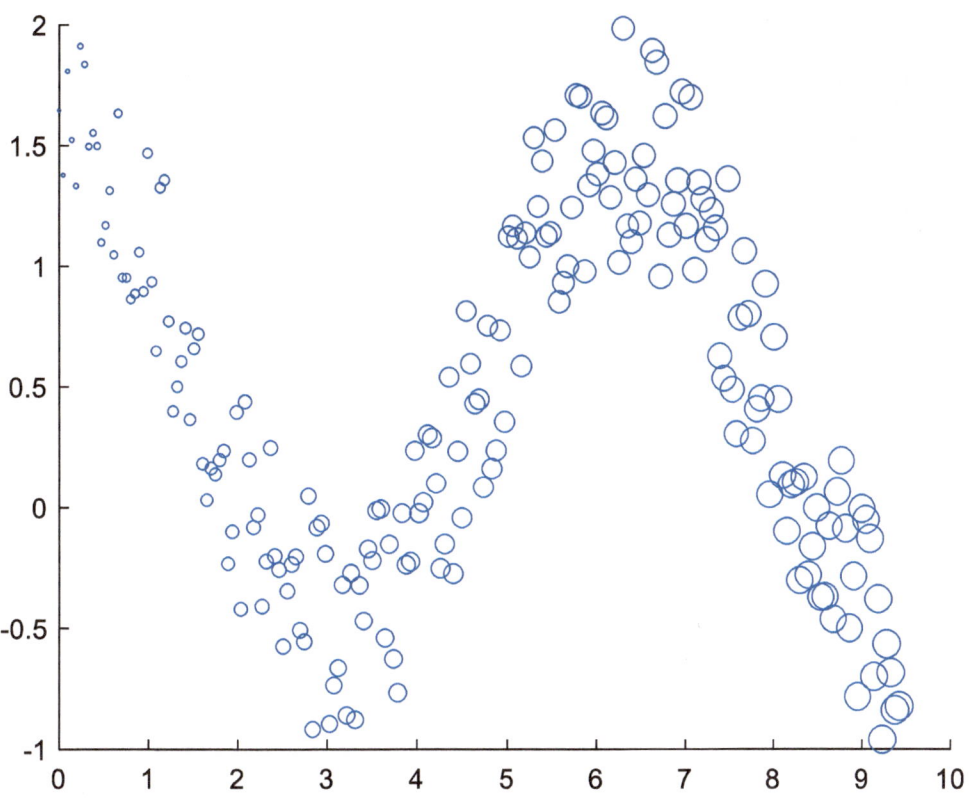

Fig. 32.3 Plotting the scatter plot with different sizes

Example

```
x = linspace(0,3*pi,200);
y = cos(x) + rand(1,200);
c = linspace(1,10,length(x));
scatter(x,y,[],c)
```

Description

Plotting the scatter plot with various colors. See Fig. 32.4.

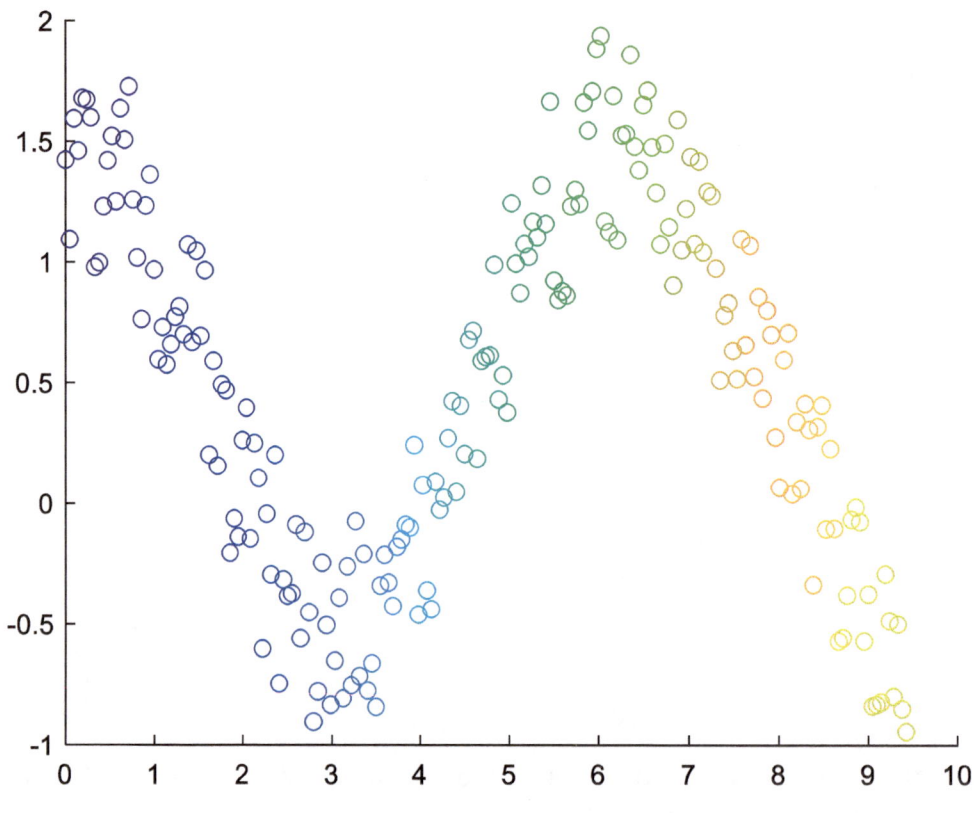

Fig. 32.4 Plotting the scatter plot with various colors

Example

x = linspace(0,3*pi,200);
y = cos(x) + rand(1,200);
c = linspace(1,10,length(x));
scatter(x,y,[],'m')

Description

Plotting the scatter plot in magenta as the chosen color (See Fig. 32.5). The list of line colors available in MATLAB is shown in Fig. 32.6.

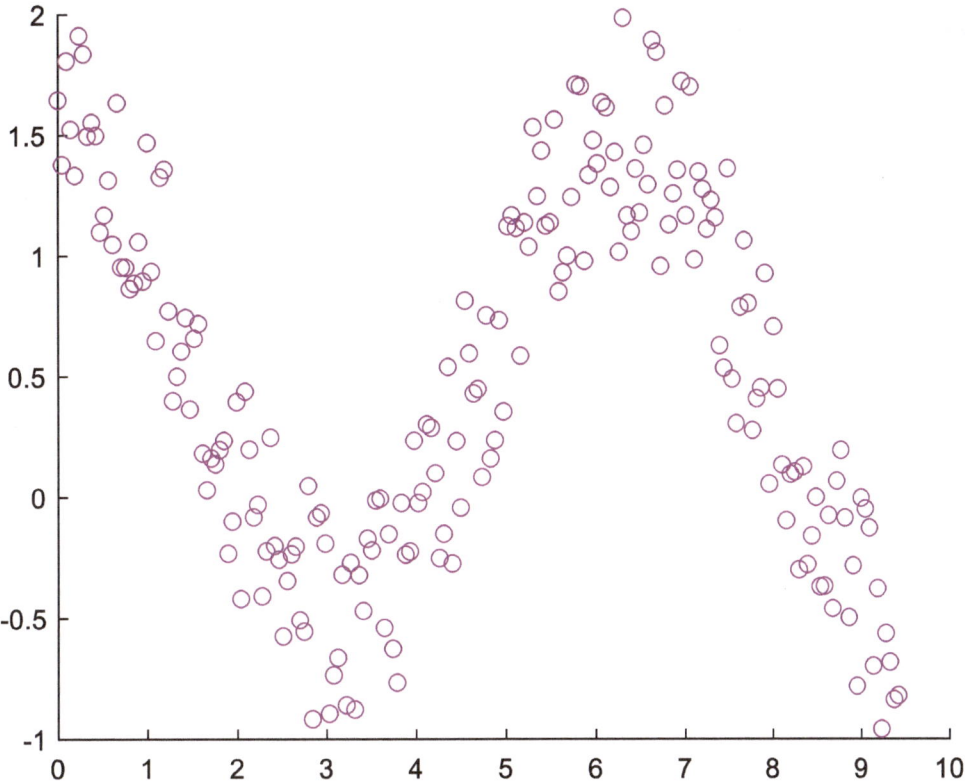

Fig. 32.5 Plotting the scatter plot with a specific color (magenta)

Color Name	Short Name	Appearance
"red"	"r"	
"green"	"g"	
"blue"	"b"	
"cyan"	"c"	
"magenta"	"m"	
"yellow"	"y"	
"black"	"k"	
"white"	"w"	

Fig. 32.6 The list of line colors available in MATLAB

```
x = linspace(0,3*pi,200);
y = cos(x) + rand(1,200);
sz = 25;
c = linspace(1,10,length(x));
scatter(x,y,sz,c,'filled')
```

Description

Plotting the scatter plot with filled markers. See Fig. 32.7.

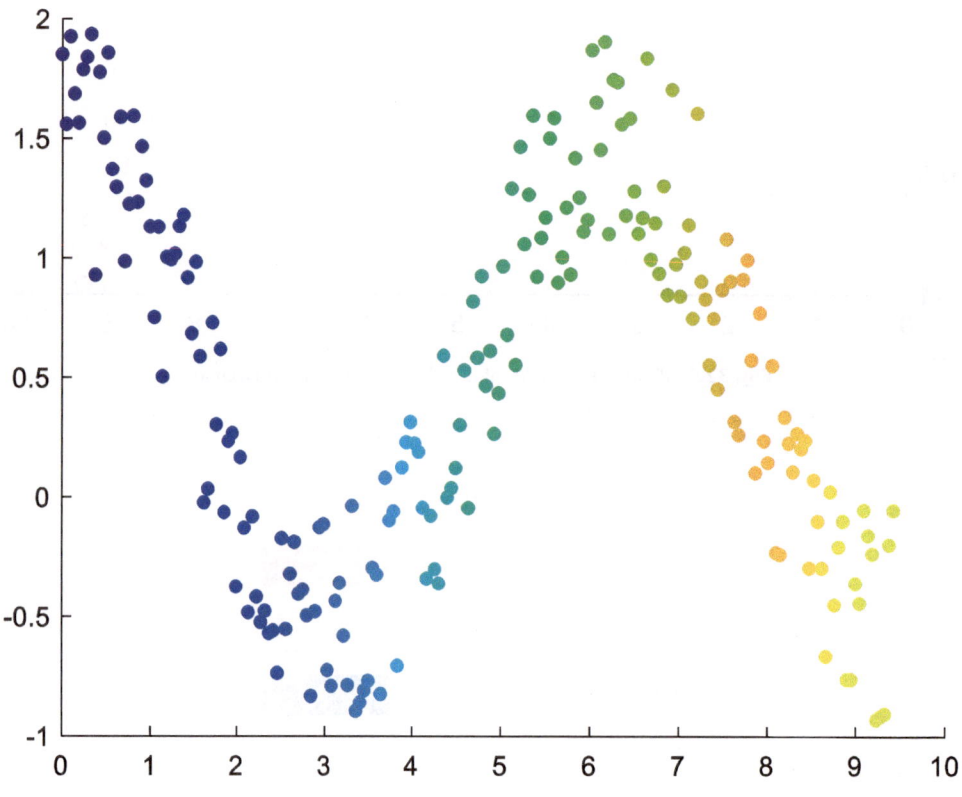

Fig. 32.7 Plotting the scatter plot with filled markers

Example

```
x = linspace(0,3*pi,200);
y = cos(x) + rand(1,200);
scatter(x,y,'^','LineWidth',1.5)
```

Description

Plotting the scatter plot with different marker and line width (See Fig. 32.8). The list of line markers available in MATLAB is illustrated in Fig. 32.9.

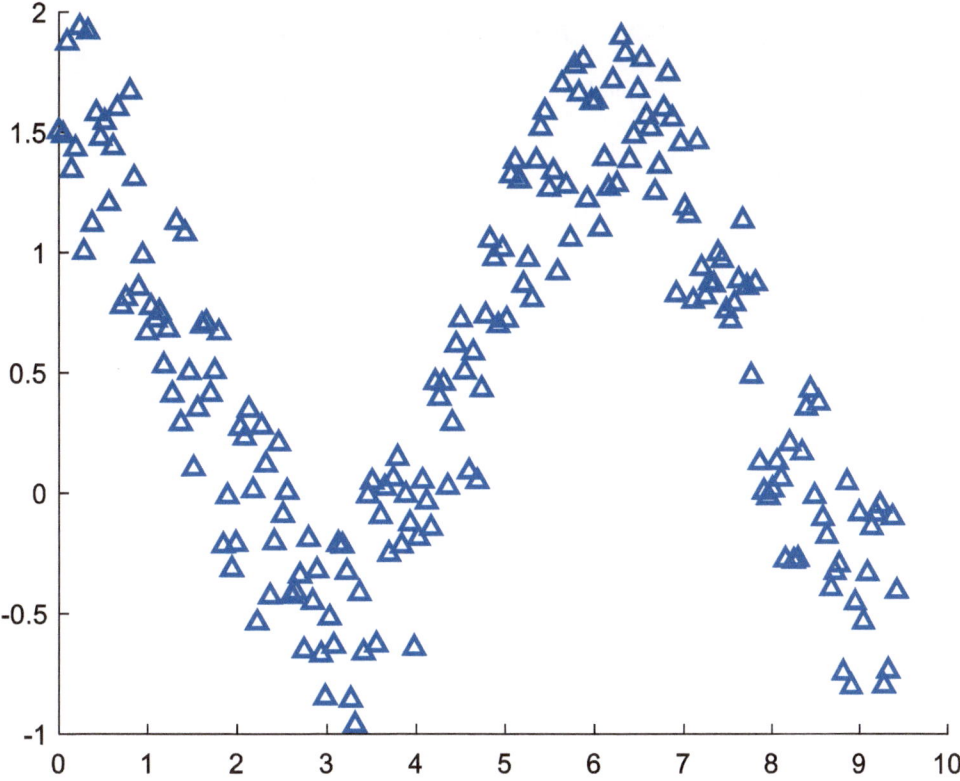

Fig. 32.8 Plotting the scatter plot with different marker and line width

Marker	Description	Resulting Marker
"o"	Circle	○
"+"	Plus sign	+
"*"	Asterisk	✳
"."	Point	•
"x"	Cross	×
"_"	Horizontal line	—
"|"	Vertical line	|
"square"	Square	□
"diamond"	Diamond	◇
"^"	Upward-pointing triangle	△
"v"	Downward-pointing triangle	▽
">"	Right-pointing triangle	▷
"<"	Left-pointing triangle	◁
"pentagram"	Pentagram	☆
"hexagram"	Hexagram	✡

Fig. 32.9 The list of line markers available in MATLAB

Demonstration

Execute the codes in your computer and observe the result(s).

```
theta = linspace(0,1,500);
x = exp(theta).*cos(100*theta);
y = exp(theta).*sin(100*theta);
c = linspace(1,10,length(x));
scatter(x,y,[],c,'filled')
```

Reference

1. MATLAB 2023a.

Abstract

In this chapter, the reference functions concerned with the programming methods in MATLAB are presented and described. In this regard, several examples and exercises for each section of the chapter are presented. The exercises that include writing the codes, executing them, and achieving the results need to be done by students to master programming skills. In this book, the codes, outputs, and descriptions are in blue, black, and green colors, respectively. To program in MATLAB, a script file can be created and saved with an appropriate name (e.g., untitled01) in the preferred directory of a computer. The program can be run by clicking on the "Run" available on the top toolbar of the script in MATLAB or calling the script by typing its name in Command Window or in the other scripts.

33.1 Defining a Function

- function [y1,...,yN] = myfun(x1,...,xM) defines a function named myfun that accepts inputs x1,...,xM and returns outputs y1,...,yN. This declaration statement must be the first executable line of the function. Valid function names begin with an alphabetic character and can contain letters, numbers, or underscores [1].
- You can save your function in a function file which contains only function definitions. The name of the file must match the name of the first function in the file. Or, you can save it in a script file which contains commands and function definitions. Functions must be at the end of the file. Note that script files cannot have the same name as a function in the file.

Example

```
function ave = calculateAverage(x)
   ave = sum(x(:))/numel(x);
end
```

Description

To define a function, open a new script file, write the above-mentioned codes in that and save it with the name calculateAverage.m in your computer's desktop. This function accepts an input vector, calculates the average of the values, and returns a single result.

Example

```
z = 1:99;
ave = calculateAverage(z)
```

Output

```
ave =
  50
```

Description

Enter the values of input vector and then call the function from the command line. The function will calculate the average of the values.

Exercise

(a) *Write the codes to define a function (with the name sum.m) to accept an input vector, calculate the sum of the values, and return a single result.*

(b) *Execute the codes in your computer to return the result(s).*

Example

```
function [m,s] = stat(x)
   n = length(x);
   m = sum(x)/n;
   s = sqrt(sum((x-m).^2/n));
end
```

Description

To define another function, open a new script file, write the above-mentioned codes in that, and save it with the name state.m in your computer's desktop. This function accepts an input vector, calculates the average and standard deviation of the values, and returns multiple results.

Example

```
values = [12.7, 45.4, 98.9, 26.6, 53.1];
[ave,stdev] = stat(values)
```

Output

ave =
    47.3400

stdev =
    29.4124

Description

Enter the values of input vector and then call the function from the command line. The function will calculate the average and standard deviation of the values.

Example

```
xmin = 0;
xmax = pi;
f = @myIntegrand;
Area = integral(f,xmin,xmax)

function y = myIntegrand(x)
    y = sin(x).^3;
end
```

Description

To define another function, open a new script file, write the above-mentioned codes in that, and save it with the name integrationScript.m in your computer's desktop. This function computes the area under the curve of the function $y(x) = (\sin x)^3$ from 0 to π.

Output

Area =
    1.3333

Description

The program can be run by clicking on the "Run" available on the top toolbar of the script or calling the script by typing its name in Command Window or in the other scripts.

Exercise

(a) *Write the codes to define a function (with the name integrationScript2.m) to accept the lower and upper boundaries and compute the area under the curve of the function $y(x) = (\cos x)^2$.*
(b) *Execute the codes in your computer to return the result(s).*

Example

```
function [m,s] = stat2(x)
    n = length(x);
    m = avg(x,n);
    s = sqrt(sum((x-m).^2/n));
end

function m = avg(x,n)
    m = sum(x)/n;
end
```

Description

To define another function, open a new script file, write the above-mentioned codes in that, and save it with the name state2.m in your computer's desktop. This function computes the average and standard deviation of the values. Herein, the function "avg" is a local function. Local functions are only available to other functions within the same file.

Example

```
values = [12.7, 45.4, 98.9, 26.6, 53.1];
[ave,stdev] = stat2(values)
```

Output

```
ave =
    47.3400

stdev =
    29.4124
```

Description

The program can be run by calling the script by typing its name in Command Window.

33.2   if, elseif, else

- The {if *expression, statements*, end} block evaluates an expression and executes a group of statements when the expression is true. An expression is true when its result is nonempty and contains only nonzero elements (logical or real numeric). Otherwise, the expression is false.
- Herein, "elseif" and "else" are optional; however, "elseif" can add new expression if needed.

Example

```
x = 10;
if x ~= 0
    disp('Nonzero value')
end
```

Output

Nonzero value

Description

A simple example of "if" block. This program determines if a given number is nonzero or not.

Example

```
x = 0;
if x ~= 0
    disp('Nonzero value')
else
    disp('Zero value')
end
```

Output

Zero value

Description

Applying "else" in an "if" block. This program determines if a given number is nonzero or zero.

Example

```
PassingGrade = 60;
Grades = [
    88
    66
    100
    91
    82
    60
    67
    75];

if any(Grades < PassingGrade)
    disp('There is at least one student who fails the course.')
else
    disp('All students pass the course.')
end
```

Output

All students pass the course.

Description

Applying a relational operator and "else" in an "if" block. This program determines if all the students have passed the course or there is at least one student who has failed the course.

Example

```
Grades = [
    88
    66
    100
    91
    82
    60
    67
    75];

if any(Grades == 100)
    disp('There is at least one student who has received the full grade.')
else
    disp('There is no student who has received the full grade.')
end
```

Output

There is at least one student who has received the full grade.

Description

Applying an equality operator and "else" in an "if" block. This program determines if there is at least one student who has received the full grade or no student has received the full grade.

Example

```
x = -1;
if x > 0
    disp('Positive value')
elseif x < 0
    disp('Negative value')
end
```

Output

Negative value

Description

Applying "elseif" in an "if" block. This program determines the sign of a given quantity.

Example

```
x = 5;
minVal = 2;
maxVal = 6;

if (x >= minVal) && (x <= maxVal)
   disp('Th value is within the specified range.')
elseif (x > maxVal)
   disp('The value exceeds the maximum value.')
else
   disp('The value is below the minimum value.')
end
```

Output

The value is within the specified range.

Description

Applying multiple conditions in an "if" block. This program determines if a given quantity is within the specified range or outside of that.

Example

```
Score = 89;
if Score >= 90
   disp('Grade: A')
elseif Score >= 80
   disp('Grade: B')
elseif Score >= 70
   disp('Grade: C')
elseif Score >= 60
   disp('Grade: D')
else
   disp('Grade: F')
end
```

Output

Grade: B

Description

Applying multiple conditions in an "if" block. This program determines a student's letter grade based on his/her score.

Exercise

(a) *Write the codes using an "if" block to determine if a person is eligible to vote based on his/her age.*
(b) *Execute the codes in your computer to return the result(s).*

Exercise

(a) *Write the codes using an "if" block to determine if a given number is a multiple of 5 or not.*
(b) *Execute the codes in your computer to return the result(s).*

33.3 for

- The {for *index = values, statements*, end} block executes a group of statements in a loop for a specified number of times.
- To programmatically exit the loop, use a "break statement."
- To programmatically skip the rest of the instructions in the loop and begin the next iteration, use a "continue statement."

Example

```
for x = 1:9
   disp(x)
end
```

Output

```
1
2
3
4
5
6
7
8
9
```

Description

Displaying the single-digit integers using a "for" block.

Example

```
M = [1 2 3; 4 5 6; 7 8 9];
    for i=1:3
        Column = M(:,i)
    end
```

Output

```
Column =
    1
    4
    7

Column =
    2
    5
    8

Column =
    3
    6
    9
```

Description

Separately displaying each column of a matrix using a "for" block.

Example

```
Sum = 0;
for i = 1:10
    Sum = Sum + i;
end
Sum
```

Output

```
Sum =
    55
```

Description

Calculating sum of the integers less than 11 by using a "for" block.

Exercise

(a) *Write the codes using a "for" block to calculate the sum of the squared value of integers less than 20.*
(b) *Execute the codes in your computer to return the result(s).*

Example

```
j=1;
for i = 1:19
   if mod(i,2) == 0
      EvenNumbers(j) = i;
      j = j + 1;
   end
end
EvenNumbers
```

Output

```
EvenNumbers =
   2   4   6   8   10   12   14   16   18
```

Description

Displaying the even numbers less than 20 using an "if" block inside a "for" block.

Exercise

(a) *Write the codes using a "for" block to display the odd numbers less than 30.*
(b) *Execute the codes in your computer to return the result(s).*

Example

```
V = [-5 -6 -11 0 2 5 1 -4 8];
for i = 1:length(V)
   if V(i)>0
      FirstPositive = V(i)
      Location = i
      break
   end
end
```

Output

FirstPositive =
2

Location =
5

Description

Applying a "break statement" in an "if" and "for" block. Herein, the first positive number and its location in the vector are displayed.

Exercise

(a) *Write the codes using a "for" block and "break statement" to display the first even number of the vector [1 3 5 7 8 9 11 12].*

(b) *Execute the codes in your computer to return the result(s).*

Example

```
for n = 1:50
    if mod(n,7)
        continue
    end
    disp(['Divisible by 7: ' num2str(n)])
end
```

Output

Divisible by 7: 7
Divisible by 7: 14
Divisible by 7: 21
Divisible by 7: 28
Divisible by 7: 35
Divisible by 7: 42
Divisible by 7: 49

Description

Applying a "continue" in an "if" and "for" block. Herein, "continue" is used to skip the "disp statement" and pass control to the next iteration of the "for" loop whenever a number is not divisible by 7. The program displays the multiples of 7 from 1 through 50.

Exercise

(a)   *Write the codes using "for" and "if" blocks to display the double-digit multiples of 8.*
(b)   *Execute the codes in your computer to return the result(s).*

33.4   while

- The {while *expression, statements*, end} block evaluates an expression and repeats the execution of a group of statements in a loop while the expression is true. An expression is true when its result is nonempty and contains only nonzero elements (logical or real numeric). Otherwise, the expression is false.
- To programmatically exit the loop, use a break statement.
- To programmatically skip the rest of the instructions in the loop and begin the next iteration, use a continue statement.
- If you inadvertently create an infinite loop (that is, a loop that never ends on its own), stop execution of the loop by pressing Ctrl+C.

Example

```
Sum = 0;
i = 1;
while i<=10
    Sum = Sum + i;
    i = i + 1;
end
Sum
```

Output

```
Sum =
   55
```

Description

Calculating sum of the integers less than 11 by using a "while" block.

Exercise

(a)   *Write the codes using a "for" block to calculate the sum of the squared value of integers less than 20.*
(b)   *Execute the codes in your computer to return the result(s).*

Example

```
Sum = 0;
i = 1;
while 1
   Sum = Sum + i;
   if i >= 10
      break
   end
   i = i + 1;
end
Sum
```

Output

```
Sum =
   55
```

Description

Calculating the sum of the integers less than 11 by using a "break statement" in an "if" and a "while" block. Herein, "while 1" means that the loop is forever until it is broken.

Example

```
j=1;
while i < 20
   if mod(i,2) == 0
      EvenNumbers(j) = i;
      j = j + 1;
   end
   i = i +1;
end
EvenNumbers
```

Output

```
EvenNumbers =
   2   4   6   8   10   12   14   16   18
```

Description

Displaying the even numbers less than 20 by using an "if" and a "while" block.

(a) *Write the codes using a "while" block to display the odd numbers less than 30.*

(b) *Execute the codes in your computer to return the result(s).*

Example

```
n = 10;
f = n;
while n > 1
    n = n-1;
    f = f*n;
end
disp(['n! = ' num2str(f)])
```

Output

n! = 3628800

Description

Calculating factorial(10) by using a "while" block.

33.5   switch case

- The {switch *switch_expression, case case_expression*, end} block evaluates an expression and chooses to execute one of several groups of statements. Each choice is a case. The full format of the "switch" block is as follows.

```
switch switch_expression
    case case_expression
        statements
    case case_expression
        statements
    ...
    otherwise
        statements
end
```

- The switch block tests each case until one of the case expressions is true. A case is true when for numbers, *case_expression* == *switch_expression*, for character vectors, strcmp(*case_expression,switch_expression*) == 1, for objects that support the eq function, *case_expression* == *switch_expression*. The output of the overloaded eq function must be either a logical value or convertible to a logical value.
- When a case expression is true, MATLAB executes the corresponding statements and exits the switch block.
- An evaluated *switch_expression* must be a scalar or character vector. An evaluated *case_expression* must be a scalar, a character vector, or a cell array of scalars or character vectors.
- The otherwise block is optional. MATLAB executes the statements only when no case is true.

Example

```
n = input('Enter a number: ');
switch n
   case 3.1416
      disp('Right Answer')
   otherwise
      disp('Wrong Answer')
end
```

Output

Enter a number: 3
Wrong Answer

Enter a number: 3.1416
Right Answer

Description

Displaying different text conditionally, depending on a value entered at the command prompt. The program can be used to identify the correct answer of a problem.

Reference

1. MATLAB 2023a.

Abstract

In this chapter, the reference functions concerned with the numerical methods in MATLAB are presented and described. In this regard, several examples and exercises for each section of the chapter are presented. The exercises that include writing the codes, executing them, and achieving the results need to be done by students to master programming skills. In this book, the codes, outputs, and descriptions are in blue, black, and green colors, respectively. To program in MATLAB, a script file can be created and saved with an appropriate name (e.g., untitled01) in the preferred directory of a computer. The program can be run by clicking on the "Run" available on the top toolbar of the script in MATLAB or calling the script by typing its name in Command Window or in the other scripts.

34.1 The Reference Function "fzero": Root of Nonlinear Function

- This reference function in the format x = fzero(fun,x0) tries to find a point x where fun(x) = 0 [1–9].

Example

```
fun = @sin;
x0 = 3;
x = fzero(fun,x0)
```

Output

```
x =
   3.1416
```

Description

Finding the zero of a function from a given point. Herein, the zero of sine function near 3 is calculated.

Exercise

(a) *Write the codes to calculate the zero of cosine function near 1.5.*
(b) *Execute the codes in your computer to return the result(s).*

Exercise

(a) *Write the codes to calculate the zero of tangent function near 3.*

(b) *Execute the codes in your computer to return the result(s).*

Example

```
fun = @cos;
x0 = [1 2];
x = fzero(fun,x0)
```

Output

```
x =
   1.5708
```

Description

Finding the zero of a function between a range. Herein, the zero of cosine function between 1 and 2 is calculated.

Exercise

(a) *Write the codes to calculate the zero of sine function between −1 and 1.*

(b) *Execute the codes in your computer to return the result(s).*

Exercise

(a) *Write the codes to calculate the zero of cotangent function between 0 and 3.*

(b) *Execute the codes in your computer to return the result(s).*

Example

```
myfun = @(x,c) cos(c*x);
c = 2;
fun = @(x) myfun(x,c);
x = fzero(fun,0.1)
```

Output

```
x =
   0.7854
```

Description

Finding the root of a function including an extra parameter.

Example

```
fun = @f;
x0 = 2;
z = fzero(fun,x0)
```

Output

```
z =
   2.0946
```

Description

Finding the zero of a function (near 2) defined by a file. Herein, the file needs to be saved with the name "f.m" with the following codes:

```
function y = f(x)
y = x.^3 - 2*x - 5;
```

34.2   The Reference Function "fminsearch": Minimum of Unconstrained Multivariable Function

- This reference function in the format x = fminsearch(fun,x0) starts at the point x0 and attempts to find a local minimum x of the function described in fun.
- This reference function is called nonlinear programming solver.

Example

```
fun = @(x)100*(x(2) - x(1)^2)^2 + (1 - x(1))^2;
x0 = [-1.2,1];
x = fminsearch(fun,x0)
```

Output

```
x =
   1.0000   1.0000
```

Description

Minimizing a function with the given starting point. Herein, the function is $100\left(x_2 - x_1^2\right)^2 + (1 - x_1)^2$ and the starting point is $[-1.2,1]$.

Example

```
f = @(x,a)100*(x(2) - x(1)^2)^2 + (a-x(1))^2;
a = 3;
fun = @(x)f(x,a);
x0 = [-1,1.9];
x = fminsearch(fun,x0)
```

Output

```
x =
   3.0000   9.0000
```

Description

Minimizing a function with the given starting point and extra parameters. Herein, the function, starting point, and the value of parameter (a) are $100\left(x_2 - x_1^2\right)^2 + (a - x_1)^2$, $[-1,1.9]$, and 3, respectively.

Exercise

Execute the codes below in your computer. Then, write the result(s) in the following.

```
f = @(x,a)100*(x(2) - x(1)^2)^2 + (a-x(1))^2;
a = 2;
fun = @(x)f(x,a);
x0 = [-1,1.9];
x = fminsearch(fun,x0)
```

Example

```
x0 = [1–3];
fun = @(x)-norm(x+x0)^2*exp(-norm(x-x0)^2 + sum(x));
[x,fval] = fminsearch(fun,x0)
```

Output

```
x =
   1.5359   2.5645   3.5932

fval =
   -5.9565e+04
```

Description

Finding both location and minimum value of a function.

Execute the codes below in your computer. Then, write the result(s) in the following.

```
f = @(x,a)100*(x(2) - x(1)^2)^2 + (a-x(1))^2;
a = 2;
fun = @(x)f(x,a);
x0 = [-1,1.9];
[x,fval] = fminsearch(fun,x0)
```

34.3 The Reference Function "fminbnd": Minimum of Single-Variable Function on Fixed Interval

- This reference function in the format x = fminbnd(fun,x1,x2) returns a value x that is a local minimizer of the scalar valued function that is described in fun in the interval x1 < x < x2.
- This reference function in the format x = fminbnd(fun,x1,x2,options) minimizes with the optimization options specified in options.

Example

```
fun = @sin;
x1 = 0;
x2 = 2*pi;
x = fminbnd(fun,x1,x2)
```

Output

```
x =
   4.7124
```

Description

Finding the point where the function takes its minimum in the range. Herein, x = 4.7124 is the same as x = 3π/2.

Example

```
fun = @sin;
x1 = 0;
x2 = 2*pi;
options = optimset('Display','iter');
x = fminbnd(fun,x1,x2,options)
```

Output

Func-count	x	f(x)	Procedure
1	2.39996	0.67549	initial
2	3.88322	-0.67549	golden
3	4.79993	-0.996171	golden
4	5.08984	-0.929607	parabolic
5	4.70582	-0.999978	parabolic
6	4.7118	-1	parabolic
7	4.71239	-1	parabolic
8	4.71236	-1	parabolic
9	4.71242	-1	parabolic

Optimization terminated:
  the current x satisfies the termination criteria using OPTIONS.TolX of 1.000000e-04

x =
  4.7124

Description

Monitoring the steps that the solver takes to minimize the function.

Exercise

Execute the codes below in your computer. Then, write the result(s) in the following.

```
fun = @cos;
x1 = 0;
x2 = 2*pi;
options = optimset('Display','iter');
x = fminbnd(fun,x1,x2,options)
```

Example

```
fun = @sin;
[x,fval] = fminbnd(fun,1,2*pi)
```

Output

x =
  4.7124

fval =
  -1.0000

Description

Finding both the location and minimum value of a function.

Exercise

Execute the codes below in your computer. Then, write the result(s) in the following.

```
fun = @cos;
[x,fval] = fminbnd(fun,1,2*pi)
```

Example

```
a = 9/7;
fun = @(x)sin(x-a);
x = fminbnd(fun,1,2*pi)
```

Output

```
x =
  5.9981
```

Description

Minimizing a function with an extra parameter. Herein, x = 5.9981 is the same as x = 3*pi/2 + 9/7.

Exercise

Execute the codes below in your computer. Then, write the result(s) in the following.

```
a = pi;
fun = @(x)cos(x-a);
[x,fval] = fminbnd(fun,1,2*pi)
```

34.4   The Reference Function "integral": Numerical Integration

- This reference function in the format q = integral(fun,xmin,xmax) numerically integrates function fun from xmin to xmax using global adaptive quadrature and default error tolerances.

Example

```
fun = @(x) exp(-x.^2).*log(x).^2;
q = integral(fun,0,Inf)
```

Output

q =
   1.9475

Description

Calculating the integral of a function from x = 0 to x = Inf.

Exercise

Choose a function from your Calculus textbook and calculate its integral. Then, write the codes below, execute them in your computer, and write the result(s) in the following.

Example

```
fun = @(x,c) 1./(x.^3-2*x-c);
q = integral(@(x) fun(x,5),0,2)
```

Output

q =
   -0.4605

Description

Calculating the integral of a parameterized function from x = 0 to x = 2 at c = 5.

Exercise

Execute the codes below in your computer. Then, write the result(s) in the following.

```
fun = @(x,c) 1./( x.^2-c);
q = integral(@(x) fun(x,0.5),1,2)
```

Example

```
fun = @(z) 1./(2*z-1);
q = integral(fun,0,0,'Waypoints',[1+1i,1-1i])
```

Output

q =

  0.0000 - 3.1416i

Description

Calculating the complex integral of a complex function in the complex plane over the triangular path from 0 to $1 + 1i$ to $1 - 1i$ to 0 by specifying waypoints.

Exercise

Choose a complex function from your Mathematics of Engineering and Science textbook and calculate its complex integral. Then, write the codes below, execute them in your computer, and write the result(s) in the following.

Exercise

Execute the codes below in your computer. Then, write the result(s) in the following.

```
fun = @(z) 1./(z-0.5);
q = integral(fun,0,0,'Waypoints',[1+1i,1-1i])
```

Exercise

Execute the codes below in your computer. Then, write the result(s) in the following.

```
fun = @(z) 1./(z-0.5);
q = integral(fun,0,0,'Waypoints',[1-1i,1+1i])
```

Exercise

Execute the codes below in your computer. Then, write the result(s) in the following.

```
fun = @(z) 1./(z+0.5);
q = integral(fun,0,0,'Waypoints',[1-1i,1+1i])
```

34.5   The Reference Function "gradient": Gradient of Scalar Field

- This reference function in the format FX = gradient(F) returns the one-dimensional numerical gradient of scalar field F. The output FX corresponds to $\partial F/\partial x$, which are the differences in the x (horizontal) direction. The spacing between points is assumed to be 1.
- This reference function in the format [FX,FY] = gradient(F) returns the x and y components of the two-dimensional numerical gradient of matrix F. The additional output FY corresponds to $\partial F/\partial y$, which are the differences in the y (vertical) direction. The spacing between points in each direction is assumed to be 1.

Example

```
x = [1   2   3   4   5];
fx = gradient(x)
```

Output

```
fx =
   1   1   1   1   1
```

Description

Calculating the gradient of a monotonically increasing scalar field.

Exercise

Execute the codes below in your computer. Then, write the result(s) in the following.

```
x = [-1   -2   -3   -4   -5];
y = [5   4   3   2   1];
fx = gradient(x)
fy = gradient(y)
```

Example

```
x = -2:0.2:2;
y = x';
z = x .* exp(-x.^2 - y.^2);
surf(x,y,z)
xlabel('x')
ylabel('y')
zlabel('z')
```

Description

Plotting the scalar field. See Fig. 34.1.

Example

[px,py] = gradient(z);
quiver(x,y,px,py)

Description

Calculating the 2D gradient of the scalar field and plotting it. See Fig. 34.2.

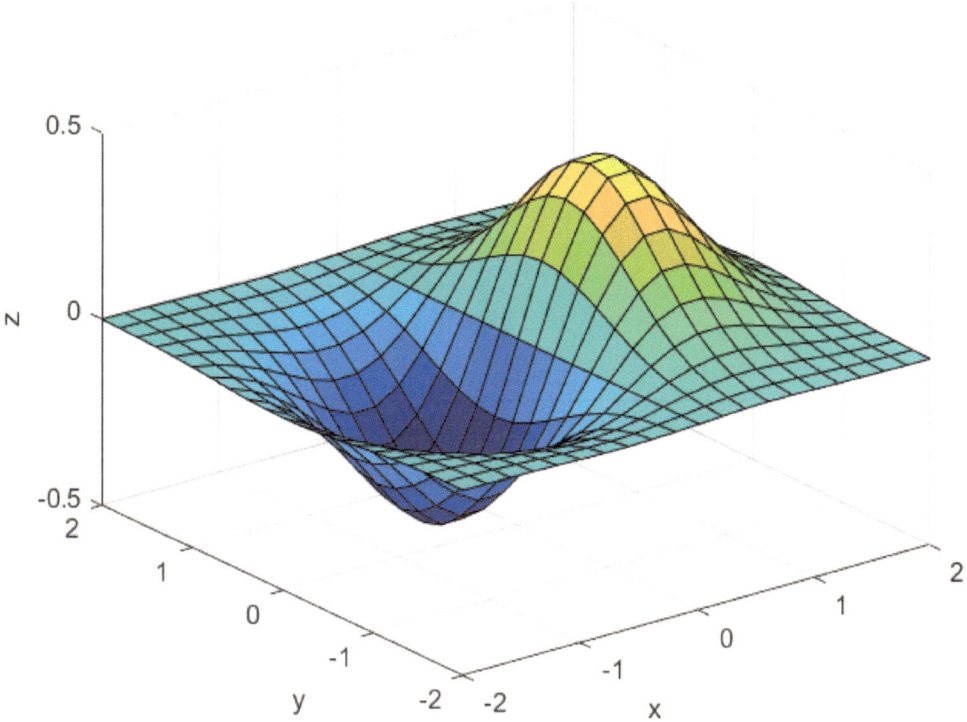

Fig. 34.1 Plotting the scalar field

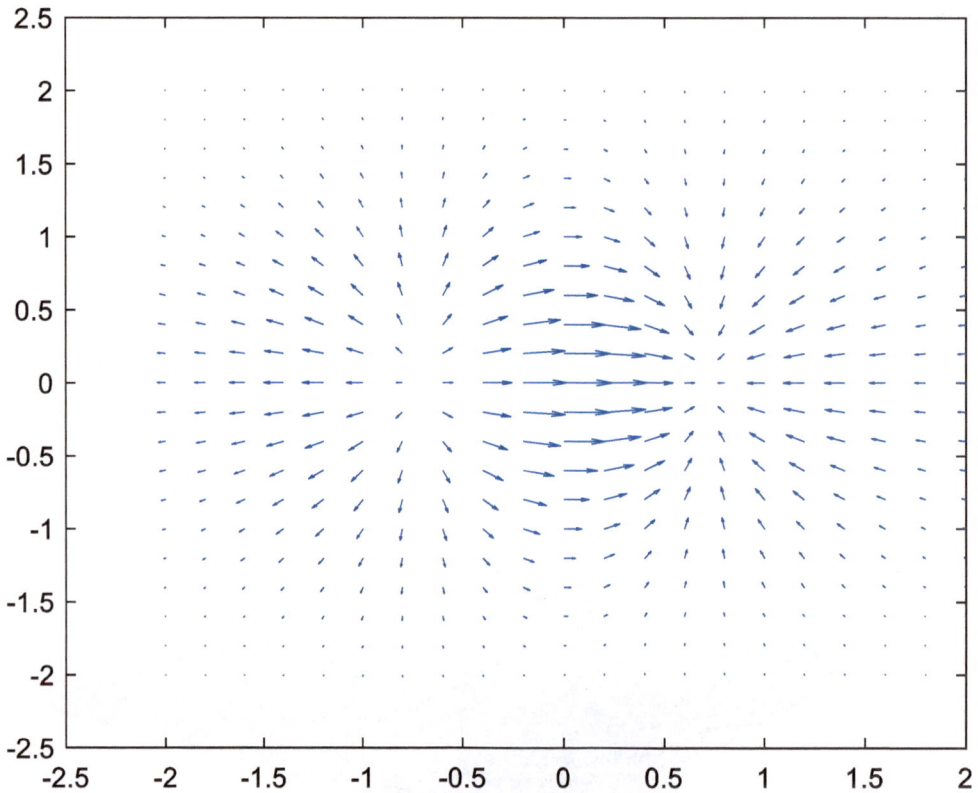

Fig. 34.2 The 2D gradient of the scalar field

Demonstration

First, execute the codes below in your computer and observe the result(s).

```
x = -2:0.2:2;
y = x';
z = x .* exp(x.^2 + y.^2);
surf(x,y,z)
xlabel('x')
ylabel('y')
zlabel('z')
```

Then, execute the codes below in your computer and observe the result(s).

```
[px,py] = gradient(z);
quiver(x,y,px,py)
```

Demonstration

First, execute the codes below in your computer and observe the result(s).

```
x = -2:0.2:2;
y = x';
z = x .* exp(-x.^(-2) - y.^(-2));
surf(x,y,z)
```

```
xlabel('x')
ylabel('y')
zlabel('z')
```

Then, execute the codes below in your computer and observe the result(s).

```
[px,py] = gradient(z);
quiver(x,y,px,py)
```

Example

```
x = -3:0.2:3;
y = x';
z = x.^2 .* y.^3;
surf(x,y,z)
xlabel('x')
ylabel('y')
zlabel('z')
```

Description

Plotting the scalar field. See Fig. 34.3.

Example

```
[px,py] = gradient(z);
quiver(x,y,px,py)
```

Description

Calculating the 2D gradient of the scalar field and plotting it. See Fig. 34.4.

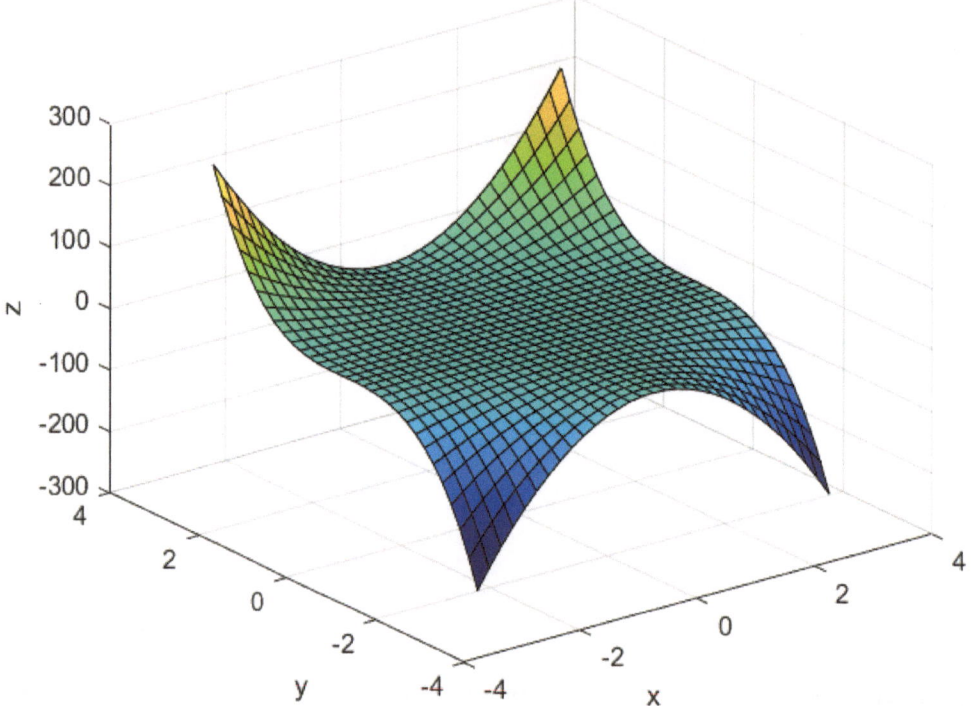

Fig. 34.3 Plotting the scalar field

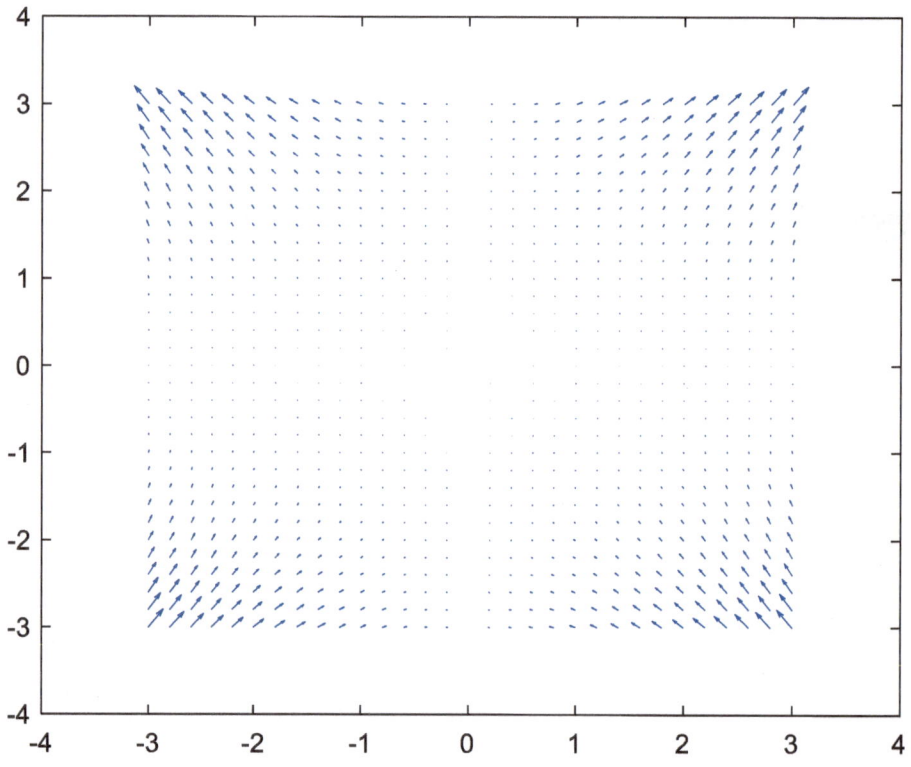

Fig. 34.4 The 2D gradient of the scalar field

First, execute the codes below in your computer and observe the result(s).

```
x = -3:0.2:3;
y = x';
z = x.^20 .* y.^30;
surf(x,y,z)
xlabel('x')
ylabel('y')
zlabel('z')
```

Then, execute the codes below in your computer and observe the result(s).

```
[px,py] = gradient(z);
quiver(x,y,px,py)
```

34.6   The Reference Function "divergence": Divergence of Vector Field

- This reference function in the format div = divergence(X,Y,Z,Fx,Fy,Fz) computes the numerical divergence of a 3D vector field with vector components Fx, Fy, and Fz.
- The arrays X, Y, and Z, which define the coordinates for the vector components Fx, Fy, and Fz, must be monotonic, but do not need to be uniformly spaced.
- X, Y, and Z must be 3D arrays of the same size, which can be produced by meshgrid.

```
[x,y] = meshgrid(-8:2:8,-8:2:8);
Fx = 200 - (x.^2 + y.^2);
Fy = 200 - (x.^2 + y.^2);
quiver(x,y,Fx,Fy)
```

Plotting the vector field. See Fig. 34.5.

```
D = divergence(x,y,Fx,Fy);
contour(x,y,D,'ShowText','on')
```

Calculating the numerical divergence of the 2D vector field and plotting it. See Fig. 34.6.

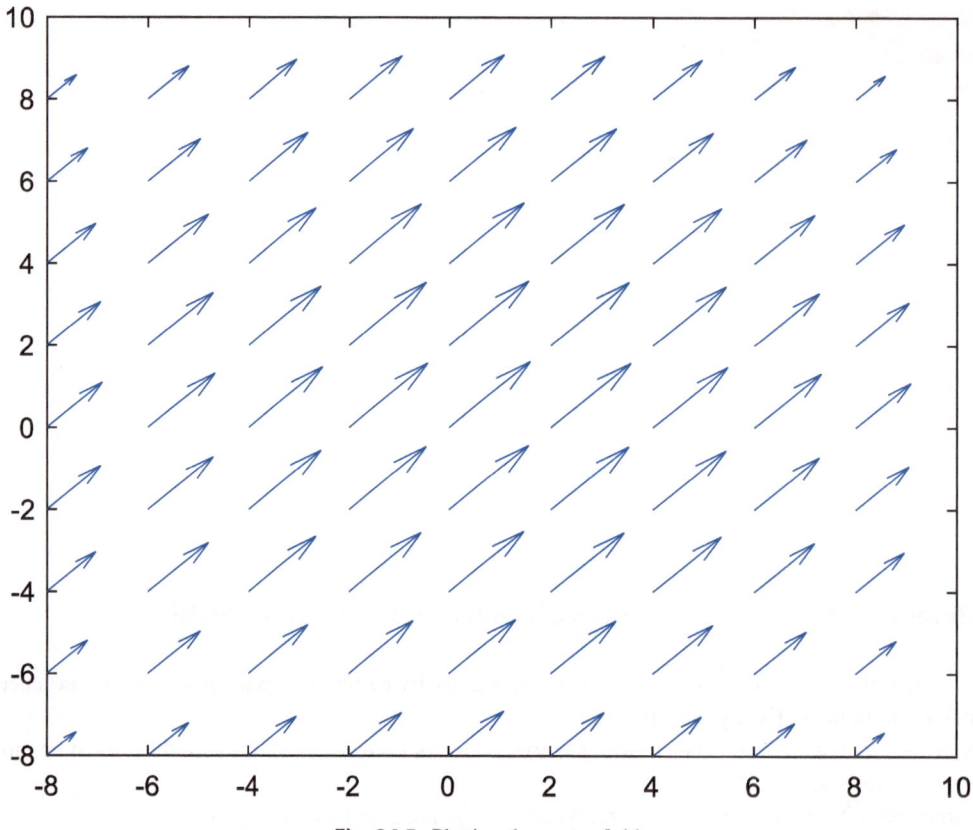

Fig. 34.5 Plotting the vector field

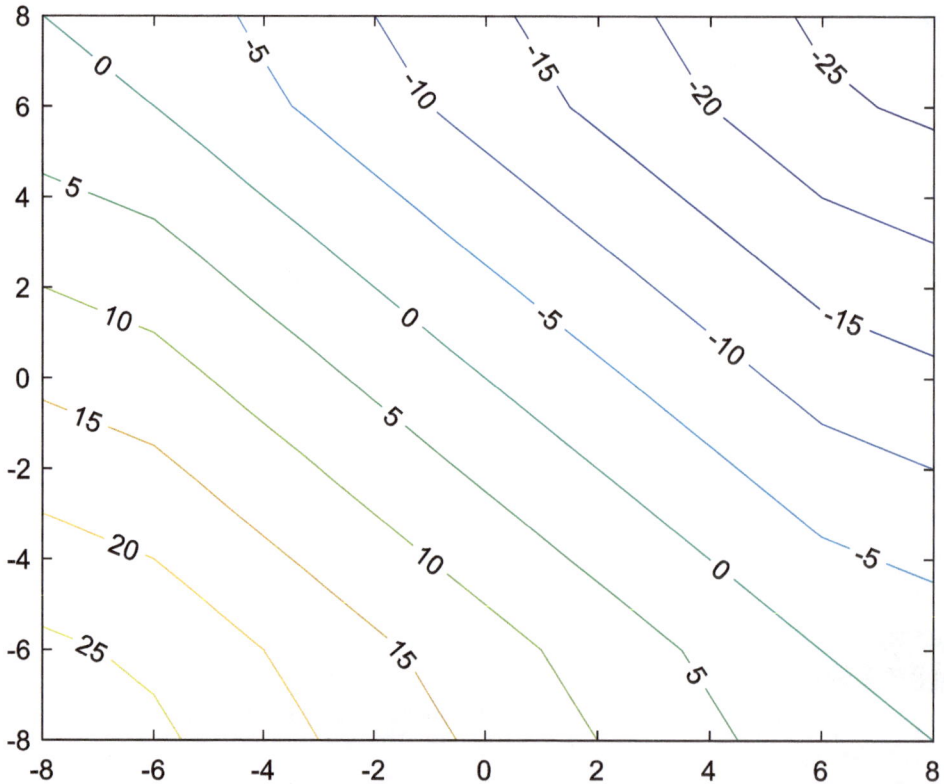

Fig. 34.6 The numerical divergence of the 2D vector field

[x,y] = meshgrid(-8:2:8,-8:2:8);
Fx = (x.^3 + y.^3);
Fy = (x.^3 + y.^3);
quiver(x,y,Fx,Fy)

Plotting the vector field. See Fig. 34.7.

D = divergence(x,y,Fx,Fy);
contour(x,y,D,'ShowText','on')

Calculating the numerical divergence of the 2D vector field and plotting it. See Fig. 34.8.

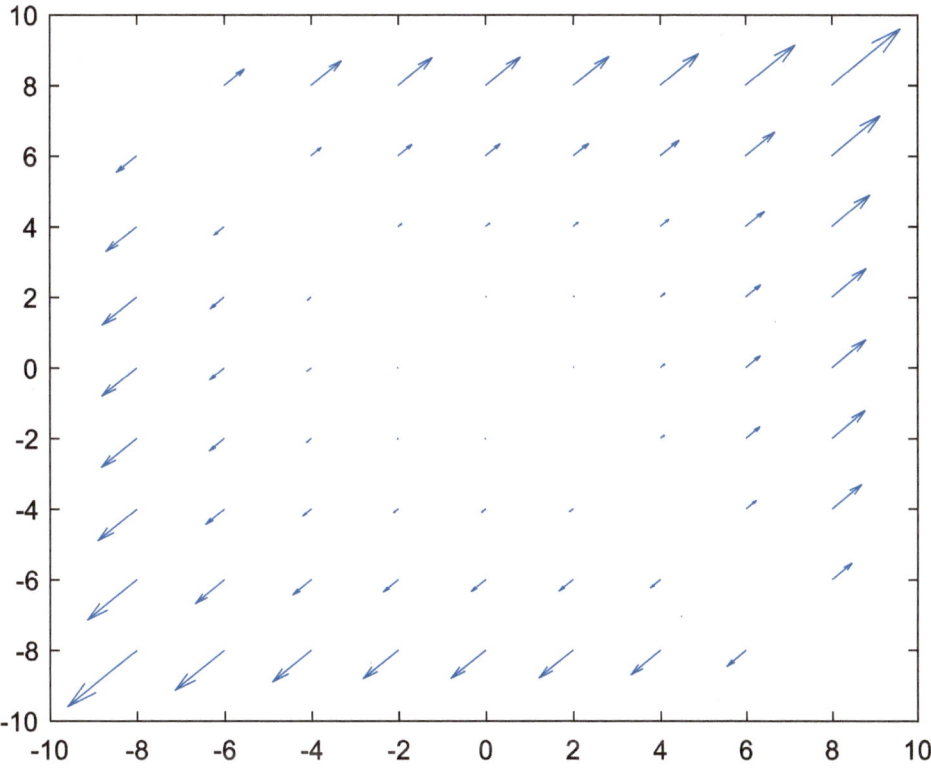

Fig. 34.7 Plotting the vector field

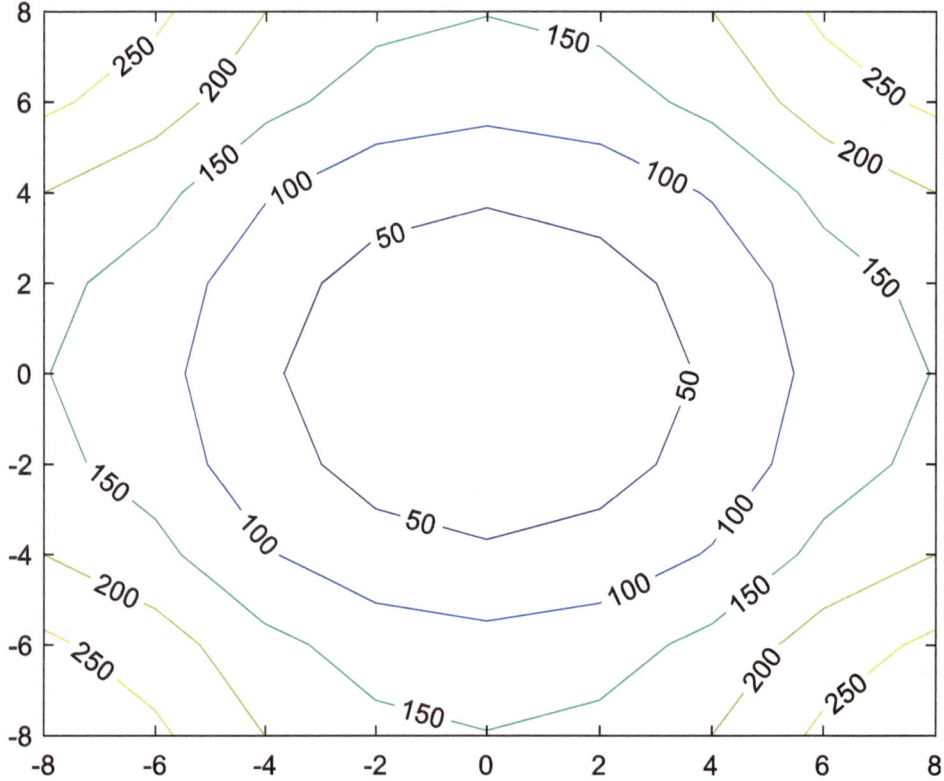

Fig. 34.8 The numerical divergence of the 2D vector field

34.7 The Reference Functions "ode45" and "ode15s": Solving Non-stiff and Stiff Ordinary Differential Equations (Medium Order Method)

- The reference function in the format [t,y] = ode45(odefun,tspan,y0), where tspan = [t0 tf], integrates the system of non-stiff ordinary differential equations (ODEs) y = f(t,y) from t0 to tf with initial conditions y0. Each row in the solution array y corresponds to a value returned in column vector t.
- The reference function in the format [t,y] = ode15s(odefun,tspan,y0) is used to solve stiff ODEs.

Example

```
tspan = [0 5];
y0 = 0;
[t,y] = ode45(@(t,y) 2*t, tspan, y0);
plot(t,y,'-.')
```

Description

Plotting the solution of an ODE with single initial condition. See Fig. 34.9. Herein, the solution of the ODE $y(t) = 2t$ with the initial condition $y(t = 0) = 0$ is plotted.

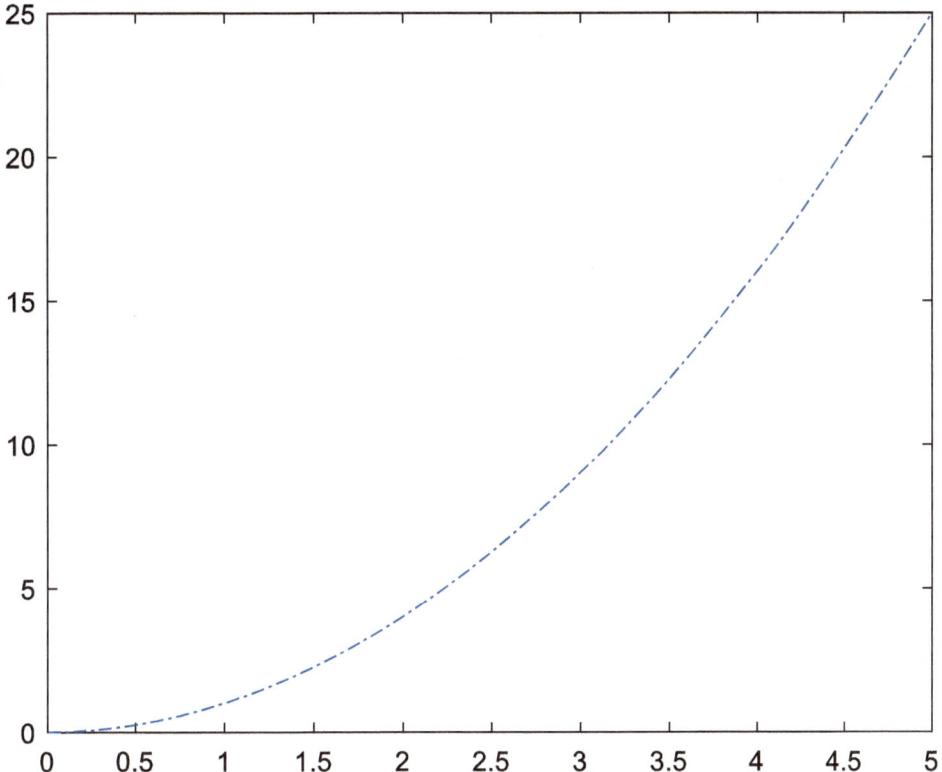

Fig. 34.9 Plotting the solution of an ODE with single initial condition

Demonstration

Execute the codes below in your computer and observe the result(s).

```
tspan = [-5 5];
y0 = 20;
[t,y] = ode45(@(t,y) t^2, tspan, y0);
plot(t,y,'--')
```

Example

```
yprime = @(t,y) -2*y + 2*cos(t).*sin(2*t);
y0 = -5:5;
tspan = [0 3];
[t,y] = ode45(yprime,tspan,y0);
plot(t,y)
grid on
xlabel('t')
ylabel('y')
```

Description

Plotting the solution of an ODE with multiple initial conditions. See Fig. 34.10. Herein, the solution of the ODE $y'' = -2y + 2\cos(t)\sin(2t)$ with the initial conditions $y(0) = -5, -4, \ldots, 4, 5$ is plotted.

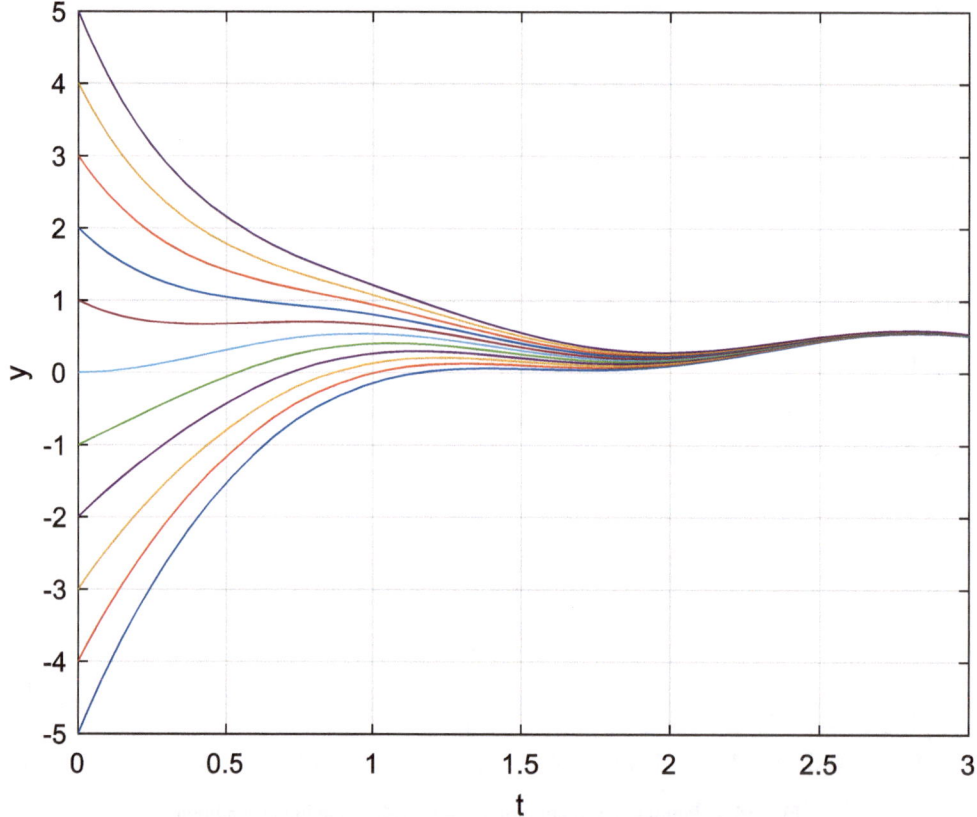

Fig. 34.10 Plotting the solution of an ODE with multiple initial conditions

34.8   The Reference Function "interp1": 1D Data Interpolation

- This reference function in the format vq = interp1(x,v,xq) returns interpolated values of a 1D function at specific query points using linear interpolation by default. Vector x contains the sample points, vector v contains the corresponding values, and vector xq contains the coordinates of the query points.

Example

```
x = 0:pi/4:2*pi;
v = sin(x);
xq = 0:pi/16:2*pi;
vq1 = interp1(x,v,xq);
plot(x,v,'o',xq,vq1,':.');
xlim([0 2*pi]);
title('Linear Interpolation (Default)');
```

Description

Linear interpolation of coarsely sampled sine function and plotting it. See Fig. 34.11.

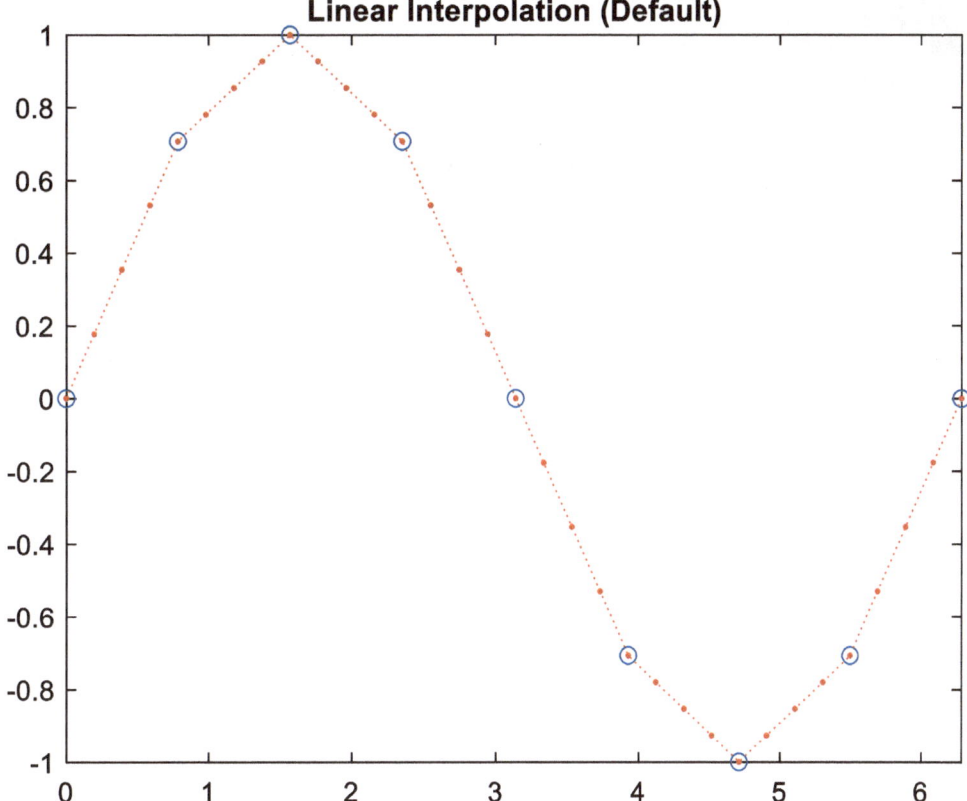

Fig. 34.11 The plot of linear interpolation of coarsely sampled sine function

---

Demonstration

Execute the codes below in your computer and observe the result(s).

```
x = 0:pi/4:2*pi;
v = cos(x);
xq = 0:pi/16:2*pi;
vq1 = interp1(x,v,xq);
plot(x,v,'o',xq,vq1,':.');
xlim([0 2*pi]);
title('Linear Interpolation (Default)');
```

Example

```
x = 0:pi/4:2*pi;
v = sin(x);
xq = 0:pi/16:2*pi;
vq2 = interp1(x,v,xq,'spline');
plot(x,v,'o',xq,vq2,':.');
xlim([0 2*pi]);
title('Spline Interpolation');
```

Description

Interpolation of coarsely sampled sine function using the spline method and plotting it. See Fig. 34.12.

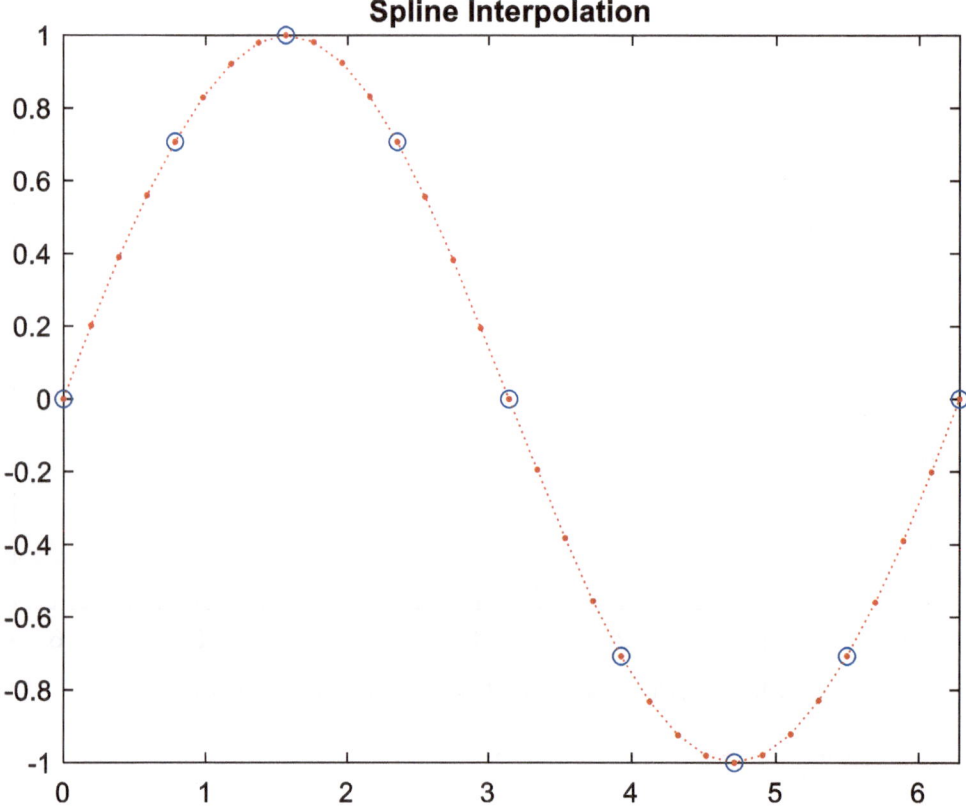

Fig. 34.12 The plot of interpolation of coarsely sampled sine function using the spline method

Demonstration

Execute the codes below in your computer and observe the result(s).

```
x = 0:pi/4:2*pi;
v = cos(x);
xq = 0:pi/16:2*pi;
vq2 = interp1(x,v,xq,'spline');
plot(x,v,'o',xq,vq2,':.');
xlim([0 2*pi]);
title('Spline Interpolation');
```

Example

```
x = 1:10;
v = (5*x)+(x.^2*i);
xq = 1:0.25:10;
vq = interp1(x,v,xq);
```

```
plot(x,real(v),'^m',xq,real(vq),'-.m');
hold on
plot(x,imag(v),'sb',xq,imag(vq),'-.b');
```

Description

Interpolation of a complex quantity and plotting its real and imaginary parts. See Fig. 34.13.

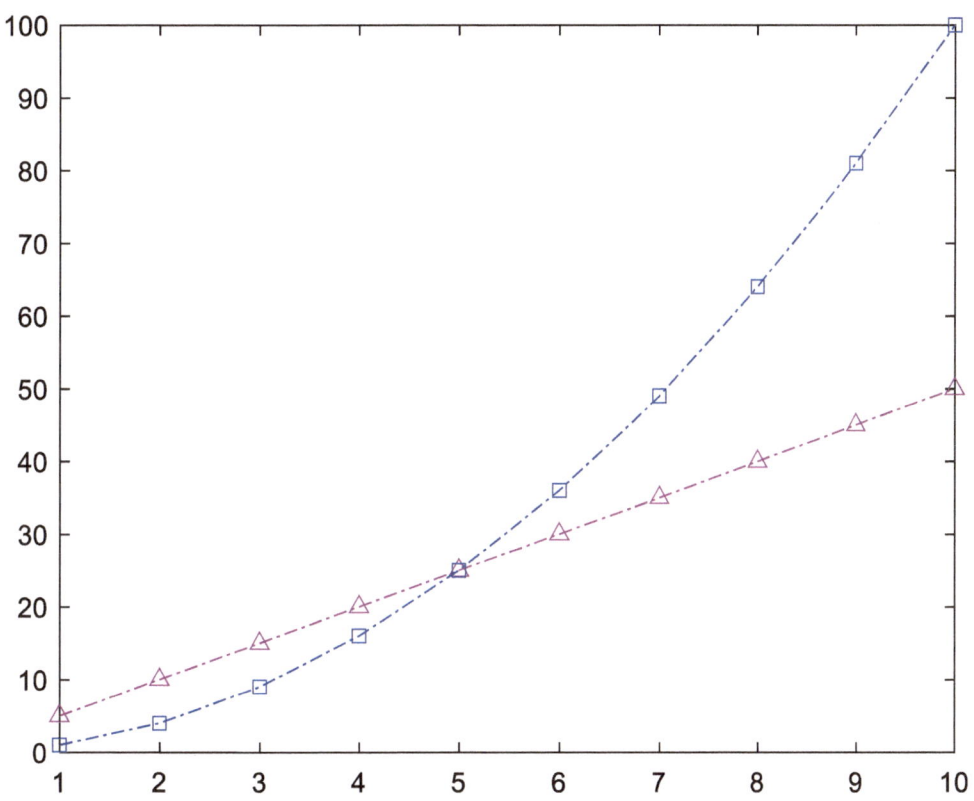

Fig. 34.13 The plots of real and imaginary parts of the interpolated complex quantity

Demonstration

Execute the codes below in your computer and observe the result(s).

```
x = 1:10;
v = (0.3*x)+(x.^0.3*i);
xq = 1:0.25:10;
vq = interp1(x,v,xq);
plot(x,real(v),'^m',xq,real(vq),'-.m');
hold on
plot(x,imag(v),'sb',xq,imag(vq),'-.b');
```

34.9   The Reference Functions "spline", "pchip", and "mkpp": Cubic Spline Data Interpolation, Piecewise Cubic Hermite Interpolating Polynomial, and Make Piecewise Polynomial

- The reference function in the format s = spline(x,y,xq) returns a vector of interpolated values s corresponding to the query points in xq. The values of s are determined by cubic spline interpolation of x and y.
- The reference function in the format p = pchip(x,y,xq) returns a vector of interpolated values p corresponding to the query points in xq. The values of p are determined by shape-preserving piecewise cubic interpolation of x and y.
- The reference function in the format mkpp(breaks,coefs) builds a piecewise polynomial pp from its breaks and coefficients. Use ppval to evaluate the piecewise polynomial at specific points, or unmkpp to extract details about the piecewise polynomial.

Example

```
x = [0 1 2.5 3.6 5 7 8.1 10];
y = sin(x);
xx = 0:.25:10;
yy = spline(x,y,xx);
plot(x,y,'o',xx,yy)
grid on
```

Description

Interpolation of the function using spline method. See Fig. 34.14.

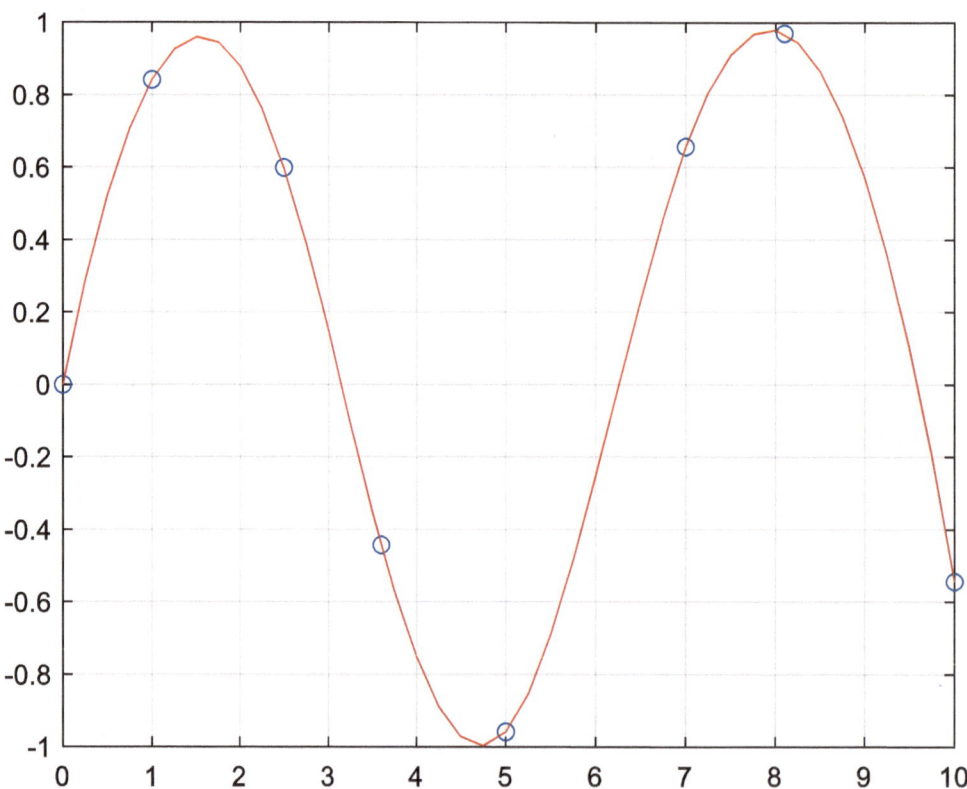

Fig. 34.14 The plot of interpolation of the function using spline method

Demonstration

Execute the codes below in your computer and observe the result(s).

```
x = [0 1 2.5 3.6 5 7 8.1 10];
y = cos(x);
xx = 0:.25:10;
yy = spline(x,y,xx);
plot(x,y,'o',xx,yy)
grid on
```

Example

```
x = -3:3;
y = [-1 -1 -1 0 1 1 1];
xq1 = -3:.01:3;
p = pchip(x,y,xq1);
s = spline(x,y,xq1);
m = makima(x,y,xq1);
plot(x,y,'o',xq1,p,'-',xq1,s,'-.',xq1,m,'--')
grid on
legend('Sample Points','pchip','spline','makima','Location','SouthEast')
```

Description

Interpolation of the data set using spline, pchip, and makima methods and plotting them. See Fig. 34.15. As can be seen, for this data set, pchip and makima have similar behavior in which they avoid overshoots and can accurately connect the flat regions.

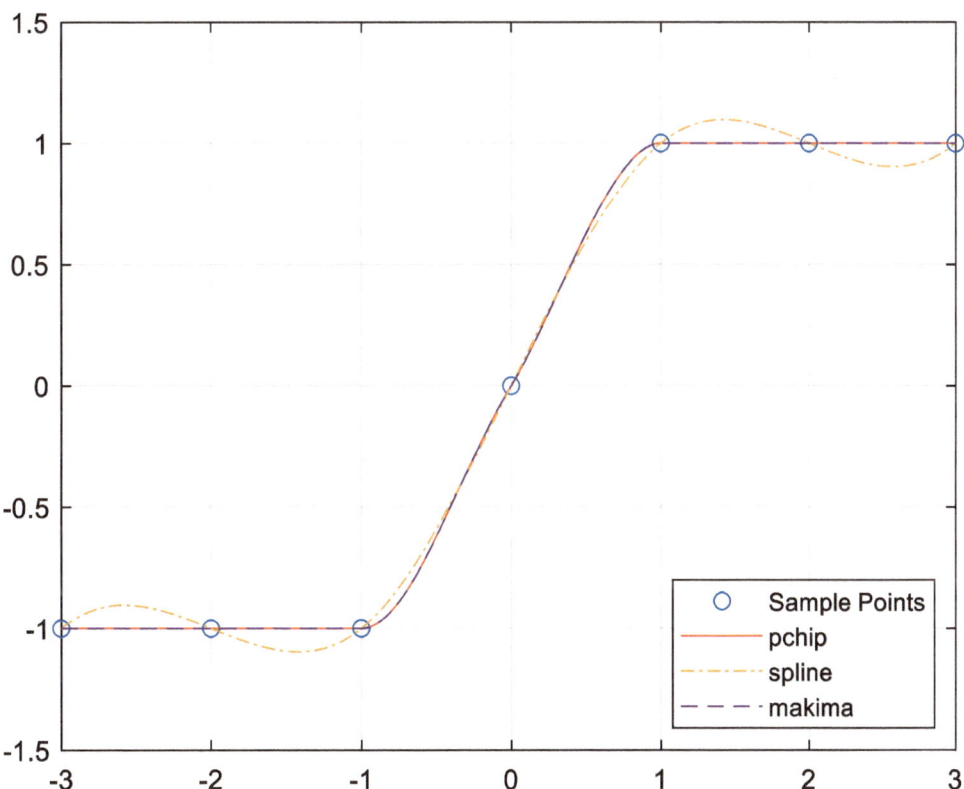

Fig. 34.15 The plots of interpolation of the data set using spline, pchip, and makima methods

```
x = 0:15;
y = besselj(1,x);
xq2 = 0:0.01:15;
p = pchip(x,y,xq2);
s = spline(x,y,xq2);
m = makima(x,y,xq2);
plot(x,y,'o',xq2,p,'-',xq2,s,'-.',xq2,m,'--')
grid on
legend('Sample Points','pchip','spline','makima')
```

Interpolation of another data set using spline, pchip, and makima methods and plotting them. See Fig. 34.16. As can be seen, for this data set, spline and makima have similar behavior. They capture the movement between points better than pchip, which is aggressively flattened near local extrema, when the underlying function is oscillatory.

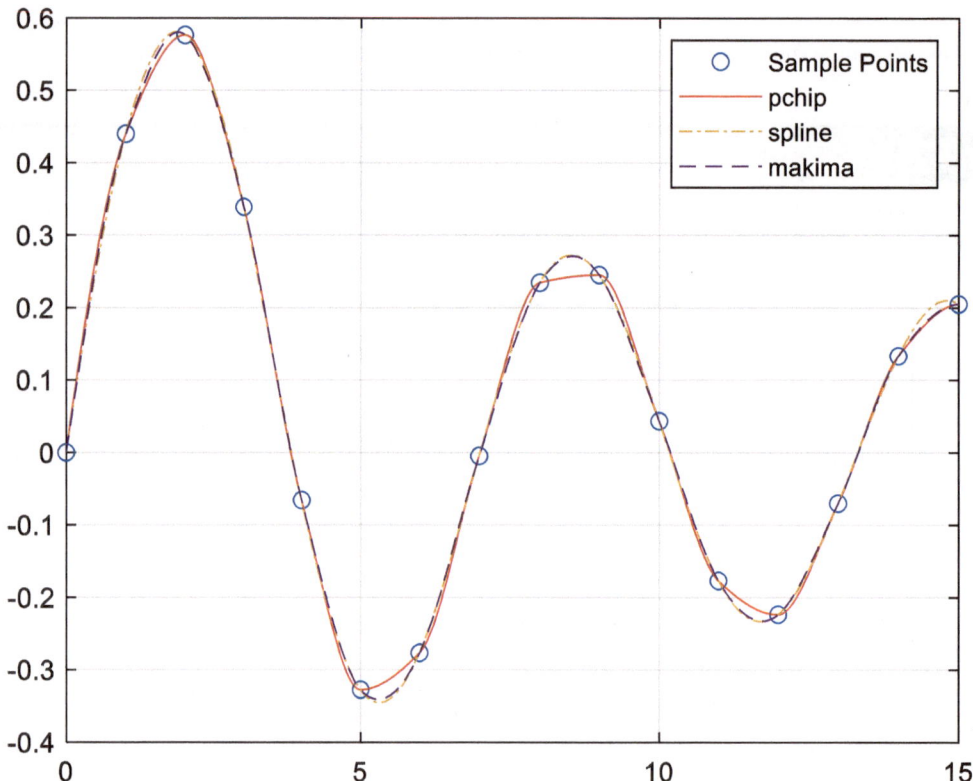

Fig. 34.16 The plots of interpolation of another data set using spline, pchip, and makima methods

34.10 The Reference Function "poly": Polynomial with Specified Roots or Characteristic Polynomial

- This reference function in the format p = poly(A), where A is an n-by-n matrix, returns the n + 1 coefficients of the characteristic polynomial of the matrix, $\det(\lambda I - A)$.
- This reference function in the format p = poly(r), where r is a vector, returns the coefficients of the polynomial whose roots are the elements of r.

Example

A = [1 8 -10; -4 2 4; -5 2 8]
p = poly(A)

Output

```
A =
   1   8  -10
  -4   2   4
  -5   2   8

p =
   1.0000  -11.0000   0.0000  -84.0000
```

Description

Calculating the characteristic polynomial of a matrix.

Example

r = roots(p)

Output

```
r =
  11.6219 + 0.0000i
  -0.3110 + 2.6704i
  -0.3110 - 2.6704i
```

Description

Calculating the roots of the characteristic polynomial of the matrix. As we know, the roots of characteristic polynomial of a matrix are the eigenvalues of the matrix.

Exercise

(a) *Write the codes to determine the characteristic polynomial of the matrix A = [1 2 3; 4 5 6; 7 8 0].*
(b) *Execute the codes in your computer. Then, write the result(s) in the assigned places.*

Output

A =

p =

Example

```
A = [1 8 -10; -4 2 4; -5 2 8]
e = eig(A)
p = poly(e)
```

Output

A =
     1   8   -10
    -4   2   4
    -5   2   8

e =
   11.6219 + 0.0000i
   -0.3110 + 2.6704i
   -0.3110 - 2.6704i

p =
    1.0000   -11.0000   0.0000   -84.0000

Description

Calculating the characteristic polynomial from the eigenvalues of a matrix. As we know, eigenvalues are the roots of characteristic polynomial of a matrix.

Exercise

(a)   *Write the codes to determine the characteristic polynomial of the matrix A = [1 2 3; 4 5 6; 7 8 0] by using the eigenvalues of the matrix.*
(b)   *Execute the codes in your computer. Then, write the result(s) in the assigned places.*

Output

A =

e =

p =

34.11 The Reference Function "polyval": Polynomial Evaluation

- This reference function in the format y = polyval(p,x) evaluates the polynomial p at each point in x. The argument p is a vector of length n + 1 whose elements are the coefficients (in descending powers) of an nth-degree polynomial $p(x) = p_1x^n + p_2x^{n-1} + \ldots + p_nx + p_{n+1}$.
- To evaluate a polynomial in a matrix sense, "polyvalm" can be use, instead.

Example

p = [3 2 1];
x = [5 7 9];
y = polyval(p,x)

Output

y =
    86   162   262

Description

Evaluating the polynomial $p(x) = 3x^2 + 2x + 1$ at the points $x = 5,7,9$.

Exercise

Execute the codes below in your computer. Then, write the result(s) in the following.

p = [1 2 3 4];
x = [0 1 2 3];
y = polyval(p,x)

Output

y =

34.12 The Reference Function "polyfit": Polynomial Curve Fitting

- This reference function in the format p = polyfit(x,y,n) returns the coefficients for a polynomial p(x) of degree n $(p(x) = p_1x^n + p_2x^{n-1} + + p_nx + p_{n+1})$ that is a best fit (in a least-squares sense) for the data in y. The coefficients in p are in descending powers, and the length of p is n + 1.

Example

x = linspace(0,4*pi,10);
y = sin(x);
p = polyfit(x,y,7);

Description

Using the reference function to fit a 7th-degree polynomial to the given points.

Example

```
x1 = linspace(0,4*pi);
y1 = polyval(p,x1);
plot(x,y,'o')
hold on
plot(x1,y1)
hold off
```

Description

Evaluating the 7th-degree polynomial on a finer grid and plotting the results. See Fig. 34.17.

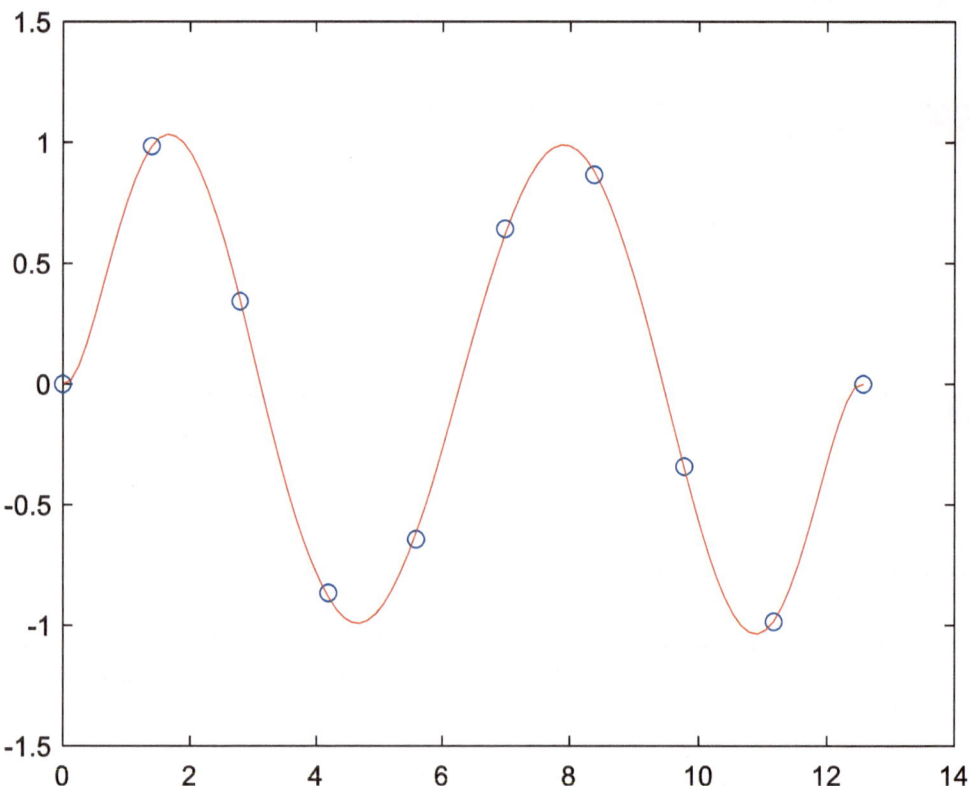

Fig. 34.17   The plot of the given points and a 7th-degree polynomial fitted to the points on a finer grid

Exercise

Execute the codes below in your computer. Then, draw the result(s) on the axes.

```
x = [0   1.3963   2.7925   4.1888   5.5851   6.9813   8.3776   9.7738   11.1701   12.5664];
y = [1   0.1736   -0.9397   -0.5000   0.7660   0.7660   -0.5000   -0.9397   0.1736   1];
p = polyfit(x,y,7);

x1 = linspace(0,4*pi);
y1 = polyval(p,x1);
plot(x,y,'o')
hold on
plot(x1,y1)
hold off
```

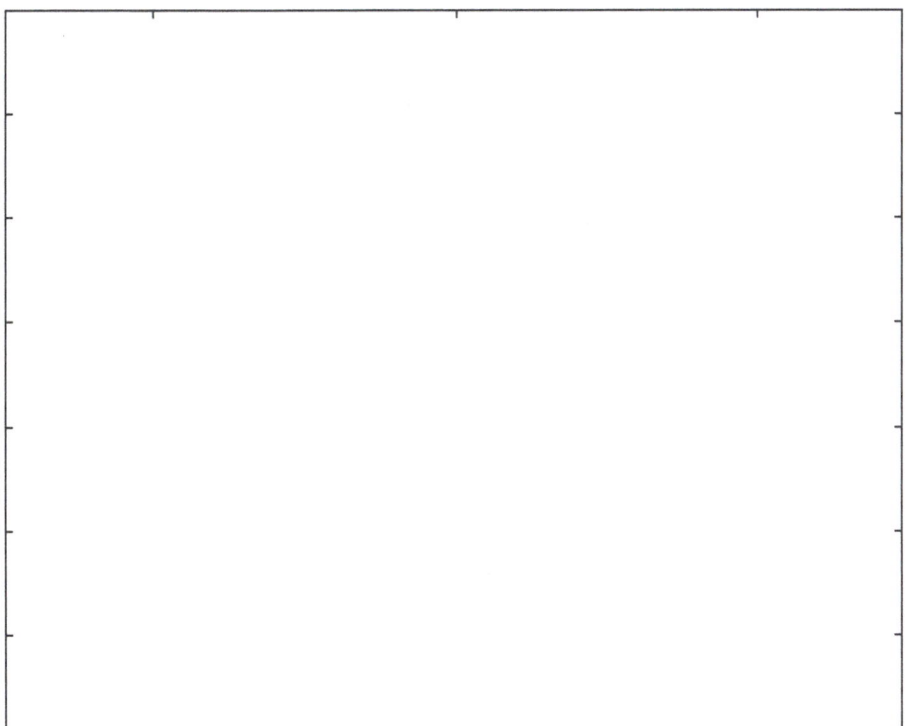

Example

```
x = linspace(0,1,5);
y = 1./(1+x);
p = polyfit(x,y,4);
```

Description

Using the reference function to fit a fourth-degree polynomial to the given points.

Example

```
x1 = linspace(0,2);
y1 = 1./(1+x1);
f1 = polyval(p,x1);
plot(x,y,'o')
hold on
plot(x1,y1)
plot(x1,f1,'r--')
legend('y','y1','f1')
```

Description

Evaluating the fourth-degree polynomial on a finer grid and plotting the results. See Fig. 34.18.

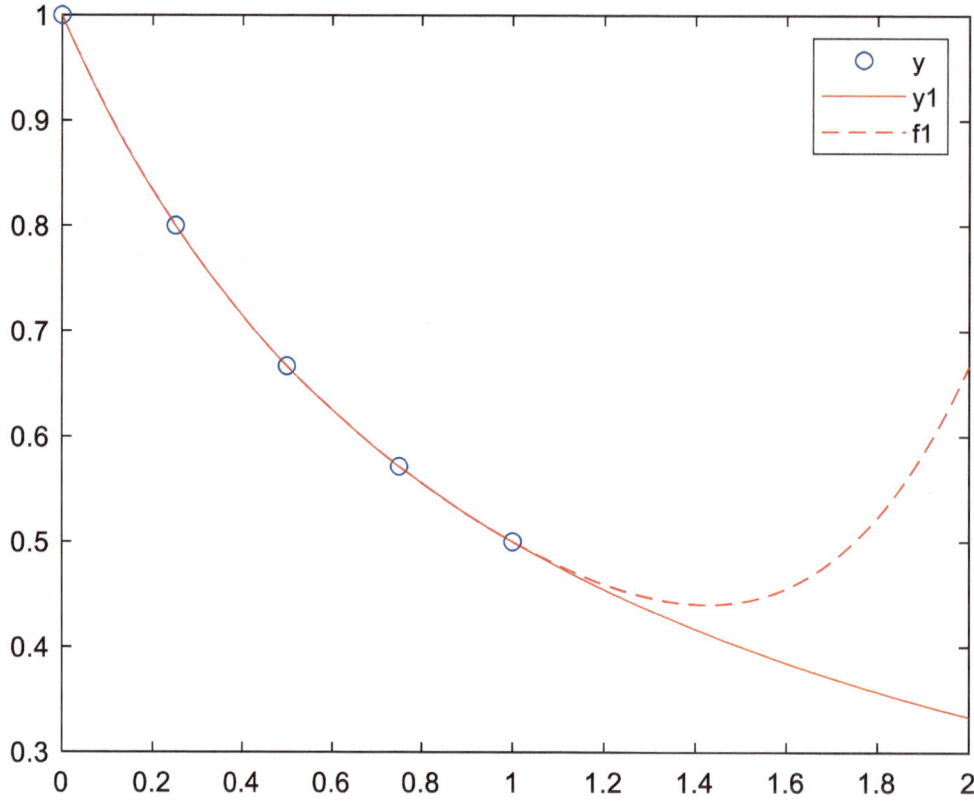

Fig. 34.18 The plot of the given points (y), a fourth-degree polynomial fitted to the points (y1), and a fourth-degree polynomial fitted to the points on a finer grid (f1)

34.13 The Reference Function "roots": Polynomial Roots

• This reference function in the format r = roots(p) returns the roots of the polynomial represented by p as a column vector. Input p is a vector containing n + 1 polynomial coefficients, starting with the coefficient of x^n. A coefficient of 0 indicates an intermediate power that is not present in the equation.

Example

```
p = [1 -2 1];
r = roots(p)
```

Output

```
r =
    1
    1
```

Description

Finding the roots of the polynomial.

Example

```
p = [1 0 1];
r = roots(p)
```

Output

```
r =
    0.0000 + 1.0000i
    0.0000 - 1.0000i
```

Description

Finding the roots of the polynomial.

Exercise

Execute the codes below in your computer. Then, write the result(s) in the following.

```
p = [1 0 -1];
r = roots(p)
```

Output

```
r =
```

Example

```
p = [1 0 0 1];
r = roots(p)
```

Output

r =
   -1.0000 + 0.0000i
   0.5000 + 0.8660i
   0.5000 - 0.8660i

Description

Finding the roots of the polynomial.

Exercise

Execute the codes below in your computer. Then, write the result(s) in the following.

p = [1 0 0 0 -1];
r = roots(p)

Output

r =

34.14   The Reference Function "residue": Partial Fraction Expansion or Decomposition

- This reference function in the format [r,p,k] = residue(b,a) finds the residues, poles, and direct term of a partial fraction expansion of the ratio of two polynomials, where the expansion is of the form below.

$$\frac{b(x)}{a(x)} = \frac{b_m s^m + b_{m-1} s^{m-1} + \ldots + b_1 s + b_0}{a_n s^n + a_{n-1} s^{n-1} + \ldots + a_1 s + a_0} = \frac{r_n}{s - p_n} + \ldots + \frac{r_1}{s - p_1} + k(s)$$

- The inputs to the reference function are vectors of coefficients of the polynomials $b = [b_m \ b_1 \ b_0]$ and $a = [a_n \ a_1 \ a_0]$. The outputs are the residues $r = [r_n \ r_1 \ r_0]$, the poles $p = [p_n \ p_1 \ p_0]$, and the polynomial k.
- This reference function in the format [b,a] = residue(r,p,k) converts the partial fraction expansion back to the ratio of two polynomials and returns the coefficients in b and a.

Example

b = [-4 8];
a = [1 6 8];
[r,p,k] = residue(b,a)

Output

r =
   -12
     8

p =
   -4
   -2

k =
   []

Description

Finding the partial fraction expansion of the ratio of two polynomials. Herein, the degree of numerator is smaller than the one of denominator. Moreover, the poles and residues are real quantities.

Exercise

Execute the codes below in your computer. Then, write the result(s) in the following.

b = [1 1];
a = [1 2 1];
[r,p,k] = residue(b,a)

Output

r =

p =

k =

Example

r = [-12; 8];
p = [-4; -2];
k = [];
[b,a] = residue(r,p,k)

Output

b =
   -4   8

a =
   1   6   8

Description

Converting the partial fraction expansion back to the ratio of two polynomials.

Exercise

Execute the codes below in your computer. Then, write the result(s) in the following.

```
r = [-1; 1];
p = [-2; -3];
k = [];
[b,a] = residue(r,p,k)
```

Output

b =

a =

Example

```
b = [2 1 0 0];
a = [1 0 1 1];
[r,p,k] = residue(b,a)
```

Output

r =
   0.5354 + 1.0390i
   0.5354 - 1.0390i
 -0.0708 + 0.0000i

p =
   0.3412 + 1.1615i
   0.3412 - 1.1615i
 -0.6823 + 0.0000i

k =
 2

Description

Finding the partial fraction expansion of the ratio of two polynomials. Herein, the degrees of numerator and denominator are equal. Moreover, the poles and residues are complex quantities.

Example

```
b = [2 0 0 1 0];
a = [1 0 1];
[r,p,k] = residue(b,a)
```

Output

r =
  0.5000 - 1.0000i
  0.5000 + 1.0000i

p =
  0.0000 + 1.0000i
  0.0000 - 1.0000i

k =
  2 0 -2

Description

Finding the partial fraction expansion of the ratio of two polynomials. Herein, the degree of numerator is larger than the one of denominator. Moreover, the poles and residues are complex quantities.

Exercise

Execute the codes below in your computer. Then, write the result(s) in the following.

b = [1 2 1];
a = [1 1];
[r,p,k] = residue(b,a)

Output

r =

p =

k =

34.15 The Reference Function "polyint": Polynomial Integration

- This reference function in the format q = polyint(p,k) returns the integral of the polynomial represented by the coefficients in p using a constant of integration k.
- This reference function in the q = polyint(p) assumes a zero value for the integration constant, that is, k = 0.

Example

p = [3 0 -4 10 -25];
q = polyint(p)

q =
   0.6000   0   -1.3333   5.0000   -25.0000   0

Calculating the integral of a polynomial with zero integration constant.

Execute the codes below in your computer. Then, write the result(s) in the following.

```
p = [5 4 3 2 1];
q = polyint(p)
```

q =

34.16 The Reference Function "polyder": Polynomial Differentiation

- This reference function in the format k = polyder(p) returns the derivative of the polynomial represented by the coefficients in p.
- This reference function in the format k = polyder(a,b) returns the derivative of the product of the polynomials a and b.
- This reference function in the format [q,d] = polyder(a,b) returns the derivative of the quotient of the polynomials a and b.

```
p = [3 0 -2 0 1 5];
q = polyder(p)
```

q =
   15   0   -6   0   1

Calculating the derivative of a polynomial.

Execute the codes below in your computer. Then, write the result(s) in the following.

```
p = [1/5 1/4 1/3 1/2 1 1];
q = polyder(p)
```

Output

q =

Example

a = [1 -2 0 0 11];
b = [1 -10 15];
q = polyder(a,b)

Output

q =
   6   -60   140   -90   22   -110

Description

Calculating the derivative of the product of two polynomials.

Exercise

Execute the codes below in your computer. Then, write the result(s) in the following.

a = [1 1];
b = [1 1];
q = polyder(a,b)

Output

q =

Example

p = [1 0 -3 0 -1];
v = [1 4];
[q,d] = polyder(p,v)

Output

q =
   3   16   -3   -24   1

d =
   1   8   16

Description

Calculating the derivative of the quotient of two polynomials.

Exercise

Execute the codes below in your computer. Then, write the result(s) in the following.

```
p = [1 2 1];
v = [1 1];
[q,d] = polyder(p,v)
```

Output

q =

d =

References

1. MATLAB 2023a.
2. Rahmani-Andebili, M. (2024). *Mathematics of engineering and science – Practice problems, methods, and solutions*. Springer Nature.
3. Rahmani-Andebili, M. (2022). *Differential equations – Practice problems, methods, and solutions*. Springer Nature.
4. Rahmani-Andebili, M. (2023). *Calculus III – Practice problems, methods, and solutions*. Springer Nature.
5. Rahmani-Andebili, M. (2023). *Calculus II – Practice problems, methods, and solutions*. Springer Nature.
6. Rahmani-Andebili, M. (2023). *Calculus I – Practice problems, methods, and solutions* (2nd ed.). Springer Nature.
7. Rahmani-Andebili, M. (2021). *Calculus – Practice problems, methods, and solutions*. Springer Nature.
8. Rahmani-Andebili, M. (2024). *Precalculus – Practice problems, methods, and solutions* (2nd ed.). Springer Nature.
9. Rahmani-Andebili, M. (2021). *Precalculus – Practice problems, methods, and solutions*. Springer Nature.

Abstract

In this chapter, the reference functions concerned with the symbolic math in MATLAB are presented and described. In this regard, several examples and exercises for each section of the chapter are presented. The exercises that include writing the codes, executing them, and achieving the results need to be done by students to master programming skills. In this book, the codes, outputs, and descriptions are in blue, black, and green colors, respectively. To program in MATLAB, a script file can be created and saved with an appropriate name (e.g., untitled01) in the preferred directory of a computer. The program can be run by clicking on the "Run" available on the top toolbar of the script in MATLAB or calling the script by typing its name in Command Window or in the other scripts.

35.1 The Reference Function "sym": Creation of Symbolic Variables, Expressions, Functions, and Matrices

- This reference function in the format A = sym('a',[n1 ... nM]) creates an n1-by-...-by-nM symbolic array filled with automatically generated elements. The generated elements do not create symbolic variables in the MATLAB workspace [1].
- This reference function in the format A = sym('a',n) creates an n-by-n symbolic matrix filled with automatically generated elements.

Example

a = sym('a',[1 4])

Output

a =
[a1, a2, a3, a4]

Description

Creating a symbolic vector with automatically generated elements.

M. Rahmani-Andebili, *MATLAB Lessons, Examples, and Exercises*, https://doi.org/10.1007/978-3-031-76177-5_35

Example

a(1)

Output

ans =
a1

Description

Calling the elements.

Example

b = sym('x_%d',[1 4])

Output

b =
[x_1, x_2, x_3, x_4]

Description

Creating a symbolic vector with automatically generated elements.

Example

b(1:2)

Output

ans =
[x_1, x_2]

Description

Calling the elements.

Example

A = sym('A',[3 4])

Output

A =
[A1_1, A1_2, A1_3, A1_4]
[A2_1, A2_2, A2_3, A2_4]
[A3_1, A3_2, A3_3, A3_4]

Description

Creating a symbolic matrix with automatically generated elements.

Example

A(1,1)

Output

ans =
A1_1

Description

Calling the elements.

Example

B = sym('B_%d_%d',3)
C = sym('C_%d_%d',3)
B + C

Output

B =
[B_1_1, B_1_2, B_1_3]
[B_2_1, B_2_2, B_2_3]
[B_3_1, B_3_2, B_3_3]

C =
[C_1_1, C_1_2, C_1_3]
[C_2_1, C_2_2, C_2_3]
[C_3_1, C_3_2, C_3_3]

ans =
[B_1_1 + C_1_1, B_1_2 + C_1_2, B_1_3 + C_1_3]
[B_2_1 + C_2_1, B_2_2 + C_2_2, B_2_3 + C_2_3]
[B_3_1 + C_3_1, B_3_2 + C_3_2, B_3_3 + C_3_3]

Description

Creating symbolic matrices and adding them.

Exercise

Execute the codes below in your computer. Then, write the result(s) in the following.

x = sym('x_%d_%d',2)
y = sym('y_%d_%d',2)
z = x + y.'

Output

x =

y =

z =

35.2   The Reference Function "factor": Factorization

- This reference function in the format F = factor(x) returns all irreducible factors of x in vector F.
- If x is an integer, factor returns the prime factorization of x.
- If x is a symbolic expression, factor returns the subexpressions that are factors of x.

Example

factor(10)

Output

ans =
    2   5

Description

Factoring an integer number.

Example

factor(13)

Output

ans =
    13

Description

Factoring an integer number. As we know, 13 is a prime number.

Example

factor(16)

Output

ans =
    2   2   2   2

Description

Factoring an integer number.

Example

factor(823429252)

Output

ans =
    2   2   59   283   12329

Description

Factoring an integer number.

Exercise

(a) *Write the codes to factor the integer numbers 30, 60, 255, 256, and 257.*
(b) *Execute the codes in your computer. Then, write the results in the following.*

Example

```
syms x
F = factor(x^2-1)
```

Output

```
F =
[x - 1, x + 1]
```

Description

Factoring a polynomial.

Example

```
syms x
F = factor(x^3-1)
```

Output

```
F =
[x - 1, x^2 + x + 1]
```

Description

Factoring a polynomial.

Exercise

(a)   *Write the codes to factor the polynomial $x^4 - 1$.*
(b)   *Execute the codes in your computer. Then, write the results in the following.*

Example

```
syms x
F = factor(x^6-1)
```

Output

```
F =
[x - 1, x + 1, x^2 + x + 1, x^2 - x + 1]
```

Description

Factoring a polynomial.

Example

syms x y
F = factor(y^2*x^2,x)

Output

F =
[y^2, x, x]

Description

Factoring a polynomial for factors containing a specific variable, that is, x.

Example

syms x y
F = factor(y^2*x^2,y)

Output

F =
[x^2, y, y]

Description

Factoring a polynomial for factors containing a specific variable, that is, y.

35.3 The Reference Function "taylor": Taylor Series

- This reference function in the format T = taylor(f,var) approximates f with the Taylor series expansion of f up to the fifth order at the point var = 0. If you do not specify var, then it uses the default variable determined by symvar(f,1). The Taylor series expansion around zero is called Maclaurin series expansion.
- This reference function in the format T = taylor(f,var,a) approximates f with the Taylor series expansion of f at the point var = a.

Example

syms x
T1 = taylor(exp(x))

Output

T1 =
x^5/120 + x^4/24 + x^3/6 + x^2/2 + x + 1

Description

Finding the Maclaurin series expansions of a function. The Taylor series expansion around zero is called Maclaurin series expansion.

Exercise

(a)   *Write the codes to find the Maclaurin series expansions of sine function.*
(b)   *Execute the codes in your computer. Then, write the results in the following.*

Example

syms x
T3 = taylor(cos(x))

Output

T3 =
x^4/24 - x^2/2 + 1

Description

Finding the Maclaurin series expansions of a function.

Example

syms x
T3 = taylor(cos(x))
sympref('PolynomialDisplayStyle','ascend');

Output

T3 =
1 - x^2/2 + x^4/24

Description

Finding the Maclaurin series expansions of a function in ascending order.

Example

syms x
T = taylor(log(x),x,1)

Output

T =
x - (x - 1)^2/2 + (x - 1)^3/3 - (x - 1)^4/4 + (x - 1)^5/5 − 1

Description

Finding the Taylor series expansions of a function around a specific value, that is, x = 1.

Exercise

(a) *Write the codes to find the Taylor series expansions of inverse cotangent function around x = 1.*
(b) *Execute the codes in your computer. Then, write the results in the following.*

References

1. MATLAB 2023a.

Index